Gene Cloning

Gene Cloning

Julia Lodge, Pete Lund & Steve Minchin

School of Biosciences
University of Birmingham
Edgbaston
Birmingham
UK

Taylor & Francis
Taylor & Francis Group

Published by:

Taylor & Francis Group

In US: 270 Madison Avenue
          New York, N Y 10016
In UK: 2 Park Square, Milton Park
          Abingdon, OX14 4RN

© 2007 by Taylor & Francis Group

First published in 2007
Reprinted in 2007

ISBN: 0-7487-6534-4

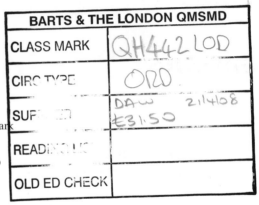
This book contains information obtained from authentic and highly regarded sources. Reprinted material is quoted with permission, and sources are indicated. A wide variety of references are listed. Reasonable efforts have been made to publish reliable data and information, but the author and the publisher cannot assume responsibility for the validity of all materials or for the consequences of their use.

A catalog record for this book is available from the British Library.

Library of Congress Cataloging-in-Publication Data

Lodge, Julia.
    Gene cloning : principles and applications / Julia Lodge, Pete Lund
& Steve Minchin.
        p. ; cm.
    Includes bibliographical references and index.
    ISBN 0-7487-6534-4 (alk. paper)
    1. Molecular cloning.   I. Lund, Peter A. II. Minchin, Steve.
III. Title.
    [DNLM: 1. Cloning, Molecular.   2. Gene Library.   3. Genomics
--methods.   QU 450 L822g 2007]
    QH442.2.L63 2007
    660.5′5--dc22

Editor: Elizabeth Owen
Editorial Assistant: Kirsty Lyons
Production Editor: Georgina Lucas
Typeset by: Phoenix Photosetting, Chatham, Kent, UK
Printed by: MPG BOOKS Limited, Bodmin, Cornwall, UK

Printed on acid-free paper

10 9 8 7 6 5 4 3 2

Taylor & Francis Group, an informa business   Visit our web site at http://www.garlandscience.com

Contents

Chapter 1   **Introduction**

1.1   The Beginning of Gene Cloning   1

1.2   How To Use This Book   3

1.3   What You Need To Know Before You Read This Book   5

1.4   A Request From the Authors   5

Further Reading   6

Chapter 2   **Genome Organization**

2.1   Introduction   7

2.2   The C-value Paradox   8

2.3   The Human Genome   9

2.4   Genomes of Other Eukaryotes   19

2.5   Bacterial Genomes   24

2.6   Plasmids   25

2.7   Viral Genomes   26

2.8   GC Content   27

2.9   Physical Characteristics of Eukaryotic Chromosomes   28

2.10   Karyotype   28

2.11   Euchromatin and Heterochromatin   30

2.12   CpG Islands   31

Questions and Answers   32

Further Reading   34

Chapter 3   **Key Tools for Gene Cloning**

3.1   Introduction   35

3.2   Vectors   36

3.3   Restriction Enzymes   38

3.4   DNA Ligase   40

3.5   Transformation   42

3.6   Purification of Plasmid DNA   45

3.7   More Restriction Enzymes   47

3.8   Alkaline Phosphatase   51

3.9   More About Vectors   53

3.10   Analyzing Cloned DNA by Restriction Mapping   58

3.11   Measuring the Size of DNA Fragments   59

3.12   The Polymerase Chain Reaction and Its Use in Gene Cloning   64

3.13   How Does PCR Work?   67

3.14   Designing PCR Primers   72
3.15   The PCR Reaction   73
3.16   Uses for PCR Products   74
3.17   Cloning PCR Products   74
3.18   Real-time PCR for Quantification of DNA   76
3.19   Advantages and Limitations of PCR   76
        Questions and Answers   78
        Further Reading   83

Chapter 4   Gene Identification and DNA Libraries

4.1   The Problem   85
4.2   Genomic Library   87
4.3   Constructing a Genomic Library   87
4.4   How Many Clones?   89
4.5   Some DNA Fragments are Under-represented in Genomic
        Libraries   90
4.6   Using Partial Digests to Make a Genomic Library   90
4.7   Storage of Genomic Libraries   92
4.8   Advantages and Disadvantages of Genomic Libraries   92
4.9   Cloning Vectors for Gene Libraries   93
4.10   Vectors Derived from Bacteriophage λ   93
4.11   Packing Bacteriophage λ *In Vitro*   95
4.12   Cloning with Bacteriophage λ   97
4.13   Calculating the Titer of your Library   98
4.14   Cosmid Libraries   98
4.15   Making a Cosmid Library   99
4.16   YAC and BAC Vectors   100
4.17   cDNA Libraries   101
4.18   Making a cDNA Library   103
4.19   Cloning the cDNA Product   105
4.20   Expressed Sequence Tags   108
4.21   What are the Disadvantages of a cDNA Library?   108
        Questions and Answers   109
        Further Reading   116

Chapter 5   Screening DNA Libraries

5.1   The Problem   117
5.2   Screening Methods Based on Gene Expression   118
5.3   Complementation   119
5.4   Immunological Screening of Expression Libraries   120
5.5   Screening Methods Based on Detecting a DNA Sequence   123
5.6   Oligonucleotide Probes   124
5.7   Cloned DNA Fragments as Probes   127

5.8   Colony and Plaque Hybridization   127
5.9   Differential Screening   132
5.10 Using PCR to Screen a Library   133
      Questions and Answers   135
      Further Reading   140

Chapter 6   Further Routes to Gene Identification
6.1   How Do We Get From Phenotype to Gene: a Fundamental
      Problem in Gene Cloning   141
6.2   Gene Tagging: A Method That Both Mutates and Marks Genes   142
6.3   A Simple Example of Transposon Tagging in Bacteria: Cloning
      Adherence Genes from *Pseudomonas*   149
6.4   Signature-tagged Mutagenesis: Cloning Bacterial Genes with
      "Difficult" Phenotypes   152
6.5   Gene Tagging in Higher Eukaryotes: Resistance Genes in
      Plants   156
6.6   Positional Cloning: Using Maps to Track Down Genes   159
6.7   Identification of a Linked Marker   161
6.8   Moving From the Marker Towards the Gene of Interest   161
6.9   Identifying the Gene of Interest   166
6.10 Cloning of the CF Gene: A Case Study   168
      Questions and Answers   169
      Further Reading   171

Chapter 7   Sequencing DNA
7.1   Introduction   173
7.2   Overview of Sequencing   174
7.3   Sanger Sequencing   175
7.4   The Sanger Sequencing Protocol Requires a Single-stranded
      DNA Template   179
7.5   Modifications of the Original Sanger Protocol   181
7.6   Strategies for Sequencing a DNA Fragment   182
7.7   High-throughput Sequencing Protocols   184
7.8   The Modern Sequencing Protocol   185
7.9   Genome Sequencing   188
7.10 High-throughput Pyrosequencing   197
7.11 The Importance of DNA Sequencing   202
      Questions and Answers   203
      Further Reading   205

Chapter 8   Bioinformatics
8.1   Introduction   207
8.2   What Does a Gene Look Like?   208

8.3   Identifying Eukaryotic Genes   214
8.4   Sequence Comparisons   217
8.5   Pair-wise Comparisons   217
8.6   Identity and Similarity   220
8.7   Is the Alignment Significant?   222
8.8   What Can Alignments Tell Us About the Biology of the
       Sequences Being Compared?   224
8.9   Similarity Searches   224
8.10   Fasta   226
8.11   BLAST   228
8.12   What Can Similarity Searches Tell Us About the Biology of
       the Sequences Being Compared?   230
8.13   Multiple Sequence Alignments   233
8.14   What Can Multiple Sequence Alignments Tell Us About the
       Structure and Function of Proteins?   234
8.15   Consensus Patterns and Sequence Motifs   235
8.16   Investigating the Three-dimensional Structures of Biological
       Molecules   237
8.17   Using Sequence Alignments to Create a Phylogenetic Tree   239
       Questions and Answers   242
       Further Reading   246

Chapter 9   Production of Proteins from Cloned Genes

9.1   Why Express Proteins?   249
9.2   Requirements for Protein Production from Cloned Genes   252
9.3   The Use of *E. coli* as a Host Organism for Protein Production   252
9.4   Some Problems in Obtaining High Level Production of
       Proteins in *E. coli*   260
9.5   Beyond *E. coli:* Protein Expression in Eukaryotic Systems   265
9.6   A Final Word About Protein Purification   274
       Questions and Answers   275
       Further Reading   277

Chapter 10   Gene Cloning in the Functional Analysis of Proteins

10.1   Introduction   279
10.2   Analyzing the Expression and Role of Unknown Genes   280
10.3   Determining the Cellular Location of Proteins   290
10.4   Mapping of Membrane Proteins   293
10.5   Detecting Interacting Proteins   297
10.6   Site-Directed Mutagenesis for Detailed Probing of Gene and
       Protein Function   304
       Questions and Answers   309
       Further Reading   312

Chapter 11   The Analysis of the Regulation of Gene Expression

11.1   Introduction   315
11.2   Determining the Transcription Start of a Gene   318
11.3   Determining the Level of Gene Expression   326
11.4   Identifying the Important Regulatory Regions   338
11.5   Identifying Protein Factors   350
11.6   Global Studies of Gene Expression   353
Questions and Answers   361
Further Reading   364

Chapter 12   The Production and Uses of Transgenic Organisms

12.1   What is a Transgenic Organism?   365
12.2   Why Make Transgenic Organisms?   367
12.3   How are Transgenic Organisms Made?   377
12.4   Drawbacks and Problems   396
12.5   Knockout Mice and Other Organisms: The Growth of
       Precision in Transgene Targeting   398
12.6   Is the Technology Available to Produce Transgenic People?   406
Questions and Answers   407
Further Reading   410

Chapter 13   Forensic and Medical Applications

13.1   Introduction   411
13.2   Forensics   411
13.3   DNA Profiling   413
13.4   Multiplex PCR   414
13.5   Samples for Forensic Analysis   415
13.6   Obtaining More Information from DNA Profiles   416
13.7   Other Applications of DNA Profiling   417
13.8   Medical Applications   418
13.9   Techniques for Diagnosis of Inherited Disorders   422
13.10 Whole Genome Amplification   435
13.11 Diagnosis of Infectious Disease   437
13.12 Diagnosis and Management of Cancer   439
Questions and Answers   441
Further Reading   444

Glossary   445

Index   453

1 Introduction

1.1 The Beginning of Gene Cloning

In November 1973, a five-page paper was published in the prestigious journal *Proceedings of the National Academy of Sciences USA* by Stanley Cohen, Annie Chang, Herb Boyer and Robert Helling from Stanford University in California. The title of the paper was "Construction of biologically functional plasmids *in vitro*", and it described for the first time the production of an organism into which a DNA molecule had been introduced which consisted of DNA sequences from two different sources, joined together in the test tube. Although this work itself built on an earlier body of research, it may justifiably be seen as the paper which marked the birth of a scientific revolution which has continued to this day.

Great changes in science come about in different ways. Sometimes, they are the result of new concepts that transform our way of looking at things, or give us new insights into areas of knowledge which had previously been obscure. Such a revolution in biology had already occurred in the two decades before Cohen's paper, with the realization that the fundamental stuff of inheritance is DNA, with the discovery of DNA's remarkable structure, and with the unscrambling of the genetic code. Other dramatic changes in science have been more technical than conceptual, and are no less important for that. Cohen's paper describes the first methods for manipulating DNA in ways that began to give the experimenters a measure of control over these molecules, hence enabling manipulation of the genetic properties of the organisms that contain them. Humans have, of course, been selectively breeding organisms for particular traits for millennia: domestication of wild plants for crops was a hugely successful early experiment in genetic engineering. But with the advent of what are commonly called recombinant DNA techniques, the degree to which we can produce predetermined genetic changes with high precision has grown to the point where now it is commonplace to make bacteria or plants that produce human proteins, to tinker with the basic structures of enzymes to alter their activity or stability, or to pull a single gene from the tens of thousands present in a human chromosome and identify within in it a single changed base that may give rise to a crippling genetic disease.

One of the hallmarks of the maturity of a technology is the extent to which it works its way through the system. It begins as the preserve of one or a few specialized laboratories; then, it becomes more widely available and used, and commercial applications begin to appear; ultimately, it makes its way onto undergraduate and even school curricula. Experiments that were once the material of Nobel prizes become the topic of routine practicals. For many years we have run, here in our own school, a practical for second-year undergraduates which is, in essence, not that dissimilar to the breakthrough experiment described by Cohen and his colleagues in 1973. Many hundreds of our undergraduates have become gene cloners by the end of their second year at university, and this is also true in thousands of institutions all over the world, including probably the one where you are studying. In addition many aspects of the biology that are taught are only known and understood today because of the incredible power that recombinant DNA methods give us to answer fundamental questions about the nature of life and the functions of cells and organisms. Many of our graduates go on to use these methods extensively in their own careers as research scientists in academia and industry.

If the basic technique described by Cohen *et al.* was all there was to recombinant DNA methods, our life as academics teaching molecular biology would be very simple. But, in fact, Cohen and his colleagues were in many ways the Wright brothers of the field, and in little more than three decades since they published their paper, we have moved from fragile biplanes to jumbo jets. Today's research laboratories have access to hundreds of different approaches to biological investigation which use aspects of recombinant DNA techniques, and whole industries are founded upon their exploitation. Methods have been introduced and refined at a dizzying pace that shows no obvious sign of relenting. Some – such as the ability to exponentially amplify vanishingly small amounts of DNA in a test tube, to manipulate the germ line of complex multicellular organisms, or to determine the complete sequences of the genomes of many organisms – have been revolutions in their own right. Others represent incremental improvements to basic techniques which have nonetheless transformed complex methods into processes that can be done using off-the-shelf kits, or (increasingly) performed by robots. This presents something of a problem for us as teachers, or rather two problems. The first is that the pace of change is such that it is difficult to know what to include and what to omit from undergraduate courses on the subject, and hard to find text books at the right level that are up to date. The second is that the essentially technical nature of the recombinant DNA revolution means it is important to present the subject in such a way that it is not just a dry list of methods, but which also conveys a sense of the excitement and insight that these approaches have brought to so many different areas, not only in the research laboratory but also in everyday applications. Hence the book that you are now reading. In it we have tried to present a selection of what we regard as the key concepts and methods that underlie gene cloning, at a level which should be easily understandable to a typical undergraduate in a bioscience or medical subject, and to illustrate these as much as possible with examples drawn from the laboratories of universities and companies around the world. Our aim throughout has been to be as comprehensive as possible both with basic methods and with their more advanced applications, subject to space constraints. Inevitably, we have had to be selective in the material that we have covered, and even during the course of writing the book we have had to go back and revise or add to early material as new methods have been published. But we believe that the major aspects of the subject are all here, presented in a form that you will find easy to understand, and which will interest and enthuse those of you that read and use it.

1.2 How To Use This Book

The layout of the book is quite traditional, with the different chapters dealing with methods and concepts of increasing complexity through the book. Although we have made the individual chapters self-contained, and used extensive cross-referencing between chapters, we expect most people will start at the beginning and work their way through the book as needed according to the course they are studying. Each chapter starts with a list of "learning outcomes" – that is, a list of the things you should be able to do once you have read and understood the material in the chapters. These should help you to assess whether you have understood what the chapter is all about. By way of an introduction to the book we present some information about the way genomes are organized in both prokaryotic and eukaryotic organisms. There follows a group of chapters which present basic details about the enzymes and reactions used in simple gene manipulations, and then go on to talk about how genes are actually cloned and identified. We have gone into the details of how clones of particular genes are found, since our experience has been that this is an area that students often find difficult to understand. The advent of high-throughput genome sequencing and the consequent availability of huge amounts of gene sequence data online means that approaches to gene cloning have changed a lot in recent years, but we feel it is still important for you to understand the "traditional" (i.e. more than 10 years old!) methods, even though the use of gene libraries is becoming less common.

It would be ridiculous in a book of this nature, however, not to give a good deal of weight to the topic of genomics (i.e. all aspects of studying organisms at the whole genome level), since this constitutes one of the more recent revolutions in the methodology of the biosciences. Two chapters describe how DNA is sequenced and how the large amounts of sequence data deposited in international databases can be mined and analyzed – although we have not gone into this latter area in too much technical detail, since this is a whole new discipline in its own right and requires skills in mathematics and computer programming which are beyond the remit of this book.

We then turn to more applied aspects of recombinant DNA methods, including how cloned genes can be used as the source of large amounts of proteins, and how genes can be manipulated and introduced into higher organisms to produce so-called transgenic organisms, the uses of some of which are described as case studies. We discuss also some of the powerful research uses of these methods, such as deepening our understanding of how genes are regulated in cells, and enabling us to functionally dissect proteins. Finally, we conclude with a chapter which discusses some further applications, mainly in a medical context, to add to those used as illustrations in earlier chapters.

Although most of the text will be self-explanatory, assuming you have a degree of basic knowledge (the things we expect you to already know are listed in the next section), there are some places where particular general concepts seemed to us to be sufficiently important that we have put them in boxes, separate from the rest of the text.

One thing that you will notice in the book is the use of large numbers of examples, based on genuine experiments and published results, to illustrate the points that we are making. One of the features of molecular biology is that the methods can be applied to all living organisms, and you will find that in some cases, our examples will be based on bacterial systems (prokaryotes), and on others they will refer to eukaryotes, ranging from single-celled organisms such as yeast all the way to humans. We (the authors) have research and teaching experience both with prokaryotes and eukaryotes, and it has been our experience that it is often best to discuss the simpler prokaryotic systems first, to introduce basic concepts, before going on to talk about the more complex eukaryotes. At the end of each chapter, we have included references for the papers which are referred to in the case studies discussed in that chapter. In most cases, these are available online; if not, they should be in your institution's library. Reading these papers should add to your understanding of the methods and their applications discussed in this book. Some of the papers are quite straightforward, while others are complex and may be tricky to follow in places. However, learning to read the scientific literature is an essential part of any undergraduate degree, and we encourage you to read as many of these papers as you are able. A key feature of the book is the questions that are included in the text of each chapter. Some of these are simply designed to make sure that you have taken in what you have just read, by (for example) setting simple problems based on the previous sections. Others do require a bit of extra thought, or the bringing together of several different topics. As it is only by trying to answer questions on it that you can really tell how good your understanding of a subject is, we encourage you to persevere with these questions, even if they appear difficult at first, before turning to our answers at the end of each chapter.

1.3 What You Need To Know Before You Read This Book

In writing this book, we have tried to pitch it roughly at a level that would be understandable by undergraduates in the UK in their second year, although some of the more advanced material would perhaps be left until the following final year, and these are the levels at which we have experience of teaching these topics. It is important to be clear, therefore, that this is not a textbook about fundamental concepts in genetics, cell biology, or biochemistry, and it is assumed that you will already know these before you start. We take it that anyone studying this book will be familiar with:

- The structure of DNA
- The nature of the genetic code
- The way in which information flows from DNA via RNA to proteins, and the basic nature of the mechanisms (transcription and translation) by which this happens
- The nature of proteins, including the way in which their structure determines their function, and their different roles in cells
- Basic cell biology of both prokaryotes and eukaryotes

If these are not areas that you are familiar with, then much of the material in this book will be hard to follow, and it would be better to study a more basic text first before trying to use the current book.

1.4 A Request From the Authors

We have tried very hard to make this book precise, informative, interesting and correct. Some of the material has been tested extensively on undergraduates here in Birmingham or has grown from material that we have been teaching for many years. Other material is relatively new, and has involved us in a great deal of research of our own, reading original papers and talking with people using methods with which we ourselves were not directly familiar. It is inevitable, however, that the book will contain flaws, and we genuinely do want to hear about these so that in the event that future editions are needed, we can incorporate any suggestions which are made by you for the benefit of other readers. If you have comments or corrections to make, do please send them by e-mail to gene_cloning@bham.ac.uk. We look forward to reading your comments, and we hope you find the book a valuable aid to studying the fascinating and important topic of gene cloning.

Further Reading
Construction of biologically functional bacterial plasmids *in vitro*. (1973) Cohen SN, Chang AC, Boyer HW and Helling RB. Proc Natl Acad Sci USA, Volume 70 Pages 3240–3244.

The first paper to describe the production of an organism that contained DNA sequences from two different sources.

2 Genome Organization

Learning outcomes:

By the end of this chapter you will have an understanding of:

- the genomic organization of prokaryotes and eukaryotes, and in particular the human genome
- the different types of sequence within the eukaryotic genome: coding, non-coding, non-repetitive and repetitive
- the physical characteristics of chromosomes
- why an appreciation of genome organization is important in the context of gene cloning

2.1 Introduction

The genome of an organism can be defined as "the total DNA content of the cell", and as such it contains all the genetic information required to direct the growth and development of the organism. For all multicellular organisms this growth and development starts from a single cell, the fertilized egg. In the case of humans the egg develops into an adult comprising approximately 10^{12} cells made up from over 200 different cell types.

As you will be aware the gene is the basic unit of biological information. Most genes code for a protein product: the gene is transcribed to RNA and this RNA messenger is then translated to the protein product. In addition to genes which encode proteins, there are many genes which encode stable RNAs such as ribosomal RNA and transfer RNA. The number of genes contained within the genome of an organism ranges from around 500 for the bacterium *Mycoplasma genitalium* to over 50,000, predicted to be present in most plants.

In bacteria the genetic information is normally carried on one circular DNA molecule referred to as the bacterial chromosome, which may be supplemented with several small self-replicating DNA molecules, also known as plasmids. Eukaryotic cells contain several linear chromosomes within the nucleus. Human cells, for example, contain 23 pairs of chromosomes. In addition to the DNA present in the nucleus, mitochondria and chloroplasts contain DNA that encodes a fraction of the functions of these organelles. For multicellular organisms the genome content is identical for all cells with only a few exceptions (such as red blood cells, which contain no nuclei and hence no nuclear DNA).

Although the genome and gene content of an organism is necessary for the development and survival of that organism, it is not sufficient. There are important proteins in the fertilized egg whose function is to control how theses genes are used. Because of this no free-living organism could be created from its DNA alone. All cells present on Earth today have arisen from pre-existing cells. Genetic engineering cannot lead to the generation of novel organisms from basic components, genetic engineering can only modify the genetic make-up of pre-existing cells by adding or removing functions from the organism's genome.

In order to understand how to manipulate DNA it is important to understand the way it is organized in different organisms. In this chapter we will discuss the main features of the genome of higher eukaryotic organisms using the human genome as our primary example. We will also discuss bacterial and viral genomes so as to understand how they differ from those of eukaryotic organisms.

2.2 The C-value Paradox

The C-value is a measure of genome size, typically expressed in base pairs of DNA per haploid genome. The use of the term haploid genome refers to a single copy of all the genetic information present in the nucleus. Diploid nuclei of organisms produced sexually will of course contain two complete,

Table 2.1 Characteristics of the genomes of example organisms

Organism	Genome Size (bp)	Chromosome Number (n)	Predicted Number of Genes
Mycoplasma genitalium	580,000	1	500
Escherichia coli K12	4,639,000	1	4,500
Saccharomyces cerevisiae (yeast)	12,069,000	16	6,000
Caenorhabditis elegans (worm)	97,000,000	6	20,000
Drosophila melanogaster (fly)	137,000,000	6	15,000
Oryza sativa (rice)	420,000,000	12	40,000
Arabidopsis thaliana (weed)	115,000,000	5	28,000
Fugu rubripes (pufferfish)	390,000,000	22	25,000
Mouse	2,500,000,000	20	25,000
Humans	3,300,000,000	23	25,000

and not quite identical, copies of a haploid genome, each derived from one of the parents. Table 2.1 gives the C-value for a range of different organisms. One surprising outcome of analyzing the C-value from different organisms is the so-called "C-value paradox" which refers to the fact that genome sizes (and hence the C-values) do not always correlate with genetic and/or morphological complexity. The C-value paradox states that the organism with the largest genome is not necessarily the most complex and that genome size cannot be used as a predictor of genetic or morphological complexity. For example humans and mice, in common with most other mammals, have a genome size of around 3 billion base pairs (3×10^9 bp). However the unicellular protozoan *Amoeba dubia* has a genome size of over 600 billion base pairs (6×10^{11} bp) about 200 times as big. The C-value paradox means that organisms with similar complexity may have very different genome sizes and conversely organisms with similar C-values may not be equally complex. The main exception to the C-value paradox is found in the prokaryotic kingdom, where genome size is a good predictor of metabolic complexity, which in turn often relates to the range of different niches within which a particular bacterium can survive.

In prokaryotic organisms like bacteria, genes are packed tightly together with very little non-coding DNA being present, although all higher eukaryotes contain a large amount of repetitive non-coding DNA. The presence of varying amounts of this non-coding DNA in different eukaryotic organisms explains how relatively simple organisms can have more DNA in their genomes than more complex ones. Remarkably, genes which encode proteins are present as oases in a desert of non-coding "junk" sequences. In the human genome, for example, there is on average only one gene for every 100 kb of sequence.

Q2.1. If the genomes of higher organisms were similar to bacteria, how many genes could be encoded by the human genome? Assume that in bacteria genes are about 1 kb in length.

2.3 The Human Genome

The human haploid nuclear genome contains 3×10^9 base pairs, a vast amount of DNA. To put this in context: if you were to begin reading the human genome sequence at 1 base per second and continue for 24 hours a day, 7 days a week, it would take you approximately 100 years to finish. This DNA is organized into linear segments, the chromosomes, which vary in length from 47×10^6 bp to 246×10^6 bp. There are two sex chromosomes (the X and Y), which determine the gender of the person carrying them. The other 22 chromosomes are called autosomes. Somatic cells in diploid eukaryotes contain pairs of each autosome plus two sex chromosomes. The pairs are not perfect pairs, however, in that they are not identical in

DNA sequence, as one originates from the father, and one from the mother, of the person in question. The number of chromosomes quoted for an organism is usually the haploid number (23 for humans). Somatic human cells therefore contain approximately 6×10^9 bp of DNA – two complete (and non-identical) copies of the genome. A recent milestone in science has been the sequencing of the entire human genome.

In addition to the nuclear genome, mitochondria also contain DNA: in humans, the mitochondrial genome is about 17×10^5 bp in length. As each of the 800 mitochondria per cell contains 10 copies of the mitochondrial genome, this is in fact the most abundant DNA molecule in the cell. The abundance of mitochondrial DNA has been exploited in the analysis of ancient DNA (aDNA), and in forensic investigations where limiting starting material can hinder genetic analysis. Mitochondria are maternally inherited, coming almost exclusively from the egg. The maternal inheritance of mitochondria can also be exploited in both population genetics studies and in forensic analysis.

How is the genome organized? As pointed out above, much of the genome in higher eukaryotes is non-coding. Initial analysis has indicated that less than 2% of the human genome actually encodes protein. The remaining non-coding DNA is often referred to as junk DNA and in most cases probably serves no purpose. However, this junk DNA, some of which has arisen due to DNA duplication, has allowed the evolution of new genes and hence the generation of genetic variation. Recent comparison of dog, mouse and human whole genome sequences has shown a surprising degree of sequence conservation in regions of non-coding DNA. Approximately 5% of the mammalian genome is under selective pressure, i.e. 5% of their genome sequences have not changed significantly since mice, dogs and humans diverged from a common ancestor; this suggests that at least 3% of the non-coding DNA is conserved. The conserved non-coding DNA will include regulatory sequences required to control gene expression to ensure that the correct proteins are made at the right time in the appropriate cells. However, some of the conserved sequences may have functions that we do not yet fully understand. The study of this junk DNA will play an important role in understanding how modern genomes have evolved. In addition, as we will see in Section 13.2, this non-coding DNA has been exploited in forensic and population studies. Repetitive non-coding DNA also presents many problems for genome sequencing projects; this will be discussed in Chapter 7.

Non-coding DNA

Non-coding DNA can be grouped into four main categories, namely: (1) introns; (2) simple sequence repeats; (3) interspersed transposon-derived repeats; and (4) non-repeat, non-coding DNA. All these are discussed below. In addition there are also pseudogenes and segmental duplication, both of which have arisen through the duplication of regions of the genome during evolution. A duplicated region can either be on the same or a separate chromosome.

Introns

Most eukaryotic genes are made up of stretches of DNA that code for amino acids, interrupted by non-coding sequences. The coding and non-coding sections are called exons and introns respectively. An example of eukaryotic gene structure is shown in Figure 2.1, where the gene indicated has three exons separated by two introns. Eukaryotic RNA polymerases transcribe both the exons and introns to yield an RNA molecule called the primary transcript; it requires further processing before it can be used as a template for translation (Figure 2.2). This processing includes removal of the introns in a process called splicing, which edits the exons into a contiguous coding sequence that can be translated by the ribosome. Introns make up 24–28% of the human genome, and the "average" human gene is encoded by nine small exons of 145 bp each separated by large introns of over 3 kb each.

The existence of introns has many implications for gene cloning and analysis. Most genetic engineering requires the manipulation of a sequence that is devoid of introns, i.e. is a single contiguous coding sequence. As you can see from Figure 2.2a, the mature messenger RNA contains an uninterrupted copy of the protein coding sequence. Therefore the easiest way to obtain a contiguous coding sequence is to isolate messenger RNA and to then create a DNA copy. A DNA copy of messenger RNA is called cDNA, and the way in which this is produced and used will be discussed in Section 4.18. Another issue that is raised by introns is the problem of identifying genes within whole genome sequences. This is made particularly difficult because of the fact that small exons are separated by very large introns and predicting exon-intron boundaries is a difficult challenge. The problem of identifying the coding sequences within genomic

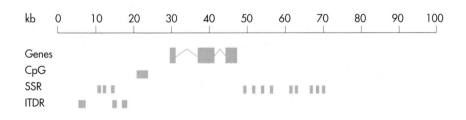

Figure 2.1 Schematic representation of 100 kb of a human chromosome. The four rows represent different classes of sequence found within the genome. This region of the genome contains only one gene, the exons are shown as blocks and the introns as lines. The position of simple sequence repeats (SSRs), interspersed transposon-derived repeats (ITDRs) and CpG islands (CpG) are indicated by the blocks.

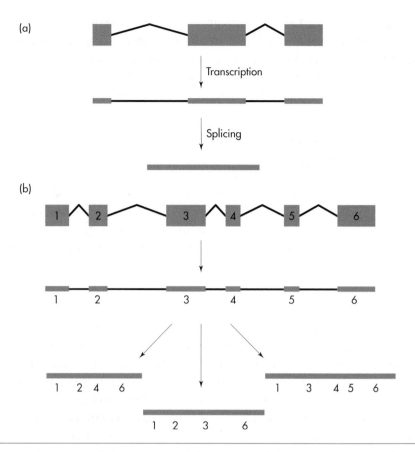

Figure 2.2 Eukaryotic genes contain introns. a) Schematic representation of a gene. Exons and introns are shown as blocks and lines respectively. RNA polymerase transcribes both the exons and the introns, the introns are then edited out in a process called splicing to generate the mRNA molecule that is exported to the cytoplasm and translated. b) A schematic representation of a gene with six exons. Three splice variants are shown.

DNA is therefore generally resolved by analysing cDNA in parallel with genomic DNA; this will be discussed in Section 8.3.

An added complexity of human genome biology is the prediction that at least 35% of human genes are alternatively spliced. This is a much higher level than for any other organism for which genomic information is available. Alternative splicing is shown in Figure 2.2b, which depicts a hypothetical gene that contains six exons. In this example the six exons could be spliced in three different ways leading to three different proteins. It should be noted that the three proteins will share common peptide sequences, i.e. all three have the same N-terminal and C-terminal sequences. Alternative splicing may actually explain how a relatively small number of genes (20–25,000) can code for an organism as complex as a human. Alternative splicing generates the potential for over 100,000 different proteins, so the relative complexity of the human proteome (the catalog of human proteins) may be due to alternative splicing rather than a high gene number. This impacts on our understanding of development, health and disease. Methods used to study the proteome (total protein profile) and transcriptome (total mRNA profile) of cells will also be discussed in Section 11.6.

Q2.2. Given a gene with the following structure:

E1	Intron	E2	Intron	E3	Intron	E4

How many alternatively spliced variants can be generated? Note alternative splicing results in one or more exons being spliced from the final message; it cannot result in the reordering of exons, i.e. a final message with the exon order E1+E3+E4 can occur, but not E3+E1+E4. You should assume that E1 and E4 are present in all spliced mRNAs.

Simple sequence repeats

A region of the genome where a DNA sequence is repeated many times in tandem is called a simple sequence repeat (SSR). These regions are described as satellite DNA, mini-satellite DNA and micro-satellite DNA depending on the number and length of the repeats (see Box 2.1 for details). SSRs, i.e. satellite, mini-satellite and micro-satellite, make up 3% of the human genome. Most SSRs are not associated with a phenotype as they fall in non-coding DNA, however a subclass of micro-satellites, often referred to as trinucleotide or trimer repeats, are within protein coding genes and as such can give rise to altered phenotypes and, in certain circumstances, disease. The rapid evolution of different dog breeds is thought, in part, to be due to changes within these trimer repeats. One example is the repeat 5'-CAG-3' found within the Huntington (hdh) gene. Most individuals have 10–35 copies of the 5'-CAG-3' repeat within each of the two copies of the hdh gene in the cell. However, if the number of repeats exceeds 35 within either copy of the gene, this leads to Huntington's disease. Several other genetic diseases are due to trinucleotide repeat expansion. It was analysis of mini-satellites that formed the original basis of DNA fingerprinting, whereas modern DNA profiling techniques are based on the analysis of micro-satellite DNA (see Section 13.2).

Interspersed transposon-derived repeats

By far the most abundant repeat sequences found within the human genome are the transposon-derived repeats or interspersed repeat sequences, which account for 45% of the human genome. These are derived from transposons, mobile genetic elements capable of being copied and inserted into new sites in the genome. The ability of transposons to spread through the genome is thought to have given rise to an

Box 2.1 Simple Sequence Repeats (SSRs)

Simple sequence repeats (SSRs) are found in all eukaryotes, although their relative abundance does vary between organisms. When genomic DNA is isolated by centrifugation on a cesium chloride gradient it forms a discrete band although smaller satellite bands are often visible. These contain simple sequence repeat DNA which is also known as satellite DNA. SSRs can be classified into three categories, satellite DNA, mini-satellite DNA and micro-satellite DNA.

Satellite DNA sequences are typically greater than 100 kb in length and composed of repeats five to 200 bp in length. An example is the alphoid DNA found in centromeric DNA, in which a 171 bp sequence is repeated many times, resulting in a region of around 500 kb.

Mini-satellite sequences are generally less than 20 kb of DNA composed of repeats less than 25 bp in length. An example of a mini-satellite sequence is one found near the insulin gene, which has the sequence 5′-ACAGGGGTGTGGGG-3′. The number of repeats of the mini-satellite sequence at each locus varies between individuals and they are therefore also referred to as variable number tandem repeats or VNTRs. It was analysis of these VNTRs that formed the original basis of DNA profiling.

Micro-satellite DNA is normally less than 150 bp in length and composed of 5 bp repeat sequences. Micro-satellites are also known as simple tandem repeats (STRs). An example is the D18S51 micro-satellite, which has the sequence 5′-AGAA-3′. Most individuals will have between seven and 27 copies of D18S51 on each copy of chromosome 18. Analysis of these simple tandem repeats has formed the basis of modern DNA profiling techniques (Section 13.2).

accumulation of these sequences in eukaryotic genomes over evolutionary time. Similarly, over time these transposons have been subject to mutations and deletions, resulting in most cases in the loss of the ability to transpose. There are four classes of transposable elements, namely: LTR retroposons (retrovirus-like elements), long interspersed elements (LINEs), short interspersed elements (SINEs) and DNA transposons; a more detailed discussion of interspersed transposon-derived repeats can be found in Box 2.2.

In addition to the multiple repeat elements mentioned above, approximately 5% of the genome appears to be a duplication of another segment of the genome. There are several mechanisms that, during evolution, result in sections of the genome being copied and pasted into the same or different location on the chromosome. Often the duplicated segment will have

Box 2.2 Interspersed Transposon-Derived Repeats

Interspersed transposon-derived repeats are found in all eukaryotes. They are typically the most abundant repeat sequences found within the genome and account for 45% of the human genome. There are four classes of transposable elements, LTR retroposons (retrovirus-like elements), long interspersed elements (LINEs), short interspersed elements (SINEs) and DNA transposons.

LTR retroposons contain long terminal repeats (LTR elements) and are relics of retroviruses. They contain the necessary signals to allow transposition within the genome, a process that requires transcription of the retroposon element followed by reverse transcription to form a DNA copy which then integrates at a random site within the genome. LTR retroposons account for approximately 8% of the human genome.

Long interspersed elements (LINEs) are retroelements, 6–8 kb in length, that do not have an LTR sequence. The LINE sequence encodes two proteins that are involved in transposition of the sequence within the genome. The inefficient mechanism used during transposition often results in truncated forms of the LINEs being transposed within the genome. LINE elements make up approximately 20% of the human genome.

Short interspersed elements (SINEs) are short DNA sequences, 100–300 bp in length, which are not capable of transposition on their own but can do so using functions provided by LINE elements. The Alu element, a 300 bp sequence, is the most abundant SINE with more than 1,000,000 copies present within the human genome representing 10% of the genome. In total, SINEs make up approximately 13% of the human genome.

Active DNA transposons are 2–3 kb in length and contain a transposase gene flanked by terminal repeat sequences; inactive transposons contain the inverted repeats but have lost the transposase gene. DNA transposons make up 3% of the human genome.

LTR retroposons, LINEs and SINEs all transpose via an RNA intermediate, whereas DNA transposons spread through the genome via a copy and paste mechanism.

contained functional genes which therefore leads to redundancy. The duplicated gene or genes can pick up mutations that either lead to change or loss of function. So gene duplication is one way in which genes with new functions can occur and is responsible for the evolution of gene families. Loss of function will result in a pseudogene, i.e. a sequence on the genome that looks like a gene but no longer gives rise to a functional protein.

Non-coding, non-repeat DNA

SSRs and interspersed transposon-derived repeats account for 50% of our genome, while the coding DNA makes up only 1–1.4% of the genome. This leaves approximately 50% of the genome made up of "unique" non-coding DNA, half of which will be present in introns. A small percentage of this non-coding DNA will contain regulatory sequences such as enhancers and promoters that are required for the control of gene transcription (see Section 11.1). It is highly unlikely that the remaining 25% arose from segmental duplication and therefore this must, at some point, have been derived from ancient transposable elements that have changed during evolution such that they are no longer recognized as transposable elements.

The impact of non-coding DNA on genome analysis

What is the relevance of interspersed repeat and simple sequence elements to gene cloning? These elements provide a fundamental problem for assembly of whole genome sequences. All current genome sequencing protocols, at some point, involve generating a random set of overlapping fragments of only a few kilobases in length (Chapter 7). Once sequenced, the fragments need to be assembled into the correct order to yield the finished genome sequence. Clearly, if the genome is littered with repeat sequences, many of which are several kilobases in length, this can lead to mis-assembly of the finished sequence. Segmental duplication gives rise to gene families, which means that it is important to differentiate between a gene that you wish to study and the similar genes within the genome. Pseudogenes must also be taken into account when trying to identify "functional" genes and when predicting the gene content of a particular genome.

The genes and other functionally important sequences

The coding and other functionally important DNA makes up only a small fraction of the human genome sequence. Because of the difficulty in analysing the genome sequence and predicting precisely where all the genes occur, we do not have a complete catalog of all human genes. This is exemplified by the fact that initial analysis of the human genome sequence by the public genome consortium predicted the existence of 30,000 genes, whereas Celera Genomics predicted the existence of 40,000 genes. Although the two groups both identified the 10,000 previously known genes, there was little overlap in the predicted novel genes within their data sets. With the publication of the "finished" human genome sequence in 2004 the number of predicted genes has fallen further. The best estimate for the number of human genes is still a figure of 20,000 to 25,000. Genes are non-randomly scattered around the genome, with gene rich and gene poor regions. Not all chromosomes have the same gene density, as you can see from Table 2.2 which summarizes data from the public genome consortium.

Table 2.2 Human chromosome size and gene content

Chromosome	Size (bp)	Number of Genes
1	246,000,000	2800
2	243,000,000	1800
3	200,000,000	1500
4	191,000,000	1100
5	181,000,000	1200
6	171,000,000	1350
7	159,000,000	1300
8	146,000,000	950
9	138,000,000	1050
10	135,000,000	1000
11	134,000,000	1600
12	132,000,000	1300
13	114,000,000	500
14	106,000,000	900
15	100,000,000	950
16	89,000,000	1050
17	79,000,000	1400
18	76,000,000	400
19	64,000,000	1550
20	62,000,000	750
21	47,000,000	300
22	50,000,000	600
X	155,000,000	1100
Y	58,000,000	130

Figures given are as predicted by the Ensembl project (www.ensembl.org). The gene number represents genes that have been identified or predicted as of November 2004.

Q2.3. Using the data in Table 2.2, determine the gene density for the human chromosomes 13, 19 and the Y chromosome in terms of genes per million base pairs. Which is the most gene-rich and which is the most gene-poor chromosome?

As genome sequencing projects reveal the sequence of other mammalian genomes, it is becoming clear that human developmental and intellectual complexity is not due to increased gene number. All vertebrates carry roughly the same number of genes and it is possible that flowering plants contain more genes than humans. The mouse genome contains an almost identical set of genes to ours; almost every human gene has a mouse equivalent. So what makes us different from a mouse? The key difference between organisms is how the genetic information is utilized, in particular when and where the genes are expressed. It appears that the DNA itself is less important than what you do with it. If we wish to understand the genetic aspects of human development, health and disease, we need to be able to monitor, on a genomic scale, both the timing and level of gene expression of all genes, as well as the generation of protein diversity due to alternative splicing of mRNAs. These topics will be discussed in Section 11.6.

Variation between individuals: single nucleotide polymorphisms

We all have copies of the same genes, so the question can be asked: What generates the diversity seen within the human population? What makes you different from your siblings and your friends? Diversity arises from the fact that many genes are polymorphic, i.e. there are many different versions of the same genes within any population. In most cases, this variation is due to base changes at a single position. Such variations are called single nucleotide polymorphisms (or SNPs, pronounced "snips"). It is important to differentiate between rare mutations and SNPs. If you were to sequence the genome of tens of thousands of individuals you would probably find very rare mutations at virtually every position within the genome. SNPs, however, are defined as single bases within the genome which differ in at least 1% of the population, so they occur at much higher frequencies than rare mutations.

SNPs are distributed non-randomly throughout the genome, occurring at a frequency of once every 1000 bases. There are two types of SNPs found within the protein coding region of genes. First are those that cause no change in the amino acid sequence of a protein. These SNPs change a base but, due to the degeneracy of the genetic code, the amino acid encoded by the codon of which this base is a part is not altered. These are therefore examples of silent mutations. The second type of SNP does alter the amino acid encoded and may therefore alter the property of the protein. SNPs do not have to occur in coding regions; they can, for example, also occur within promoter regions (sequences that control if, when and to what level a gene is expressed). And, in fact, most SNPs will occur within non-coding regions of the genome, and because of this it has been predicted that less than 1% of SNPs will have a significant phenotypic effect. These SNPs that do have an effect on the phenotype are responsible for almost all genetic variation between members of a species, such as variation in response to drug treatment and variation in susceptibility to, or protection from, all kinds of diseases (Box 13.2).

Even the 99% of SNPs that are predicted to have no direct impact on phenotype are targets of genome biologists since many will be positioned close to other genes of interest, and they will therefore be linked to these genes. SNPs are also useful markers in population genetics and evolutionary studies. Because of this, SNP study has become a major focus of genome analysis.

Q2.4. Below is a section of a hypothetical gene with the sequence of the encoded protein. Two SNPs are marked that are found in this sequence. Would both SNPs result in an altered protein sequence? Would you expect two SNPs to be this close together?

```
          A                                                        A
          ↑                                                        ↑
          |                                                        |
GGGCTAGCCAACAGGCCAAGGATTTGGGAGGAGAGACTAAACATGGTGGGTGACTGTGAT
GlyLeuAlaAsnArgProArgIleTrpGluGluArgLeuAsnMetValGlyAspCysAsp
```

2.4 Genomes of Other Eukaryotes

In addition to the human genome, the genomes of many other organisms have been sequenced, including those of a number of model organisms (Box 2.3). With some exceptions, most eukaryotes have a similar genome organization to humans, in that they contain a vast amount of non-coding DNA that contains simple sequence repeats and transposon-derived repeats. A degree of relationship between the complexity of the organism and gene content is observed, so that, for instance, the yeast *Saccharomyces cerevisiae*, the fruit fly, *Drosophila melanogaster* and the thale cress *Arabidopsis thaliana* are predicted to contain around 6000, 15,000 and 25,000 genes, respectively – increasing as the complexity of the organism increases. However, as already pointed out above, this relationship does not always hold true and genome size, or C-value, is not always well correlated with the complexity of the organism.

In eukaryotic organisms the genome is organized into a number of linear chromosomes. The number of chromosomes is essentially arbitrary and sometimes even quite closely related organisms can have different chromosome numbers, with no obvious evolutionary reason or biological consequences. This is classically exemplified by the small muntjac deer, which has a genome almost as large as the human genome. The Chinese subspecies contains 23 pairs of chromosomes, whereas the Indian subspecies contains just three pairs. Both subspecies contain the same genes but in the Indian subspecies the 23 chromosomes present in its Chinese relative have become joined into three very large chromosomes. To date it appears that the Indian muntjac not only contains the smallest number of chromosomes observed in a mammalian species but also the largest chromosome, with one of its chromosomes containing more than 1 billion base pairs of DNA.

The biological significance, if any, of the way in which the genome is organized into different numbers of chromosomes in different species is

Box 2.3 Model Organisms

Throughout the history of biology many experimental scientists have chosen to work on a limited number of "model" organisms that range from bacteria to the chimpanzee. Model organisms are useful to scientists because they provide a standard within each area of research, and by concentrating research on these organisms a large body of information is accumulated. Because fundamental biological processes such as metabolism, development and genetics are conserved by evolution it is possible to extrapolate findings based on studies of relatively primitive model organisms such as the fruit fly or the nematode worm to other more complex organisms such as humans.

The main model organisms tend to be organisms which are easy to grow and to manipulate and for all of the examples listed below extremely powerful genetic techniques have been developed and the genomes have been sequenced. Many experiments discussed in the following chapters have been completed using one or other of these model organisms.

- The bacterium *Escherichia coli*
  E. coli is very easy to grow and has a very short doubling time. Virtually all cloning experiments use *Escherichia coli* at some point.
- The yeast *Saccharomyces cerevisiae*
  This single-celled organism is the most primitive model eukaryotic organism. It is very easy to grow and to manipulate.
- The weed *Arabidopsis thaliana*
  This small plant has a very short life cycle and is easy to grow. It has a relatively small genome for a plant.
- The fruit fly *Drosophila melanogaster*
  This organism is easy to grow in large numbers and over many years of research a large number of mutants have been identified and characterized.
- The nematode worm *Caenorhabditis elegans*
  These are easy to grow in large numbers and, because they are transparent, every cell in the body can be viewed in live organisms. There is a strictly defined number of cells in the mature organism and the development pathways are well characterized.
- The zebra fish *Danio rerio*
  This small fish is a relatively new model organism, it has a transparent embryo and is easy to grow in large numbers. As a vertebrate it is used as a model for some aspects of human biology as it is easier to breed than mice and rats.
- The mouse *Mus musculus*, and the rat *Rattus norvegicus*
  These mammals have essentially the same gene content as humans. Powerful genetic techniques have been developed for their manipulation, including, in the case of the mouse, the ability to "knock out"

individual genes. However, these animals are relatively slow to breed and expensive to keep.
- The chimpanzee
  This is our closest relative and has been used for experiments which would not be ethical if performed on humans. However, chimpanzees are not cheap to maintain or easy to breed and are not used for routine purposes.

not known. Similar organisms have similar gene sets but these can be arranged differently at the chromosome level. During evolution this genetic information has been moved around, but it does appear to have been reorganized in "blocks". For example, two chromosomes which are separate in one species may have been joined (fusion) in another, or a single chromosome from one species may have become two in another (fission), but the overall gene order is essentially the same. When the orders of genes are rearranged in closely related organisms, it is often by large inversions or similar events such that the order of genes within the inversion is unchanged but they are reversed with respect to those around them. Therefore if you look at a particular gene homologue in the genomes of related organisms, for instance within the mammals, it is often found with the same neighboring genes. This phenomenon is called synteny. In most cases not only are the genes found together but also they are arranged in the same order, a phenomenon termed colinearity.

Because of these phenomena, people have begun to ask the following question: how many rearrangements, such as large inversions, fusions, fissions or translocations (where a chunk of DNA is moved from one chromosome to another), would it take to get from the genome of one species to that of a related species? The answer is often surprisingly few. For example, it is possible to take the human genome, fragment it into approximately 200 pieces containing at least two genes and then rearrange these 200 pieces to "create" the mouse genome. (This "created" mouse genome is not identical to the actual mouse genome, as sequences of individual genes differ between mice and men, but the order of genes is essentially the same.) The relationship between human chromosome 11 and several mouse chromosomes is shown in Figure 2.3. Thus there are at least two levels of variation to consider when comparing genomes: variation at the DNA sequence level between individual genes, and variation in the order and arrangement of gene homologues along the chromosomes. Synteny and colinearity can be exploited in DNA analysis, and this will increase in future genome sequencing projects where comparative maps will help with the assembly of the genome sequences; for example, the human sequence has been used to help assemble the mouse sequence.

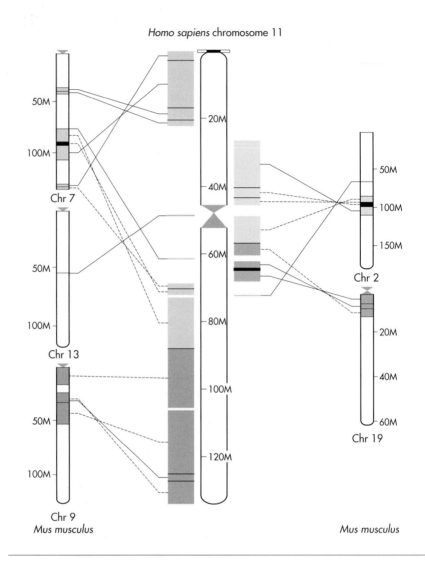

Figure 2.3 Synteny between human and mouse. A representation of the relationship between human chromosome 11 and mouse chromosomes 2, 7, 9, 13 and 19. Shown are the regions of the mouse chromosomes that would need to be rearranged to form human chromosome 11.

You may at this point be asking: since we have sequenced the human genome what is the value of sequencing other genomes, such as that of the mouse? There is still much to learn about the human genome, we still have not identified all of the regions which potentially encode genes, and once a gene has been "discovered" determining its function is another major undertaking. These problems are much more tractable if you take a "comparative biology" approach. Evolutionary pressure has meant that genes and other important sequences, such as control regions, change at a much slower rate than junk DNA. (To be more accurate, the rate of change at the DNA level is probably roughly constant for all bases, but changes in control and coding regions are much more likely to be deleterious to the organism than changes in the junk DNA and so these changes will be selected against.) The net effect of this will be that changes in the control and coding regions will be less common than those in the junk DNA. Therefore when you compare genomes from different organisms the sequences of interest are highlighted as regions of sequence similarity against the junk DNA background. An example of the power of comparative biology comes from studies on the pufferfish, *Fugu rubripes*. Fugu contains the same number of genes as humans but has a relatively compact genome of 365 million base pairs when compared to other vertebrates, i.e. it is one eighth the size of the human genome. The compact size is due to the presence of very little repetitive DNA and the small size of its introns. Fugu contains one gene for every 10 kb of genomic sequence. Comparison of the Fugu and human sequences has shown that even in these two vertebrates that are separated by 450 million years of evolution, there is some synteny. More importantly, comparative analysis has led to the identification of 900 novel putative human genes.

By comparing the human genome sequence with the genome sequence of more distantly related animals including other mammals, we can catalog important sequences, including genes and regulatory regions, because they have been conserved throughout evolution. In contrast, by comparing the human genome with the recently sequenced chimpanzee genome it is hoped that we will be able to identify the important differences that make us human. The human and chimpanzee genomes that have been sequenced only differ by 1.23%. However, some of the differences are within SNPs that are present in both genomes so at these positions some human and chimpanzee individuals will have the same sequence; the "real" difference between the human and chimpanzee genomes may be less than 1.06%. The fact that most human genes have homologues in all other mammals means that we can use animal models for most human genetic diseases to develop treatments that can then be used in human medicine and, in many cases, veterinary medicine.

As with animals, the size of plant genomes varies considerably: from around 50 megabases to over 100 gigabases. Plants thus exemplify the C-value paradox particularly well. For example, the wheat genome is over five times the size of the human genome. The main reason for the differences in genome size is related to the number of copies of repeat sequences within the genome: organisms such as wheat that have large genome sizes tend to have genomes littered with genome-wide repeats. For example, retrotransposons make up over 70% of the maize genome. The regular occurrence of transposons within plant genomes has been exploited in mapping studies. Another aspect of plant genetics is the phenomenon of polyploidy. Many plant species, especially flowering plants, have more than two sets of chromosomes per cell. This means that many plants have hundreds of, and in some cases over a thousand, chromosomes per cell. This has an impact on many aspects of gene cloning and analysis. Genome sequencing projects have to sequence the complete genome including all the "junk" and then identify the genes and "useful" DNA. As we will see in Section 4.20, an alternative to sequencing the complete genome is to sequence only the transcribed DNA in the form of expressed sequenced tags (ESTs).

In addition to the mitochondrial genome present in all eukaryotes, plants carry genetic information within their chloroplasts. The circular chloroplast genome encodes approximately 100 genes.

Most plants contain 40,000 to 50,000 genes and it is likely that there is little difference in gene function between different plants, i.e. plants contain virtually all the same genes. As is seen in animals, it appears that related plant species not only share the same genes but that these genes are arranged in the same order on the DNA: there is synteny and, in most cases, colinearity.

> **Q2.5.** What accounts for the difference in size of the genomes of different eukaryotes?

2.5 Bacterial Genomes

Most bacterial genomes are single circular DNA molecules that are 0.5–10 Mb in size. Many bacterial genomes have been sequenced including that of *Escherichia coli*, which is not only a model organism but is the organism which is by far the most widely used for gene cloning. The genomes of many pathogenic bacteria have been sequenced including *Mycobacterium tuberculosis* (which causes TB), *Treponema pallidum* (which causes syphilis), *Rickettsia prowazekii* (which causes typhus), *Vibrio cholerae* (which causes cholera) and *Yersinia pestis* (which causes plague). Comparative analysis of bacterial genomes is proving to be a powerful tool in enhancing our understanding of genome organization in bacteria.

The size of bacterial genomes varies considerably. However, for bacteria, genome size correlates well with gene number, which in turn correlates with morphological, physiological or metabolic complexity. Bacteria with small genomes encode a small number of genes and tend to be restricted to growth in relatively few specialized niches; they are often parasites. Bacteria with large genomes encode many genes and have much higher metabolic diversity. So, for example, the 580 kb genome of the intracellular human parasite *Mycoplasma genitalium* is thought to encode just 500 genes, whereas the 7000 kb genome of the nitrogen-fixing bacterium *Mesorhizobium loti*, which colonizes leguminous plants, encodes

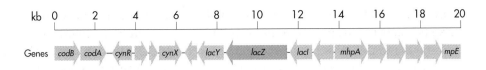

Figure 2.4 Schematic representation of 20 kb of the *Escherichia coli* chromosome. The genes are represented as blocks and the non-coding DNA as lines.

approximately 6800 genes. The average size of a gene in all bacteria is around 1000 bp; therefore, given the examples mentioned above a simple calculation tells you that bacterial genomes are densely packed with genes, i.e. the 500 genes of *M. genitalium* account for 500 kb of its 580 kb genome. The fact that bacterial genomes are packed with genes is clearly demonstrated in Figure 2.4, which shows a schematic representation of 20 kb of the *E. coli* genome (compare this with Figure 2.1). In bacteria there is very little wastage of DNA and little repetitive DNA as seen in higher eukaryotes. Only about 10% of the genome is non-coding and even this 10% in most cases plays a critical role to the cell since it contains the control signals required to co-ordinate patterns of gene expression.

The relatively small genome size of bacterial genomes, together with the fact that they contain very little non-coding or repeat DNA and that they do not contain introns, has made bacterial genomes ideal candidates for whole-genome sequencing projects. This is because, as will be discussed in Chapter 7, a relatively quick and successful strategy for whole-genome sequencing involves sequencing randomly generated overlapping fragments of only a few kilobases. These fragments are then assembled into the correct order to yield the finished genome sequence. As bacterial genomes contain very few repeat sequences assembly is relatively routine. Identification of potential genes within bacterial genomes is also much more reliable, because firstly bacterial genes do not contain introns and secondly the genes are much more closely packed. Potential genes are readily identified by computer analysis (Section 8.2).

2.6 Plasmids

Plasmids are nonessential extrachromosomal elements that control their own replication. They are found mainly in bacteria and in some eukaryotic microbes such as yeast and algae. Plasmids are normally circular double-stranded DNA molecules that range in size from as little as 1 kb to over 100 kb although linear plasmids have been observed in *Borrelia burgdorferi*, the causative agent of Lyme disease. Relative to the size of the bacterial chromosome plasmids are very small. The number of copies of a plasmid in each cell is tightly controlled with a general rule that small plasmids tend to have a high copy number, sometimes over 100 copies per cell, whereas larger plasmids may be present in one or a few copies per cell. Certain classes of plasmids are capable of horizontal transmission from one bacterium to another. In most cases the recipient bacterium does not have to be the same species as the donor bacterium and can, in fact, be distantly related. By definition, the coding potential of a plasmid is non-essential for the bacterial life cycle but plasmids frequently encode functions that give the bacteria a selective advantage in certain environments, the most infamous example being the class of genes which encode antibiotic resistance. The fact that plasmids can confer resistance to antibiotics and that horizontal transmission can occur between distantly related bacteria has major implications for medicine, since it has allowed rapid spread of antibiotic resistance throughout the microbial world. As you will see in the next chapter, plasmids have been absolutely key to the development of gene cloning.

Q2.6. What are the major differences between the genomes of bacteria and eukaryotes?

2.7 Viral Genomes

For all free-living organisms the genome content is double-stranded DNA. However, viral genomes consist of either RNA or DNA which can be either single- or double-stranded. Since viruses are obligate intracellular parasites they do not need to encode all the functions required by a free-living organism and accordingly they typically have very small genomes. The genome sizes of viruses, which may infect bacteria or eukaryotic cells, vary from 2 kb to 700 kb. It is beyond the scope of this book to discuss the complexities of viral genomes. However, viruses have been exploited throughout the history of gene cloning and as such two key viruses will be discussed below.

Bacteriophage λ is a virus that infects various bacteria including *E. coli*. It has a 48 kb double-stranded DNA genome packaged into an icosahedral head. Its genome is linear. When the λ DNA enters *E. coli* the ends of the linear genome join together to produce a circular genome. Upon infection λ can either follow a lysogenic or lytic pathway. During the lysogenic pathway, the viral genome is integrated into the host genome and is then replicated as part of the host chromosome: no expression of viral structural proteins occurs in this case. If λ enters the lytic pathway (which it can do either directly after infection or by induction from the lysogenic state), proteins required for the formation of progeny virions are made and the viral genome is replicated. Progeny virions are formed, the cell lyses, and about 100 new virions are released. The virus contains a very compact genome, the 48 kb coding for 46 genes. λ has been exploited as a vector for DNA cloning predominantly because of the efficiency with which it inserts its DNA into *E. coli* (Section 4.10).

Retroviruses are a class of viruses which infect eukaryotic cells that contain a single-stranded RNA genome in a protein capsid surrounded by a membrane envelope. The simplest retroviruses contain an 8 kb genome encoding just three genes, *gag*, *pol* and *env*. The *gag* gene encodes the proteins required to form the viral capsid, the *pol* gene encodes three enzymes, a protease, reverse transcriptase and an integrase, and the *env* gene encodes proteins that are present in the outer membrane. Upon entry into the host cell a double-stranded DNA copy is made of the RNA genome in a process termed reverse transcription, which requires the reverse transcriptase function encoded by the *pol* gene. The integrase then integrates the DNA copy of the viral genome into the host genome at a random position; integration is permanent. The host genes are transcribed and translated, and the immature polypeptide chains are cleaved by the viral protease to yield mature protein products. The exploitation of reverse transcriptase in gene cloning and analysis will be discussed in Chapters 4 and 11.

2.8 GC Content

One characteristic of an organism's genome which we have not considered so far is the DNA base composition; that is, the relative abundance of the bases guanine (G), cytosine (C), adenine (A) and thymine (T). As a result of base pairing there are always equal amounts of G and C, and equal amounts of A and T, and the base composition is usually expressed either as the ratio of (G+C) to (A+T) or as the percentage GC content. The GC content of related species, and of strains within a species, is constant. However, there is considerable variation between different species. The percentage GC content can range from 25–80% in micro-organisms, and 40–60% for higher eukaryotes. In Section 8.2 we will discuss how the GC bias can be used during sequence analysis when trying to identify potential protein coding genes.

During evolution bacteria have often acquired DNA via horizontal transfer and as this DNA may not have originated from the same species, it may have a different percentage GC content to that of the other DNA in the organism. Genome-wide analysis of GC content can thus often highlight regions of DNA that have been acquired via horizontal transfer. In pathogenic bacteria (i.e. bacteria which cause disease) these regions may represent "pathogenicity islands", the name given to regions of the genome which are distinct from the rest of the genome and which contain some of the genes which render the organism harmful. The human genome on average has a GC content of 40%; however, it does contain GC-rich and GC-poor regions which may have different properties such as gene density and repeat sequence content. In addition, differences in GC content correspond with the occurrence of bands which can be seen on stained preparations of chromosomes. For example, low GC content corresponds with dark G-bands which are gene poor, whereas high GC content corresponds with the gene-rich interband regions.

> **Q2.7.** When DNA is heated the two strands come apart, in a process termed melting, and the DNA becomes denatured. What effect would the GC content of DNA have on the temperature at which DNA melts, and why?

2.9 Physical Characteristics of Eukaryotic Chromosomes

Chromosomes can be observed by microscopy at metaphase when the DNA becomes condensed prior to cell division. When the metaphase chromosomes of eukaryotes are observed under a microscope it is clear that they are not all the same. Not only do they vary in size, but the position of the centromere is different in, and hence characteristic of, each pair of chromosomes (Figure 2.5). Further detail can be revealed by staining the chromosomes, which produces characteristic banding patterns. Depending on the stain used these bands relate to differences in chromatin structure, such as the degree to which the DNA is condensed.

2.10 Karyotype

As we have already seen, metaphase chromosomes of eukaryotic organisms vary in appearance. Chromosomes such as human chromosome 1, which contain a centromere close to the middle of the chromosome, are referred to as metacentric, whereas in acrocentric chromosomes (e.g. human chromosome 13) the centromere is close to one end of the chromosome. If the centromere is off-centre (e.g. human chromosome 4) the chromosome is called sub-metacentric. Each chromosome has two arms, one either side of the centromere. The short arm is labeled p, for petite, and the long arm q, because it follows p alphabetically.

Metaphase chromosomes can be treated enzymatically, fixed and stained with one of several dyes. A karyotype can then be produced by arranging the chromosome content of the cell in homologous pairs of decreasing size (Figure 2.5). The human male karyotype consists of 46 chromosomes, 22 pairs of autosomes, one X chromosome and one Y chromosome. Historically, the chromosomes have been numbered according to their apparent size, from the largest (chromosome 1) to the smallest (chromosome 22 in humans). This numbering is not always correct; for example, with humans further analysis revealed that chromosome 21 is actually smaller than chromosome 22 and as such is the smallest chromosome. The dye most commonly used is Giemsa stain (often following treatment with trypsin) which produces a distinctive banding pattern comprising alternating dark and light bands, often referred to as G-bands and interbands respectively. Depending on the resolution of the technique (a)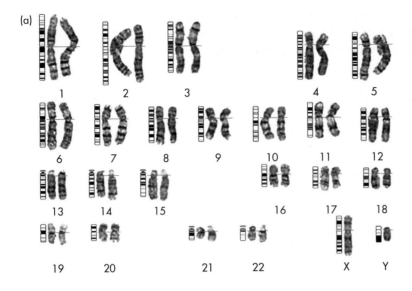

(b) Chromosome 4

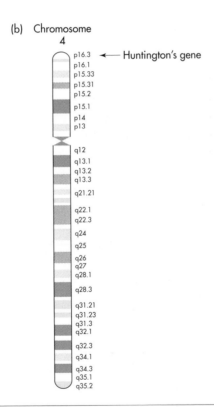

Figure 2.5 The human karyotype and example ideograms. a) The karyotype shows all human chromosomes together with a ideogram indicating the main chromosomal G bands. Note that chromosome 1 is metacentric, chromosome 13 is acrocentric, whereas chromosome 4 is submetacentric (courtesy of Lionel Willatt, Addenbrookes NHS Trust, Cambridge, UK & Digital Scientific, Sheraton House, Castle Park, Cambridge, UK). b) Ideogram representing chromosome 4.

several hundred G-bands and interbands can be visualized for each chromosome. This is often shown as an idealized diagram (ideogram) (Figure 2.5). Bands are numbered consecutively away from the centromere. So, for example, band 4p16, identifies chromosome 4 short arm, region 1, band 6. This low resolution band can be further sub-divided, for example, 4p16.3 refers to sub-division 3 of band 4p16. Often the location of a gene will be described relative to the G-band or interband designation; for example, the gene linked to Huntington's chorea (Huntington's disease) has the chromosomal location 4p16.3 (see Figure 2.5).

Examination of chromosome structure can play an important part in genetic analysis. Changes that are visible to the eye using a light microscope are an indicator of gross changes within the genome. These aberrations can either be inherited from one of the parents or can occur during gametogenesis. Often changes in chromosome structure are not tolerated and hence the fertilized egg containing such changes does not implant or the fetus is miscarried. Changes that are tolerated and lead to fetal development may, however, lead to disease, birth defects or mental retardation. The most common chromosomal change is alteration in chromosome number, referred to as aneuploidy. Monosomy is where there is one missing chromosome, and trisomy refers to a condition where there is one extra chromosome. In humans, for example, in trisomy 21 (Down syndrome) there is an additional copy of chromosome 21. In most cases aneuploidy results in failure of the fertilized egg to implant or, where implantation occurs, spontaneous miscarriage. The only numerical chromosomal aberrations that survive to term in humans are trisomy 21, trisomy 18, trisomy 13, sex chromosome aneuploidy, and triploidy (three complete sets of chromosomes).

In addition to yielding information concerning gross changes in chromosome structure linked to disease, banding gives information about chromatin structure and genome compartmentalization. When the chromosome is stained with Giemsa stain the dark bands (G-bands) have a tendency to be AT-rich and contain relatively few genes whereas the interbands tend to be gene-rich.

Q2.8. How many chromosomes would there be in a human cell displaying (a) monosomy, (b) trisomy, (c) triploidy?

2.11 Euchromatin and Heterochromatin

During mitosis, eukaryotic chromosomes become much shorter due to the DNA and protein in the chromosome taking up a highly structured form, whereupon it is said to be condensed. In eukaryotes, interphase chromosomes are decondensed compared with metaphase chromosomes, but they still contain significant structure. When cells in interphase are treated with a variety of dyes that stain the nucleus, two distinct classes of DNA are observed: lightly staining regions called euchromatin and densely staining regions called heterochromatin. Euchromatin contains most of the transcriptionally active genes, whereas heterochromatin contains few genes and is composed of repetitive non-coding DNA. Certain regions of the genome are packaged as heterochromatin in all cell types and at all stages of development, and are referred to as constitutive heterochromatin, whereas other regions of the genome may be packaged into heterochromatin in one cell type and euchromatin in others, and are therefore referred to as facultative heterochromatin regions. One method employed by the cell to silence genes permanently is to package the genes in heterochromatin. This has implications for the generation of transgenic organisms since current techniques rely on introducing the gene of interest into the fertilized egg or embryonic stem cells and then looking for transgenic organisms where the gene has been incorporated into the host genome. If the gene inserts in a region of the genome that is destined to be heterochromatin then the gene will be silenced leading to a lack of expression of the transgene and no phenotype for the transgenic organism.

2.12 CpG Islands

A characteristic of mammalian genomes is the occurrence of so-called "CpG islands". The dinucleotide CpG is unusual because it is under-represented within the genome, occurring at a frequency of less than 1%. Given that the average GC content of the human genome is around 40%, you would expect that the dinucleotide CpG would occur at a frequency of approximately 4%, a figure derived by multiplying the fraction of Cs and Gs within the genome (0.2×0.2). The explanation for this discrepancy is believed to be due to the fact that most CpG dinucleotides are methylated on the cytosine base, i.e. they contain 5-methylcytosine. Spontaneous deamination of 5-methylcytosine or cytosine generates thymine or uracil respectively. Any uracil residues generated by deamination will be repaired back to cytosine by the cell, since uracil should not be present in DNA. However, the product of 5-methylcytosine deamination, thymine, is a normal component of DNA and is therefore not removed. This means that over time methylCpG dinucleotides will slowly mutate to TpG, and so will be lost from the genome. The genome does contain "CpG islands" where the CpG dinucleotide content is approximately 4%, a frequency that matches that predicted from the GC content. These CpG islands contain CpG dinucleotides that normally contain unmethylated cytosines. These CpG islands are important because they occur at the 5′ end of genes and, as you will learn in Chapter 8, have played an important role in identifying genes and predicting the number of genes within the genome.

> **Q2.9.** Draw a cartoon of a section of a human chromosome containing a gene. Identify the location of exons, introns and CpG islands. For this gene to be transcribed and translated in a cell would it need to be in euchromatin or heterochromatin?

Much of our understanding of the way in which the genomes of animals, plants and bacteria are organized comes from studies that have only become possible because of the many ways in which we are able to manipulate DNA. We can isolate particularly interesting DNA sequences, sequence them, even change them and look to see what the effect is. It is now an almost routine proposition to sequence the whole of a bacterial genome, and we have the entire genome sequence of an increasing number of plants and animals. Gene cloning and the techniques that allow us to manipulate DNA, have driven developments in genomics, and in turn our increased understanding of genomes informs the way in which we manipulate DNA. In the following five chapters we will look in detail at the basic tools required for the manipulation of DNA.

Questions and Answers

Q2.1. If the genomes of higher organisms were similar to bacteria, how many genes could be encoded by the human genome? Assume that in bacteria genes are about 1 kb in length.

A2.1. From Table 2.1 we can see that the size of the human genome is 3,300,000,000 bp, long enough to encode 3,300,000 genes of 1 kb. In fact the human genome encodes somewhere between 20,000 and 25,000 genes, most of them bigger than bacterial genes.

Q2.2. Given a gene with the following structure:

| E1 | Intron | E2 | Intron | E3 | Intron | E4 |

How many alternatively spliced variants can be generated? Note alternative splicing results in one or more exons being spliced from the final message; it cannot result in the reordering of exons, i.e. a final message with the exon order E1+E3+E4 can occur, but not E3+E1+E4. You should assume that E1 and E4 are present in all spliced mRNAs.

A2.2. There are four possible alternatively spliced products: E1+E2+E3+E4, E1+E3+E4, E1+E2+E4, E1+E4.

Q2.3. Using the data in Table 2.2, determine the gene density for the human chromosomes 13, 19 and the Y chromosome in terms of genes per million base pairs. Which is the most gene rich and which is the most gene poor chromosome?

A2.3. Chromosome 13 encodes 500 genes and is 114 million base pairs long. The gene density is therefore 4.4 genes per million base pairs. This is the least gene rich of the autosomal chromosomes. The Y chromosome only encodes 130 genes and has a very low gene density of 2.2 genes per million base pairs. Chromosome 19 has the highest gene density at 24.2 genes per million base pairs.

Q2.4. Below is a section of a hypothetical gene with the sequence of the encoded protein. Two SNPs are marked that are found in this sequence. Would both SNPs result in an altered protein sequence? Would you expect two SNPs to be this close together?

```
        A                                                       A
        ↑                                                       ↑
GGGCTAGCCAACAGGCCAAGGATTTGGGAGGAGAGACTAAACATGGTGGGTGACTGTGAT
GlyLeuAlaAsnArgProArgIleTrpGluGluArgLeuAsnMetValGlyAspCysAsp
```

A2.4. The first SNP is in the third position of the codon, a change of C to A will not change the amino acid encoded. The second SNP is in the first position of the codon, the change of a T to an A results in a change from cysteine to serine. Although SNPs occur at a frequency of one every 1000 bases, chance can result in two SNPs very close together.

Q2.5. What accounts for the difference in size of the genomes of different eukaryotes?

A2.5. Differing amounts of non-coding or "junk" DNA accounts for the variation in the size of the haploid genome, or C-value.

Q2.6. What are the major differences between the genomes of bacteria and eukaryotes?

A2.6. Bacterial genomes are small with densely packed genes and very little non-coding DNA.

Q2.7. When DNA is heated the two strands come apart, in a process termed melting, and the DNA becomes denatured. What effect would the GC content of DNA have on the temperature at which DNA melts, and why?

A2.7. There are three hydrogen bonds in a GC base pair and only two in an AT base pair, this means that GC base pairs are more stable and hence require a higher temperature to break the hydrogen bonds, as a result DNA with a high GC content has a higher melting temperature than DNA with a high AT content.

Q2.8. How many chromosomes would there be in a human cell displaying (a) monosomy, (b) trisomy, (c) triploidy?

A2.8. (a) Monosomy refers to the condition where one chromosome of a pair is missing; the cell would only have 45 chromosomes. (b) Trisomy is where there is an extra chromosome resulting in a total of 47. (c) Triploidy refers to a condition where there is a complete set of extra chromosomes, in a human this would mean a total of 69 chromosomes.

Q2.9. Draw a cartoon of a section of a human chromosome containing a gene. Identify the location of exons, introns and CpG islands. For this gene to be transcribed and translated in a cell would it need to be in euchromatin or heterochromatin?

A2.9. *Euchromatin contains most of the transcriptionally active genes.*

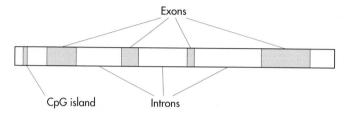

Further Reading

Genomes 3 (2006) Brown T, Garland Science, London.

Human Molecular Genetics (2003) Strachan T and Read A, Garland Science, London.
These are two good textbooks that discuss the genome organization and function

Our genome unveiled (2001) Baltimore D, Nature, Volume 409 Pages 814–816.
A commentary on the publication of the human genome sequence with links to more in-depth articles

The dog has its day (2005) Ellegren H, Nature, Volume 438 Pages 745–746.
A commentary on the publication of the dog genome sequence with links to more in-depth articles

The chimpanzee and us (2005) Li WH and Saunders MA, Nature, Volume 437 Pages 50–51.
A commentary on the publication of the chimpanzee genome sequence with links to more in-depth articles

The mouse that roared (2002) Boguski MS, Nature, Volume 420 Pages 515–516.
A commentary on the publication of the mouse genome sequence with links to more in-depth articles

Single nucleotide polymorphisms ... to a future of genetic medicine (2001) Chakravarti A, Nature, Volume 409 Pages 822–823.
A commentary on the publication of a paper reporting the identification and mapping of SNPs with links to more in-depth articles

3 Key Tools for Gene Cloning

Learning outcomes:

By the end of this chapter you will have an understanding of:

- *what gene cloning is*
- *the nature of and need for cloning vectors*
- *what restriction enzymes are and how they can be used to manipulate DNA*
- *the mode of action of DNA ligase and how it is used to join together different fragments of DNA*
- *ways of detecting the presence of cloned DNA in a vector*
- *how PCR can be used to amplify DNA*

3.1 Introduction

The discovery of the structure and role of DNA and the unravelling of the genetic code has led over the last 50 years to an explosion in our understanding of organisms and how they work. In the forefront of this revolution came gene cloning. The term "gene cloning" covers a wide range of techniques that make it possible to manipulate DNA in a test tube and also to return it to living organisms where it functions normally. The importance of this technology is that it allows us to isolate any piece of DNA from among the millions of base pairs that make up the genome of an organism. This first step is essential for a whole range of scientific and technological studies, ranging from the study of a gene that is instrumental in causing an inherited disease, to the bioengineering of a strain of yeast that produces a useful pharmaceutical product.

Gene cloning involves taking a piece of DNA from the organism where it naturally occurs and putting it into a cloning host such as the bacterium *Escherichia coli*. It is then possible to study the cloned DNA or produce the protein encoded by the gene. For many applications you may want subsequently to transfer the cloned DNA into another organism, but the initial cloning steps are almost always performed in *E. coli*.

Before looking in detail at how to clone genes, it is important to have an understanding of what the key steps are. DNA is cut into fragments and introduced into a new host, usually *E. coli*, where it is copied. However, you cannot simply introduce fragments of DNA into a cell or organism as they will probably be degraded, and even if it is not it will not be replicated and passed on when the cell divides. To make sure that the piece of cloned DNA is copied and passed on it is necessary to put it into a vector which will ensure that it is copied every time that the cell copies its own DNA and that a copy is passed on to each daughter cell at cell division. This involves cutting the vector and joining in the piece of DNA that you want to clone. This cutting and joining of DNA fragments is done using enzymes. The new molecule that you have thus created is introduced into your host cell by a process called transformation. Once in the host it will be copied and passed on every time the cell divides making many copies or clones of the original fragment.

3.2 Vectors
The most commonly used vectors for gene cloning are plasmids (Section 2.6). Figure 3.1 shows some examples. These are small circular DNA molecules found in many types of bacteria. Plasmids have an "origin of

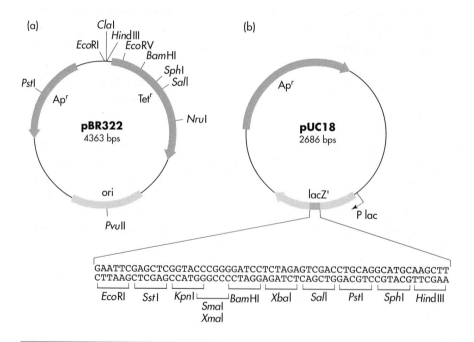

Figure 3.1 a) A map of pBR322 showing the positions of the ampicillin resistance (Ap^r) and tetracycline resistance (Tet^r) genes, the origin of replication (ori) and some of the unique restriction enzyme sites. b) A map of pUC18 showing the position of the ampicillin resistance (Ap^r) gene and the *lacZ'* gene; the sequence of the multiple cloning site encoding 10 unique restriction sites is also shown.

replication" which directs the replication of the plasmid and ensures that the cell contains many copies of the plasmid which are distributed between the daughter cells when the cell divides (Box 3.1). The exact number of copies varies according to the particular plasmid. As long as the gene that you have cloned is part of a DNA molecule with an origin of replication, that is, cloned into a plasmid, it will also be copied when the plasmid is copied.

There are a number of other features of plasmids that are useful in gene cloning. Naturally occurring plasmids can be quite large: some are more than 100 kb in size, but the ones used in routine gene cloning tend to be less than 10 kb. This makes them easy to purify and to manipulate. Plasmids commonly used in cloning contain a selectable marker, usually an antibiotic resistance gene. This means that you can tell which bacteria contain the plasmid simply by spreading them onto an agar plate containing the antibiotic. Those that contain the plasmid will grow and eventually

Box 3.1 Plasmid Origins of Replication

Naturally occurring plasmids have been modified by molecular biologists to produce the vectors that we use in gene cloning. Plasmids fall into a number of groups called incompatability groups, depending on where the origin of replication of the plasmids is derived from. Plasmid replication uses many proteins, some encoded by the host bacterium, and some encoded by the plasmid; genes for the latter are often clustered near the origin of replication in the so-called *ori* region. Sequences in the *ori* region also control how many copies of the plasmid there are for each copy of the host chromosome.

Many of the plasmids used in gene cloning, such as pBR322, are based on the ColE1 origin of replication, so-called because they were originally derived from a plasmid called ColE1. Plasmids based on the ColE1 origin of replication are usually medium copy number plasmids with between 12 and 20 copies of the plasmid per cell. The plasmid pUC18 has the ColE1 origin but also has a mutation in the *ori* region which disrupts the mechanism that controls the copy number. The consequence of this is that there can be hundreds of copies of pUC18 for each copy of the chromosome. The ColE1 origin of replication only functions in *E. coli* and closely related species; these plasmids are said to have a narrow host range.

Plasmids based on origins which function in a range of organisms are described as broad host range plasmids. The two main examples here are the R (resistance) and the F (fertility) plasmids. In addition to having a broad host range, these plasmids have a low copy number with between two and five copies for the R plasmids and as few as one copy of the F plasmid per copy of the chromosome.

form a visible colony, and all the cells within that colony will carry copies of the plasmid. Any bacteria that do not contain the plasmid will be killed by the antibiotic, and so cannot give rise to a colony.

Q3.1. If you had a culture of *E. coli*, some containing only pBR322 and some only pUC18, how would you select only those with pBR322?

Q3.2. Name two features of pBR322 that make it useful as a cloning vector.

3.3 Restriction Enzymes

If you are going to clone DNA you need a way of cutting it up. While DNA is fairly easy to damage (just shaking a tube containing large DNA molecules will soon reduce them to smaller ones) for gene cloning you need to be able to cut DNA up in a precise and repeatable way. This can be done using enzymes, which are naturally produced by bacteria, and which cut DNA whenever a particular sequence of bases occurs. These are called restriction enzymes or restriction endonucleases, a name that derives from the normal function of these enzymes in the bacteria from which they are isolated (Box 3.2). One of the most commonly used restriction enzymes is

Box 3.2 Restriction Enzymes

The discovery of restriction enzymes arose from work with bacteriophage (viruses which infect bacteria). It had been known since the 1950s that bacteria are more susceptible to infection by bacteriophage which have been grown on the same strain, than bacteriophage which have been grown on another strain. This phenomenon was known as host-controlled restriction, because the bacteriophage was restricted in which host strain it could infect. It was later discovered that this host-controlled restriction was due to the production by the bacterial host of enzymes that degrade phage DNA. These enzymes recognize a specific DNA sequence and cut the DNA at that sequence. The bacterium's own DNA is protected from being cut by these enzymes by being methylated at the site where the enzyme binds. There are three classes of restriction enzymes, but it is primarily the type II restriction enzymes that are used in gene cloning. These enzymes were called restriction enzymes because they are the enzymes responsible for host-controlled restriction. The DNA sequence they recognize is called a restriction site and the fragments of DNA produced by cutting with these enzymes are called restriction fragments. In 1978 Werner Arber, Hamilton Smith and Daniel Nathans, were awarded a Nobel Prize for the discovery of restriction enzymes and their application to problems of molecular genetics.

Box 3.3 How Do Restriction Enzyme Get Their Names?

Restriction enzymes are named after the bacteria they are isolated from. *Eco*RI was isolated from *Escherichia coli* strain RY13. The first part of the restriction enzyme name is derived from the name of the organism it was isolated from, and it is made up of the first letter of the genus name (E for *Escherichia*) and the first two letters of the species name (co for *coli*). This part of the name is usually written in italics or underlined, following the same rules that apply to writing scientific names of bacteria. The rest of the name refers to the strain and if more than one restriction enzyme is isolated from the same strain they are numbered sequentially with roman numerals. This results in a complex name for most restriction enzymes involving a combination of upper and lower case letters, italicization or underlining and roman numerals. Some scientific publications have recently decided not to italicize the first part of the name of restriction enzymes, at the moment this has not been generally accepted; we will use the established nomenclature for restriction enzymes in this book.

produced by the bacterium *E. coli*; it is called *Eco*RI (Box 3.3). It cuts DNA whenever the sequence GAATTC occurs. This sequence occurs once in the plasmid pBR322 (Figure 3.1).

Have a look at Figure 3.2: there are a number of points that you should notice about the sequence cut by *Eco*RI. The sequence, which is recognized and cut by the enzyme, is 6 bp long; we say that *Eco*RI has a six base pair recognition site. This sequence is an inverted repeat. This means if you read the sequence on the top strand from 5′ to 3′ it is the same as the sequence on the bottom strand also read from 5′ to 3′. *Eco*RI cuts both strands of the DNA and produces a staggered break (Figure 3.2b); this produces sticky or cohesive ends. You will see why this is important when we consider how to join pieces of DNA together. The pieces of DNA produced

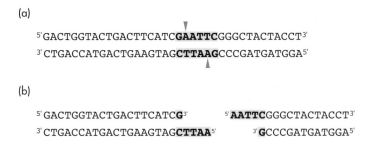

Figure 3.2 a) A section of double stranded DNA showing the *Eco*RI recognition site; the positions where the enzyme cuts are indicated with arrows. b) The result of cutting of the molecule by the restriction enzyme.

by treatment of DNA with restriction enzymes are often called restriction fragments.

> **Q3.3.** If you have a circular plasmid containing a single *Eco*RI site, and you cut it with *Eco*RI, how many pieces of DNA will be formed? What about a circular plasmid containing two *Eco*RI sites? What about a linear piece of DNA, containing one *Eco*RI site?

Many bacteria have been examined for the presence of restriction enzymes and a large number of these enzymes have been isolated. They have a range of different properties, and recognize a wide variety of different sequences, many of which are useful in gene cloning. We shall examine some key restriction enzymes in more detail after we have considered how to join DNA molecules together.

3.4 DNA Ligase
Having described a way of cutting DNA molecules up we now need to consider how to join them together in a new combination. The new molecule is called a recombinant. If the DNA has been cut up using a restriction enzyme like *Eco*RI, which produces sticky ends, then when two molecules with the same sticky ends come into contact, hydrogen bonding between the complementary bases will cause the molecules to stick together. This is in fact why these molecules are said to have sticky ends. This is not a very stable arrangement and the two molecules will soon drift apart again. For gene cloning, you need to be able to covalently link the two molecules. The enzyme that is capable of doing this is called DNA ligase.

When two restriction fragments with sticky ends are transiently held together by hydrogen bonding there are in effect two single-stranded breaks in a double-stranded molecule (Figure 3.3); DNA ligase repairs these single-stranded breaks. DNA ligase catalyzes the formation of a covalent phosphodiester bond between the 5′ phosphate on one DNA strand and a 3′ hydroxyl on another. This process requires energy. The most commonly used DNA ligase is a protein produced by a bacteriophage (a virus that infects bacteria) called T4. It uses ATP as an energy source.

A basic cloning experiment involving the cloning of genomic DNA fragments into the plasmid vector pBR322 is outlined in Figure 3.4. The first step is to cut the plasmid vector at a unique restriction site; this will produce a linear molecule. The genomic DNA, from which a fragment is to be cloned, is also cut with the same restriction enzyme to produce linear fragments, which will be of many different sizes depending on where the *Eco*RI sites occur in the DNA. After inactivating the restriction enzymes, the plasmid and restriction enzyme fragments are mixed in the presence of T4 DNA ligase. As shown in Figure 3.4, several possible events can occur in this

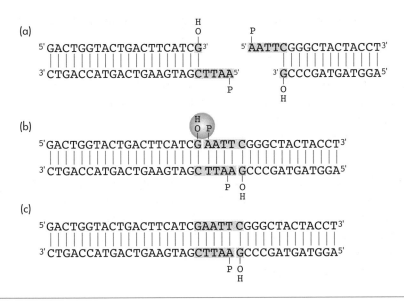

Figure 3.3 a) Two DNA molecules with sticky ends generated by cutting with *Eco*RI, the bases making up the *Eco*RI restriction site are indicated in blue. b) Hydrogen bonding between complementary bases causes the molecules, transiently, to stick together. DNA ligase (indicated by gray shading) catalyzes the formation of a phosphodiester bond between the 5′ phosphate on one molecule and the 3′ hydroxyl on the other. c) The two molecules are now covalently linked by the top strand. The nick in the bottom strand may also be sealed by DNA ligase, or may be repaired by the host bacterium.

mixture. Any molecule with an *Eco*RI sticky end can anneal to any other molecule with the same sticky end, so many fragments of genomic DNA will anneal to each other, in random order. However, these molecules will have neither an origin of replication nor an antibiotic resistance marker and so even if they can be introduced into *E. coli*, they will not lead to the formation of a colony. Another possibility is that the ends of the vector molecules may anneal with each other either reforming the original plasmid (Figure 3.4d) or forming larger molecules with more than one copy of the plasmid. A third possibility, that one vector molecule will be joined to one of the genomic DNA fragments and will circularize to form a new recombinant molecule, is the desired outcome from the cloning experiment (Figure 3.4c). It is possible to arrange the conditions in the ligation reaction so that the likelihood of the formation of such a recombinant molecule is favored. In a dilute solution, the chances of the two ends of the vector molecule coming into contact with each other are higher than the chances of an interaction between two different molecules. Ligation reactions are carried out at high DNA concentrations, typically with a molar ratio of 3:1 of insert to vector (i.e. three times as many insert molecules as

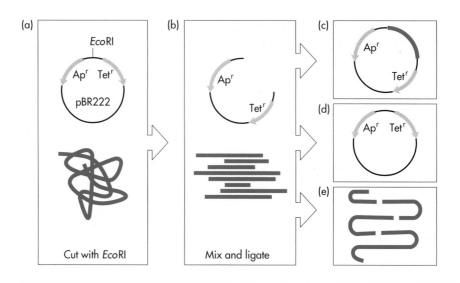

Figure 3.4 A basic cloning experiment designed to clone genomic DNA into the vector pBR222, shown uncut in a). The vector DNA is cut with *Eco*RI into a single linear fragment, the genomic DNA is also cut with *Eco*RI producing a range of fragments (shown in b)). These are mixed and DNA ligase is used to join them together. There are several possible types of recombinant molecule that can result from this. A fragment of genomic DNA can be successfully cloned into the *Eco*RI site in the vector c); this is the desired outcome. Alternatively, the sticky ends of the vector may be rejoined without any genomic DNA insert d), or the genomic DNA fragments may be joined together in a random order e). Only products containing the vector will be able to form colonies on selective media; those containing the vector alone may be distinguished from those containing vector plus insert by further analysis of the size of the plasmid and presence of the insert.

vector molecules), to increase the likelihood of the correct recombinant being formed.

Q3.4. In a DNA ligation, why is it important to inactivate the restriction enzymes before adding the DNA ligase?

Q3.5. List the steps necessary to cut a large DNA molecule up into several restriction fragments with *Eco*RI sticky ends and ligate them into pBR322. What temperature would you use for each step and why?

3.5 Transformation

The final step required in gene cloning is to introduce the new recombinant plasmid into *E. coli*. This process is called transformation, and it involves two steps. First, we need to get the DNA into the bacterial cell and then, because this is an inefficient process, we need to select those cells which contain the plasmid. Some species of bacteria such as *Neisseria gonorrhoea* naturally take up DNA from their environment, and they are described as being naturally competent for transformation. *E. coli*, however, is not naturally competent, and *E. coli* cells need to be treated in a special way to enable them to take up DNA. There are two basic methods for introducing DNA into *E. coli*: chemical treatment and electroporation.

Chemical treatment for the preparation of competent E. coli

To prepare competent *E. coli* a culture is grown and then harvested when it is in log phase, at which stage the bacteria are dividing rapidly. The cells are harvested by centrifugation and washed several times in a chilled buffer containing divalent cations, typically $CaCl_2$. The bacteria are finally suspended in a small volume of the buffer so that they are present at a high density. To introduce DNA (for instance, a recombinant plasmid molecule) into these cells, a small sample of this suspension is mixed and incubated on ice with the ligation mixture; it is then heat-shocked at 42°C for about 1 min (Figure 3.5). This technique has been widely used for many years, although the precise mechanism by which it causes *E. coli* cells to take up DNA is only now becoming clear. During transformation the DNA associates with the lipopolysaccharide on the outer surface of the competent

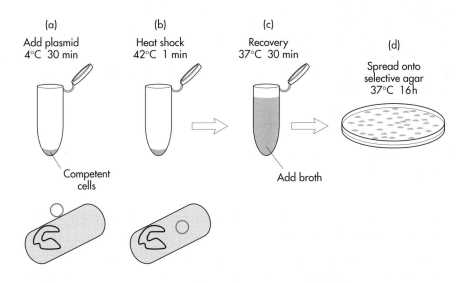

Figure 3.5 Bacteria can be made competent to take up DNA by washing in ice cold $CaCl_2$. a) A very dense suspension of competent bacteria is then mixed with plasmid DNA, during this stage the plasmids become attached to the outer surface of the bacteria. b) After heat shock the plasmids are taken up by some of the bacteria. c) Broth is added to the suspension and the bacteria are allowed to recover before d) being plated onto selective agar to detect those which have taken up the plasmid.

cells, uptake of this DNA is associated with damage to the cell walls caused, in part, by the Ca^{2+} ions and with the heat shock.

Electroporation

An alternative to chemical treatment of *E. coli* to make competent cells is called electroporation. In this process the bacteria are harvested and washed as before but in cold distilled water or a buffer with a very low ionic strength. A small sample of a dense suspension of these bacteria is then mixed with the DNA in a special cuvette and a short pulse of a very high voltage current is passed through the bacterial suspension. Again, it is not entirely clear why this technique works but it is thought that the high voltage pulse makes the bacterial membrane more permeable and possibly moves the DNA into the bacterium by a process akin to electrophoresis (Section 3.11). Electroporation is a much more efficient process than chemical transformation and it can be used with bacteria other than *E. coli*. Electroporation can also be used to transform mammalian and plant cells (see Box 11.4).

Unfortunately, transformation, even by electroporation, is a very inefficient process: only about one in a million of the bacterial cells will successfully take up the plasmids. This is one of the reasons why it is important to have a selectable marker on cloning vectors. Whichever of the two methods above is used, after the DNA has been introduced into the cells bacterial growth medium is then added to the sample and the culture allowed to recover at 37°C for 30 to 60 min. During this time expression of the antibiotic resistance gene will begin. The culture is then spread onto agar plates, containing an antibiotic, and incubated at 37°C until colonies have formed: typically, this will be overnight. Only those bacteria which have been transformed (i.e. have taken up the plasmid DNA) and which are expressing the antibiotic resistance marker will be able to grow on the agar containing the antibiotic.

After incubation at 37°C for a suitable length of time, colonies will be seen on the agar plate. Each of these colonies results from a single bacterium that has divided many times. Because each colony results from a single bacterium and has the same genetic makeup it is called a clone (Figure 3.6). If a given colony results from a single *E. coli* transformed with a recombinant plasmid, each individual bacterium in the colony will contain a copy of the same plasmid. You will thus have successfully cloned your DNA fragment and *E. coli* will have made many exact copies of it.

Q3.6. Transformation is a very inefficient process, only a very small percentage of bacteria take up plasmid DNA. In a typical gene cloning experiment how do you avoid having to screen each colony on the plate to look for those that have taken up DNA?

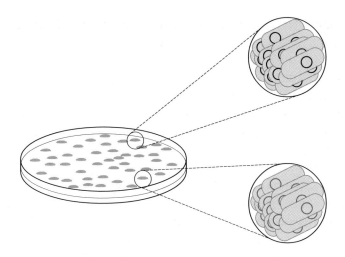

Figure 3.6 Each colony growing on an agar plate is derived from a single bacterium that has grown and divided many times. Each bacterium within a single colony is a clone because, barring mutation, it has exactly the same genetic makeup as the other members of the colony. If a mixed population of plasmids are introduced into bacteria, and the resulting transformants are plated onto agar, all of the cells in a single colony will contain the same plasmid, but this may be different from the plasmids contained in the cells in another colony.

3.6 Purification of Plasmid DNA

It is often necessary to purify plasmid DNA from *E. coli*. For example, after doing a ligation and transformation experiment, you will need to check the plasmid to ensure that it contains the DNA fragment that you are interested in. Further manipulations of the plasmid DNA often follow a successful cloning experiment, and these will also require plasmid purification. Fortunately, the purification of plasmid DNA from bacteria is a routine procedure. It involves growing a culture of cells from a single colony containing the plasmid, harvesting these cells and then breaking them open (referred to as "cell lysis"), removing non-nucleic acid components, and then selectively recovering the nucleic acid by precipitation with ethanol. For some purposes, it is also necessary to separate the plasmid DNA from other nucleic acids (RNA and chromosomal DNA). Traditionally, this second step was achieved by centrifugation on a cesium chloride gradient in the presence of ethidium bromide, a technique sometimes still used for the purification of chromosomal DNA. Which technique you choose depends on a number of factors such as how pure the DNA needs to be, how much of it you need, and how many different samples you need to isolate plasmid from. One of the most straightforward and reliable techniques for purification of plasmid DNA is called alkaline lysis. This technique can be used as a quick method for isolating plasmid DNA for analytical purposes or can be refined to produce large amounts of high quality DNA for cloning. This procedure is often described as a "miniprep" (Figure 3.7), and yields about 1 µg of DNA, which is enough for many analytical purposes such as restriction mapping (Section 3.10).

The "miniprep" method of purification of plasmid DNA can be modified to produce large quantities of DNA of high quality. A larger volume of overnight culture is used and the whole procedure is scaled up accordingly. In addition, a greater effort is made to ensure that there are no contaminating nucleic acids in the preparation. The crude lysate is usually further purified by passing it through an ion exchange column. The column binds nucleic acids, so contaminating protein and carbohydrate are removed. It is then washed with a buffer to remove any RNA and then eluted: the plasmid DNA is in the eluate (see Figure 3.8).

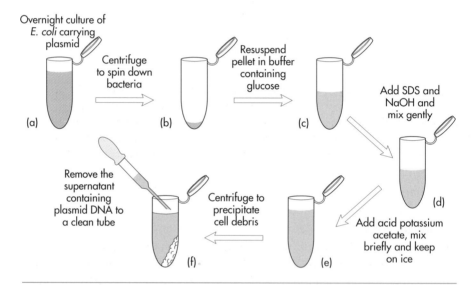

Figure. 3.7 Procedure for "miniprep" of plasmid DNA. The procedure is done on a small volume of cells (typically 1.5 mL of an overnight culture) and yields enough DNA for several restriction digests. a) A single colony is selected from an agar plate and is inoculated into growth medium, with appropriate antibiotics, and grown up overnight. A small sample (usually 1.5 mL) is used in the preparation. b) The bacteria are harvested from the culture by centrifugation, c) resuspended in buffer and d) then are lysed by addition of a solution containing sodium hydroxide and a detergent (sodium dodecyl sulfate or SDS). The DNA is denatured in these alkaline conditions, and the sample becomes very viscous as the bacteria lyse and the DNA is released. e) The addition of ice-cold potassium acetate causes cell debris, including pieces of bacterial membrane, to aggregate; f) these impurities are then removed by centrifugation to produce a "cleared lysate" containing mostly plasmid DNA, RNA, and protein. The bacterial chromosome, which is often attached to the membrane, is also removed in this step. Although plasmid DNA prepared in this way is heavily contaminated with RNA, mainly from ribosomes, the RNA does not interfere with many of the subsequent manipulations that you may wish to do, and indeed helps the DNA to precipitate.

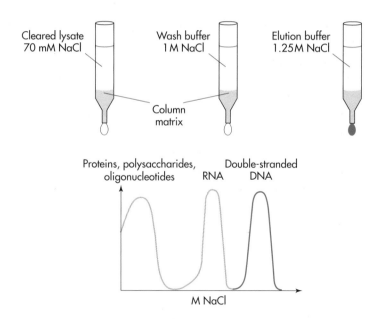

Figure 3.8 Purification of plasmid DNA on an ion exchange column.
The cleared lysate is passed down an ion exchange column. Contaminating
protein and carbohydrate do not bind to the column. The column is then washed
with a solution containing 1M salt, which removes RNA. The double-stranded
plasmid DNA is then eluted from the column with a buffer containing 1.25 M
salt.

3.7 More Restriction Enzymes

In addition to *Eco*RI there is a whole range of restriction enzymes that recognize different base sequences and cut DNA in different ways. Together
these provide an immense amount of flexibility in manipulation of DNA.
We will consider some of these additional enzymes, which exhibit a range
of properties useful in gene cloning.

First, there are a number of restriction enzymes, which, like *Eco*RI, recognize different 6 bp sequences in the DNA. The recognition sequences for
a selection of these are shown in Figure 3.9. Because these enzymes recognize and cut DNA at different sites they can be used in combination to cut
DNA up in very precise ways.

Q3.7. What restriction enzymes would you use to cut the tetracycline
resistance (tet^r) gene out of pBR322?

Restriction enzymes with a 6 bp recognition sequence are often referred
to in shorthand as "six cutters". On average a restriction enzyme with a six
base pair recognition site will cut a DNA molecule every 4^6 bp or once every

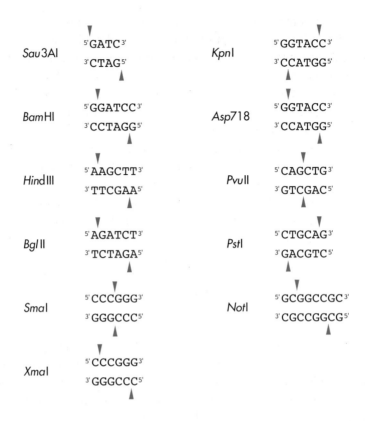

Figure 3.9 DNA sequences recognized by some of the many restriction enzymes commonly used in gene cloning. The arrowheads indicate the position where the enzyme cuts the DNA. Restriction enzymes *Bam*HI, *Bgl*II, *Hind*III, *Pst*I and *Xma*I have six base pair recognition sites and cut to give a staggered break in the DNA resulting in sticky or cohesive ends. *Pvu*II and *Sma*I, also with six base pair recognition sequences cut in the same position on both strands to give blunt or flush ends. *Sau*3AI has a 4 bp recognition site but cuts to give sticky ends.

4096 bp. This means that it is likely to cut a plasmid like pBR322 once but will cut the *E. coli* chromosome, which is 4,639,221 bp long, over 1000 times. This does not mean that the *E. coli* chromosome will be cut by *Eco*RI into restriction fragments of exactly 4096 bp but that there will be over 1000 fragments, varying in size from very small to many kilobases in length. Their average length will be about 4 kb. Some restriction enzymes have longer recognition sequences and some have shorter ones. Six cutters are particularly useful in gene cloning because they cut DNA molecules into fragments in the size range needed for gene cloning. Imagine, for example, that you had cloned a 10 kb piece of genomic DNA from human chromosome 3. You might expect that it would contain unique restriction sites for a number of restriction enzymes with six base pair recognition sites. Using these in combination you would be able to cut it up in a number of different ways, either to isolate different parts of it or to create a restriction map (Section 3.10). Similarly, a typical cloning vector like pBR322 would be expected to have unique restriction sites for a range of six cutters. In fact, pBR322 has 22 unique 6 bp restriction sites. It is important that the restriction sites you want to clone into are unique so that you only cut the vector in one place. If the restriction enzyme cuts at several sites you might loose an important part of the vector.

The problem with most of the unique restriction sites in pBR322 is that they are scattered around the plasmid and some of them are in parts of the plasmid not usually used for cloning. Modern cloning vectors are often engineered so that they have several unique restriction sites close together in one part of the plasmid; pUC18 is an example of this kind of vector (Figure 3.1). A small piece of synthetic DNA, which has on it 10 restriction enzyme recognition sites, has been cloned into the plasmid; this is called a multiple cloning site or polylinker.

Some restriction enzymes have recognition sites that are shorter than 6 bp. *Sau*3AI is an example of one of these; it has a four base pair recognition site and cuts on average every 4^4 bp or 256 bp. There are a number of important applications for restriction enzymes with four base pair recognition sites including cloning control regions like promoters (Chapter 11) and when you need to do a partial digest (Chapter 4).

Q3.8. How many restriction fragments would you predict you would get if you digest pBR322 with *Sau*3AI? Why does the predicted number in this question not exactly match with the actual number?

The problem with trying to clone *Sau*3AI fragments is that any vector will be cut into many pieces by a restriction enzyme with a 4 bp recognition site, as these sites occur more frequently than those for six base cutters. A simple way round this problem is to cut the vector with a six base cutter restriction enzyme which produces compatible sticky ends to those of *Sau*3AI. Both *Bam*HI and *Bgl*II cut to give the same sticky ends as *Sau*3AI (Figure 3.10). Restriction fragments generated by cutting genomic DNA with *Sau*3AI can thus be cloned into a vector that has been cut with either *Bgl*II or *Bam*HI.

Q3.9. Suppose you isolate a DNA fragment which has been produced by cutting a DNA molecule with *Sau*3AI, and ligate it with a vector which has been cut with *Bam*HI. Would you always expect to be able to cut the new plasmid with *Bam*HI and recover the *Sau*3AI fragment?

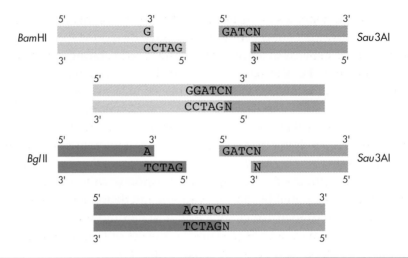

Figure 3.10 Ligation of sticky ends produced by the four cutter *Sau*3AI with those produced by six cutters *Bam*HI and *Bgl*II. Notice that the recombinant molecule can be recut with *Sau*3AI but not necessarily with *Bam*HI or *Bgl*II.

Q3.10. Why is it important that a cloning vector should have a number of unique restriction enzyme recognition sites?

Most restriction enzymes cut DNA to give sticky ends with a 5′ overhang, which means that the single-stranded piece of DNA left after cutting has a phosphate at the end. Others, like *Kpn*I, leave a 3′ overhang with a hydroxyl group at the end (Figure 3.11). The enzymes which result in a 5′ overhang are more commonly used in gene cloning because they are a better substrate for DNA ligase.

All of the enzymes which we have talked about so far cut DNA to give sticky or cohesive ends. There are also restriction enzymes that cut in the

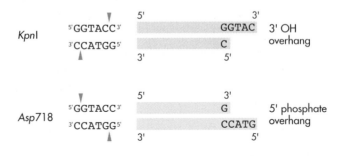

Figure 3.11 The isoschizomers *Kpn*I and *Asp*718 cut the same site but give different single-stranded extensions.

same place on both strands of the DNA, and hence give blunt or flush ends (see *Sma*I in Figure 3.9). These blunt or flush ends are more difficult to join together with ligase because there is no transient base pairing to hold the molecules together. T4 DNA ligase is capable, however, of ligating blunt-ended molecules if used at a higher concentration and there are a number of situations in which these restriction enzymes are particularly useful. Restriction enzymes, which cut to give blunt ends, are in fact very versatile. Because there is no transient base pairing of compatible sticky ends required for a successful ligation, you do not have to use the same restriction enzymes to generate the fragments that you want to ligate together. Thus a fragment generated by cutting with any restriction enzyme giving blunt ends can be ligated to a fragment cut with any other blunt-ended cutter. Also, as we will see later, it is possible to convert restriction fragments from blunt- to sticky-ended by adding linkers (Box 4.3).

> **Q3.11.** You have a restriction fragment generated by digestion with *Pvu*II which you want to clone into the multiple cloning site in pUC18. Which restriction enzyme would you cut the vector with?

Bacteria sometimes produce restriction enzymes that recognize the same sequence of bases as those from other bacteria; these enzymes are called isoschizomers. Some pairs of isoschizomers, although they recognize the same sequence, cut in different positions. *Sma*I and *Xma*I are isoschizomers, which both recognize the sequence GGGCCC, but whereas *Sma*I cuts to give blunt ends, *Xma*I cuts to give sticky ends. In some situations it would not matter which of these enzymes you use but in others you might choose one in preference to another. Another pair of isoschizomers is *Asp*718 and *Kpn*I, and whereas cutting with *Kpn*I leaves a 3′ overhang with a hydroxyl group at the end (Figure 3.11) *Asp*718 leaves a 5′ overhang which, as discussed above, is preferable in gene cloning because it is a better substrate for DNA ligase.

Some restriction enzymes cut DNA much less frequently than the ones we have looked at so far; these are often referred to as "rare cutters". Some "rare cutters" cut infrequently because they have a long recognition site for example *Not*I which recognizes the 8 bp sequence GC^GGCCGC. Others cut rarely because they recognize a sequence which is not commonly found in DNA. These restriction enzymes are useful when you want to cut whole chromosomes or whole genomes up into a few large fragments for applications like RFLP mapping (Box 6.3) and genome sequencing (Chapter 7).

3.8 Alkaline Phosphatase

In Section 3.4 we discussed the ways in which the conditions under which the ligation reaction is carried out can increase the chances of forming a

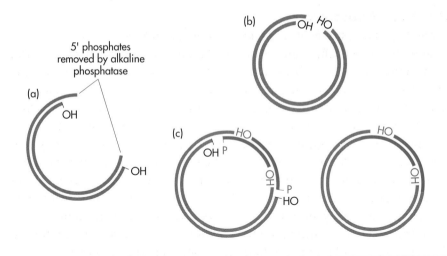

Figure 3.12 Using alkaline phosphatase to prevent vector molecules recircularizing without an insert. a) Treatment of the vector molecule with alkaline phosphatase removes the 5′ terminal phosphate from each end, b) even if the vector molecule transiently recircularizes, DNA ligase will not be able to join the ends. The DNA insert to be cloned into the vector is not treated with alkaline phosphatase, it has its 5′ terminal phosphates intact. c) DNA ligase can form one covalent bond at each join, leaving a single-stranded break in each strand. The molecule is held together by hydrogen bonding until the nicks are repaired in the bacterial host.

recombinant. These can be further improved by treating the cut vector in such a way that the two ends cannot be ligated to each other, but can still be ligated with an insert. Remember from Section 3.4 that DNA ligase requires a 3′ OH and a 5′ phosphate before it can covalently link two DNA molecules. By removing the 5′ phosphates from the cut vector it can be prevented from ligating with itself. This can be achieved using the DNA-modifying enzyme alkaline phosphatase, the most common form of which is calf intestinal alkaline phosphatase (CIAP). Alkaline phosphatase removes the terminal 5′ phosphate from DNA molecules; treating lin-earized vector with this enzyme will remove a 5′ phosphate from each end resulting in a vector which cannot recircularize (Figure 3.12b). As long as the insert molecules have their 5′ phosphates intact they can be ligated with the 3′ OH on the vector molecules (Figure 3.12c). The resulting molecule will still have two single-stranded breaks, one in each strand, but these will be repaired by the bacterial host's DNA repair mechanism once they have been transformed.

Q3.12. It is very important to inactivate the alkaline phosphatase before you mix the vector and fragment. Can you see why this is?

3.9 More About Vectors

Detecting successful ligation by insertional inactivation

There are a few useful features of pBR322 and pUC18 which we have not considered yet. Although restriction enzyme digestion is a fairly reliable process, neither ligation nor transformation is very efficient. We have discussed the ways in which the conditions under which the ligation reaction is carried out can increase the chances of forming a recombinant. In the example shown in Figure 3.4, selection for either ampicillin or tetracycline resistance will allow only the growth of clones containing either the original plasmid pBR322 or the recombinant plasmid with a fragment of genomic DNA cloned into the *Eco*RI site. These two possibilities can only be distinguished by examining the plasmids themselves, generally by preparing them and cutting them with a restriction enzyme or enzymes to compare sizes and number of fragments with the original vector.

Plasmids like pBR322, which have more than one antibiotic resistance gene in them, offer an alternative cloning strategy, called insertional inactivation, which makes it possible to distinguish between recircularized vector and recombinant plasmids. If you clone into the *Bam*HI site of pBR322 (see Figure 3.1), the tetracycline resistance gene will be interrupted, such that the plasmid will still code for ampicillin resistance but not for tetracycline resistance. You can distinguish a clone carrying recircularized vector, which is resistant to both ampicillin and tetracycline, from a clone carrying a recombinant plasmid, which is ampicillin resistant and tetracycline sensitive (Figure 3.13a).

To use insertional inactivation of the tetracycline resistance gene in pBR322 to identify recombinant clones you need to identify tetracycline sensitive clones, looking for antibiotic sensitivity is less straightforward than selecting for antibiotic resistance. First, you need to plate your transformation mix onto agar containing ampicillin and incubate overnight. All the clones that have been successfully transformed with either recircularized pBR322 or recombinant pBR322 will be able to grow on this plate. You now need to find out which of these ampicillin resistant colonies are sensitive to tetracycline. To do this it is necessary to plate colonies onto separate agar plates containing ampicillin or tetracycline. This can be done by streaking individual colonies, or by replica plating (Figure 3.13b). Any of the colonies, which do not grow on the agar containing tetracycline, can be recovered from the ampicillin plate (Figure 3.13b). DNA can now be prepared from these and checked for the presence of the correct sized insert in the vector.

Insertional inactivation of antibiotic resistance genes is a very useful technique, but because you are looking for a sensitive phenotype (ampicillin resistant, tetracycline sensitive as opposed to ampicillin resistant) a second selection involving an additional overnight incubation of the plates is required. To avoid this, insertional inactivation has been further refined

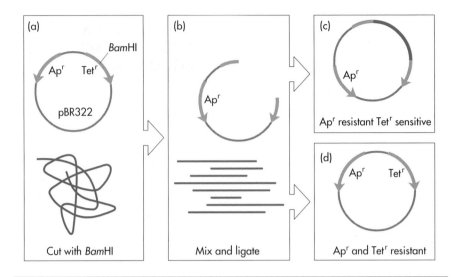

Figure 3.13a In a basic cloning experiment using insertional inactivation of the tetracycline gene of pBR322, the plasmid and genomic DNAs a) are cut with *Bam*HI, mixed and ligated b). In successful recombinants the tetracycline gene will be non-functional because it has been interrupted by the cloned fragment c); however, if the vector recircularizes the tetracycline gene will be functional d).

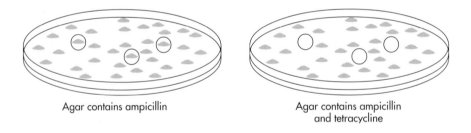

Figure 3.13b Recombinant clones which are ampicillin resistant and tetracycline sensitive can be identified by replica plating. In replica plating a sterile velvet pad is applied to the surface of the plate to be replicated, small amounts of the material from each bacterial colony adhere to the pad, the pad is then applied to a fresh agar plate and the bacterial material is transferred to the new plate, effectively inoculating it in the same way as the original. After overnight incubation the new plate is a copy of the original. If a different antibiotic selection regime is used on the second plate it is possible to detect colonies which are sensitive to an antibiotic, as they will not grow on the replica plate, and also to recover them from the original plate. The colonies circled on the figure are sensitive to tetracycline but can be recovered from the original plate containing only ampicillin.

in the plasmid pUC18 and its many derivatives to allow identification of the recombinants in a single step. These plasmids make use of a gene called *lacZ*, which encodes the enzyme β-galactosidase. The normal substrate for this enzyme is lactose, which it breaks down into glucose and galactose. However, it can also break down a compound called 5-bromo-chloro-3-indolyl-β-D-galactoside or X-gal for short. The useful thing about X-gal is that when it is broken down it produces a blue color. This means that clones with a functional *lacZ* gene can easily be identified because they are blue in color. The family of plasmids to which pUC18 belongs in fact only contains a part of the *lacZ* gene that encodes a fragment of the enzyme called the α peptide and this partial gene is called *lacZ'*. These plasmids are used in conjunction with specially engineered host strains, which supply

---

Box 3.4 Host Strains and Plasmids for Blue–White Selection

Blue–white selection is based on the *lacZ* gene, which encodes β-galactosidase. The normal function of this enzyme is to break down lactose into its components, glucose and galactose. It is also capable of breaking down the artificial substrate X-gal (5-bromo-chloro-3-indolyl-β-D-galactoside) to produce a product with a blue color. In gene cloning experiments a gratuitous inducer of the *lacZ* gene is often included. This compound, IPTG (isopropylthiogalactopyranoside), switches on expression of the gene but is not broken down by the β-galactosidase.

Insertional inactivation of *lacZ* is used in many cloning vectors as a way of identifying recombinant clones. This selection requires the use of special *E. coli* host strains. *E. coli* are naturally *lac⁺* (i.e. they can use lactose as a sole carbon source) in fact, this is one of the biochemical characteristics routinely used in the identification of *E. coli*. For blue–white selection to work special host strains, which have had the chromosomal β-galactosidase gene inactivated, are needed. These *lac⁻* strains are white when grown on agar containing X-gal and IPTG.

Incorporating the whole of the rather large *lacZ* gene into cloning vectors would be impractical. However, the β-galactosidase protein can be provided in two parts, which can associate with each other to form a functional β-galactosidase inside the cell. The part of the gene coding for the smaller N-terminal part of the β-galactosidase protein, the α peptide, is incorporated into the cloning vector and the part of the gene encoding the remaining C-terminal portion of the protein is provided by the host strain, usually encoded on a large stable plasmid called the F' plasmid. Host strains used for blue–white selection therefore have a deletion of their chromosomal *lacZ* gene but do carry the part of the gene encoding the C-terminal part of the protein.

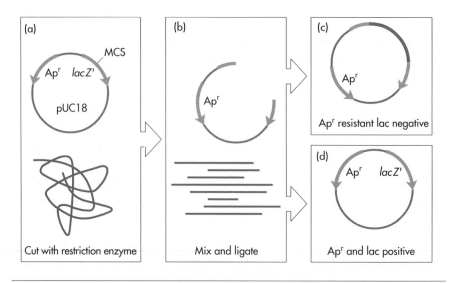

Figure 3.14a In a cloning experiment using insertional inactivation of *lacZ'* in pUC18 the plasmid and genomic DNAs a) are cut with a restriction enzyme which cuts in the multiple cloning site (MCS), and then are mixed and ligated b). In successful recombinants, *lacZ'* will be non-functional because it has been interrupted by the cloned fragment c); if the vector recircularizes the *lacZ'* gene will be functional and β-galactosidase will be produced.

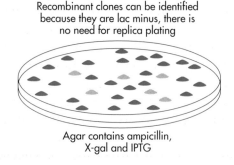

Figure 3.14b Recombinant clones can be identified because they do not express β-galactosidase, and hence they will be white on plates containing X-gal and IPTG. Colonies containing recircularized vector will be blue.

the rest of the enzyme (Box 3.4). Both the α peptide and the rest of the β-galactosidase enzyme are required for activity. The multiple cloning site in pUC18 is in the middle of the *lacZ'* gene so cloning into it will result in insertional inactivation of the gene and the resulting clones will be unable to break down X-gal. These recombinant clones will be white, and are easily distinguishable from those containing the recircularized vector, which will be blue (Figure 3.14a and b).

Q3.13. The shorthand way of describing the antibiotic resistance pheno-type of bacteria which have been transformed with a plasmid carrying an antibiotic resistance gene is to write ap^r (ampicillin resistant) or tet^r (tetra-cycline resistant), or if they are sensitive to the antibiotics ap^s, tet^s. If bacte-ria are expressing β-galactosidase they are said to be lac positive, which can be written as lac$^+$, if not then lac negative or simply lac$^-$. Write down the shorthand for the phenotype of: (a) *E. coli* transformed with pBR322; (b) *E. coli* transformed with a recombinant plasmid derived from pBR322 but with an insert cloned into the tetracycline gene; (c) A strain of *E. coli*, with a deletion of the portion of the *lacZ* gene which encodes the α peptide of β-galactosidase, transformed with pUC18.

Q3.14. What color would the colonies of the ap^r lac$^+$ transformants described in (c) above be if plated on ampicillin and X-gal? What color would they be if a fragment of DNA had been cloned into the multiple cloning site of pUC18?

Vectors for specific purposes

One other major difference between pBR322 and pUC18, which we have not yet mentioned, is the copy number. This is the number of copies of the plasmid in each bacterium, and is normally expressed as the number of copies of the plasmid per copy of the bacterial chromosome. The number of copies of a plasmid in each bacterium is regulated by the origin of repli-cation (Box 3.1). Whereas pBR322 is a medium copy number plasmid with between 12 and 20 copies per bacterial chromosome, pUC18 is a high copy number plasmid with as many as 100 copies per chromosome. For some applications it may be advantageous to have as many copies of your recombinant plasmid as possible in each bacterial cell, in which case a high copy number plasmid such as pUC18 would be a good choice of cloning vector. In other cases, for instance where the genes encoded on the recom-binant plasmid are expressed and the protein produced places a burden on the host bacterium, a plasmid with a copy number closer to pBR322 may be a better choice. If the genes you are trying to clone place a very heavy bur-den on the host or if they encode a toxic product you may choose a very low copy number plasmid based on either the F or R plasmids (Box 3.1) or you may use a plasmid which facilitates control of expression of the cloned gene (see below). Plasmids which are based on the ColE1 origin of replica-tion are described as having a narrow host range as they can only be repli-cated in *E. coli* and closely related bacterial species. Plasmids based on the F and R plasmid origins, however, have a broad host range and can func-tion in a range of Gram negative bacteria.

There is a bewildering array of vectors available for gene cloning. They are derived from naturally occurring plasmids but have been engineered to fulfill many specialized functions. Two of the features of pUC18, the multiple cloning site and the facility to use blue–white selection, have been included in many different vectors. There are many different multiple cloning sites offering a wide range of unique restriction sites for cloning into and these are incorporated into a wide range of vectors. The α peptide of the *lacZ* gene is also incorporated into many vectors, either for blue–white selection for DNA inserts as in the pUC vectors, or as a reporter gene for monitoring gene expression (Chapter 10).

In constructing DNA libraries from eukaryotic organisms with large genomes (Chapter 4) it is often necessary to be able to clone very large DNA fragments. Standard plasmid cloning vectors tend to become unstable with very large inserts; however, plasmids have been constructed, based on the F plasmid origin of replication which can accommodate up to 300 kb of insert DNA, these so called bacterial artificial chromosomes (BACs) are discussed in detail in Section 4.16.

Many applications of gene cloning involve cloning a gene in order to express the gene and to produce the protein which it encodes. This is true of many biotechnological and medical applications of gene cloning where commercially useful or therapeutic proteins are produced from cloned genes (Chapters 9 and 13) but is equally true in a research context where gene cloning is an important tool in the functional analysis of proteins (Chapter 10). A whole range of cloning vectors are available where precise control of expression of the cloned gene is possible so that there is no background expression, but when the gene is induced gene expression is rapid and large quantities of protein are produced. These are discussed in detail in Section 9.3.

3.10 Analyzing Cloned DNA by Restriction Mapping

Plasmid vectors that allow you to identify recombinant clones by looking for insertional inactivation of either antibiotic resistance genes or *lacZ* are very useful. However, potential recombinant clones still need to be analyzed by restriction mapping. This means purifying the plasmid DNA from individual clones, cutting it with restriction enzymes, and analyzing the sizes of the fragments produced. To take a simple example, if you have cloned a 1 kb *Eco*RI fragment of genomic DNA into the *Eco*RI site in the multiple cloning site of pUC18 you would expect clones to be resistant to ampicillin and lac negative. If you then purify the recombinant plasmid and cut it with *Eco*RI you would expect to get two fragments, one of 2.7 kb representing the vector and one of 1 kb representing the genomic DNA insert. Equally, if you used one of the other restriction enzymes from the multiple cloning site, for instance *Hin*dIII, you would expect to get one band of 3.7 kb, provided there were no *Hin*dIII sites present in the cloned genomic fragment. However, if *Hin*dIII cuts in the genomic DNA insert you would get more than one fragment, but the sizes of the fragments should always add up to 3.7 kb. This process can be extended using a wide range of restriction enzymes, sometimes alone and sometimes in combination, to build up a picture of the recombinant clone showing where all the restriction sites are located; this is called a restriction map. To do this, a way of measuring restriction fragment sizes is required.

Q3.15. What would be the sizes of the fragments that you would expect to find if you cut this plasmid with (a) BamHI, (b) EcoRI, (c) both BamHI and EcoRI together?

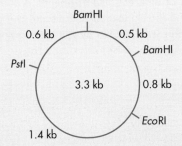

3.11 Measuring the Size of DNA Fragments

The most widely used technique that allows you measure the size of DNA fragments is called gel electrophoresis. In this technique, DNA molecules are separated according to their size in a gel, and actual size is estimated by comparison with marker DNAs of known size. The main material used for gels is called agarose, although polyacrylamide is also often used, especially when the DNA fragments are small. DNA molecules are charged at neutral pH because each phosphate in the DNA backbone contributes one negative charge. In solution, if an electric current is applied, they will move towards the positive electrode or anode. However, because the charge to mass ratio is the same for all DNA molecules, this will not result in separation of different size fragments. However, if instead of a solution the current is applied to a gel with the DNA molecules at one end, the gel will impede the movement of the molecules. Under these conditions the size of the DNA molecules becomes the most important factor as they move through the gel. A gel is essentially a tangled network of polymeric molecules with a series of pores in it; smaller molecules are able to thread their way through the gel at a faster rate than larger molecules so the molecules will be separated according to their relative sizes (Figure 3.15).

Agarose gels are used for separating DNA fragments from 100 bp to 20 kb. Agarose, a carbohydrate polymer, is a highly purified form of agar; it is supplied as a powder. The powder is dissolved in a buffer containing ionic compounds, by heating to boiling point. The molten agarose solution is poured into a gel-forming tray and a comb inserted (Figure 3.16). As the agar cools it forms a gel. The comb is then removed to reveal wells in the

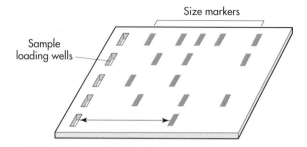

Figure 3.15 An agarose slab gel showing the sample loading wells and the bands formed by DNA samples of different size fragments. A sample containing fragments of known size is usually loaded to calibrate the gel. The distance the band has migrated from the well (shown by the arrow) is proportional to the size of the DNA fragment.

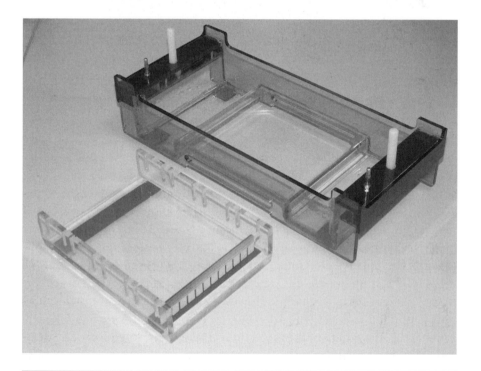

Figure 3.16 Photograph showing submarine apparatus for agarose gel electrophoresis. The molten agarose is poured into the gel forming tray (front) and once the gel is set the gel is placed in the tank (rear).

gel into which samples can be loaded. The gel is placed into a horizontal electrophoresis tank (see Figure 3.16) and submerged in buffer. DNA samples are heated briefly in loading buffer to ensure that there are no inter-molecular interactions. Loading buffer contains dye, which helps with

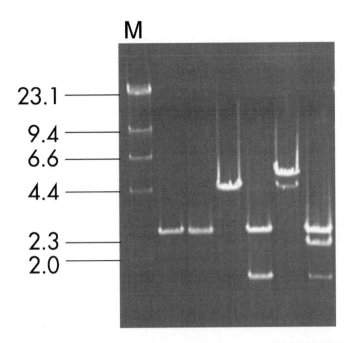

Figure 3.17 Photograph of agarose gel. Lane M shows marker DNA in this case the 6 largest bands produced by cutting bacteriophage λ DNA with *Hind*III. The other lanes show DNA fragments ranging in size from 2 to 7 kb.

monitoring the progress of electrophoresis, and a dense liquid such as glycerol, which ensures that the sample settles into the bottom of the well. The DNA samples are loaded into the wells and a current applied. The DNA molecules thread their way through the gel and are separated according to size. The DNA is visualized by staining with ethidium bromide, which fluoresces under ultraviolet light. This dye intercalates between the bases of DNA, which concentrates it in the gel where the DNA is. As a result these bands fluoresce brightly under ultraviolet illumination. This staining technique is very sensitive; as little as 0.05 μg of DNA in one band can be detected. Despite the fact that ethidium bromide is a mutagen and ultraviolet light can cause severe burns this is still the most commonly used technique for visualization of DNA.

The distance that the band has migrated from the well is a measure of the size of the DNA fragments. Gels are calibrated by running a sample of DNA, which contains a series of restriction fragments of known size, in parallel to the experimental samples. A commonly used DNA size marker is DNA from the bacteriophage lambda, cut with *Hind*III, which gives a recognizable pattern of eight bands ranging in size from 125 bp to 23.13 kb (Figure 3.17). The relationship between the size of the DNA fragments and the distance the bands migrate is not a linear one; rather, the distance migrated through the gel is proportional to the $\log_{10}$ of the molecular weight. If you plot the $\log_{10}$ of the size of the DNA fragments against the distance migrated, you will obtain a straight line (with some deviation at the extreme ends of the size range, owing to other factors). The sizes of bands from other samples can then be read off the line.

Q3.16. The table shows the distance migrated in an agarose gel by each of the eight fragments produced when λ DNA is cut with HindIII. This can be used as a size marker.

Distance Migrated (mm)	Size of Fragment (kb)
8.5	23
23.5	9.4
29	6.6
36	4.4
46.5	2.3
49	2.9
69.5	0.56
94	0.13

Draw a graph of the $\log_{10}$ of the size of each fragment against the distance migrated. Work out the size of the DNA fragments that would give rise to a band that migrated 60 mm.

A number of factors, such as the composition of the buffers, the size and thickness of the gel and the voltage applied affect the rate of migration of DNA in gels, but they do not affect the *relative* mobility of molecules of different sizes. Most laboratories use a standard set of conditions for agarose gels. The concentration of agarose in the gel affects the average pore size of the gel; this not only affects the rate of migration but also the range of molecule sizes that can be effectively separated on the gel. A 1% agarose gel would be used for molecules ranging from 200 bp to 2 kb, whereas a 0.5% gel would be more suitable for separating larger fragments, say 1 kb to 20 kb.

The final factor that affects the mobility of DNA in a gel is its conformation. Plasmid DNA can exist in one of three forms (Figure 3.18). Plasmid DNA in a bacterial cell exists as a closed circular supercoiled molecule. In this form single molecules are tightly wound up. If these molecules are damaged during extraction, single-stranded breaks may be introduced into the molecules, this causes relaxation of the supercoils to what are referred to as relaxed or open circles, where the DNA is much less highly

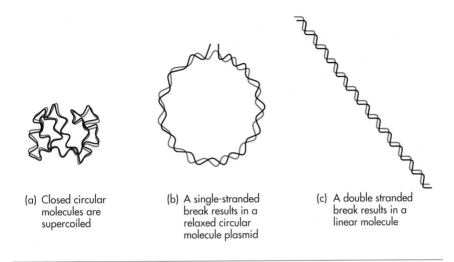

(a) Closed circular molecules are supercoiled

(b) A single-stranded break results in a relaxed circular molecule plasmid

(c) A double stranded break results in a linear molecule

Figure 3.18 The three forms of plasmid DNA: a) closed circular supercoiled; b) relaxed or open circular; and c) linear.

condensed. After cutting with a restriction enzyme with a unique restriction site, the plasmid will be linearized. Because the shapes of these three conformations are different their electrophoretic mobility will be different. It is difficult to predict where the different conformational isomers will run relative to each other on a gel and it is usual to run a sample of uncut plasmid on the gel when trying to determine whether a particular restriction enzyme cuts a plasmid. For some applications such as examining plasmid profiles of different bacterial strains, uncut DNA is analyzed; in this situation it is important to use size markers made from uncut plasmids rather than from cut DNA.

Agarose gels can be used preparatively to purify DNA as well as to analyze it. Sometimes it is necessary to purify a linear restriction fragment of DNA for further manipulation. For example, if you wish to clone a fragment from a complex digest (i.e. one that has several different sized fragments in it), it is sometimes a good idea to purify the fragment of interest first, to avoid a high frequency of colonies after ligation and transformation that contain clones with the fragments which are not of interest. One of the most straightforward ways of purifying a restriction fragment is to cut a plasmid with restriction enzymes and to run the sample on an agarose gel. The gel is run until the fragment to be purified is well separated from the other fragments. DNA is visualized under UV after staining the gel with ethidium bromide, and the section of agarose containing the desired band is cut out from the gel and the DNA extracted from the agarose. Once the restriction fragment has been purified from the agarose it can be ligated with vector in the normal way.

Q3.17. Imagine you have a 3kb *EcoRI* fragment of DNA cloned into the *EcoRI* site in the multiple cloning site of pUC18, and that you know that there is a *BamHI* site in the fragment, which divides it into unequal pieces of 1 and 2 kb. You want to subclone the 1 kb *EcoRI BamHI* fragment into pBR322. Outline the steps you would use to perform this operation. *Hints.* (i) First draw out the plasmid you want to subclone from, and (ii) there are at least two sensible approaches to this problem: in one you will need to use preparative agarose gel electrophoresis and in the other you may use a combination of antibiotic selection and restriction mapping to identify the clone of interest.

Q3.18. Suppose that in cloning fragments of genomic DNA generated by *EcoRI* digestion into the plasmid pBR322, the sticky ends of two genomic fragments annealed to each other before annealing to the sticky ends of the vector. What would be the consequence of this in terms of what would be recovered after transformation? Would it be a problem? How would you detect this?

3.12 The Polymerase Chain Reaction and Its Use in Gene Cloning

One further technique that needs to be considered in this section is the polymerase chain reaction, or PCR. This technique is more recent than the others but has rapidly evolved from being a specialist technique into a routine cloning tool, as well as having many other important applications particularly in forensics and medicine (Chapter 13).

Earlier we saw how bacteria could be used to make many copies of a piece of cloned DNA. PCR is an *in vitro* technique for synthesizing many copies of DNA. It uses the same enzyme that bacteria use, DNA polymerase, but the reaction is performed in a test tube. There are two key points about PCR. The first is that it allows you to amplify DNA. Starting with a very small sample (as little as one molecule), you can make large quantities of DNA. The second is that the technique allows you to be selective; you can start with the entire genome of an organism but amplify only the part of the DNA that codes for one particular gene.

DNA polymerase (Box 3.5) is the enzyme used by cells to copy their chromosomes prior to cell division. During DNA replication the DNA strands separate and each acts as a template for a newly replicated strand. DNA polymerase makes a complementary copy of each template strand. The reaction requires, in addition to DNA polymerase, a single-stranded DNA template and all four deoxyribonucleotides (Figure 3.19), the biological building blocks of DNA. DNA polymerase adds deoxyribonucleotide units to the 3' hydroxyl terminus of a DNA chain. Synthesis cannot begin unless there is already a short double-stranded region at the beginning of the

Box 3.5 DNA Polymerases

DNA polymerases add deoxyribonucleotide units to the 3′ hydroxyl terminus of a DNA chain; synthesis being directed by a single stranded template. DNA polymerase I from *E. coli* has been used extensively in gene cloning. In addition to its polymerase activity it also has 3′ to 5′ exonuclease activity and 5′ to 3′ exonuclease activity. The 3′ to 5′ exonuclease activity has an important role in proofreading; the terminal 3′ nucleotide can, if it is incorrectly base paired, be removed from the growing DNA chain, and the polymerase then has a second chance to add the correct nucleotide. The form of this enzyme commonly used in gene cloning is called the Klenow or large fragment of DNA polymerase I; it lacks this 5′ to 3′ exonuclease activity of the native enzyme and is useful when you wish to avoid DNA degradation.

PCR reactions require repeated cycles of heating and cooling, which would destroy enzymes like Klenow. To avoid having to add fresh enzyme for each cycle thermostable enzymes like *Taq* DNA polymerase are used. The first thermostable DNA polymerase was isolated from the bacterium *Thermus aquaticus*; because this organism lives in hot springs its enzymes are adapted to high temperatures. *Taq* polymerase has an optimum temperature for activity of 72°C and is not denatured at temperatures in excess of 100°C so it is ideal for use in PCR. One of the disadvantages of *Taq* polymerase is that it has no proofreading capability; for some applications high fidelity DNA polymerases such as Vent™ and Pwo are preferred as they are less prone to error.

stretch to be copied. In the PCR reaction this is provided by the addition of a short synthetic single-stranded DNA molecule, which will anneal to the template DNA providing a double-stranded region with a 3′ hydroxyl

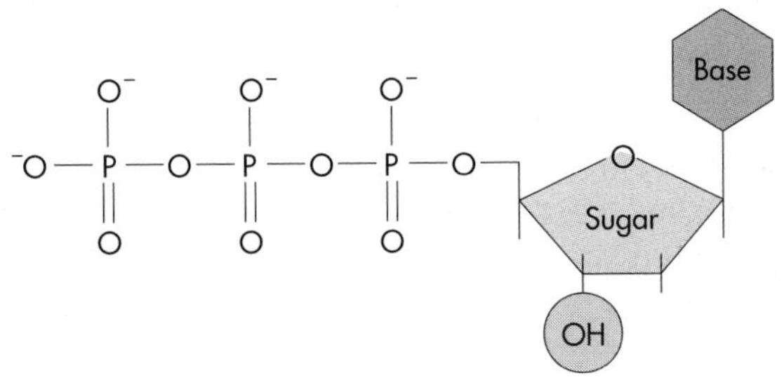

Figure 3.19 Diagram of a nucleoside triphosphate. This basic building block of DNA consists of a 5 carbon ribose sugar with a base, which can be adenine, guanine, cytosine or thymine, at the 1′ position, a hydroxyl group at the 3′ position and a triphosphate at the 5′ position.

group. This synthetic molecule primes DNA synthesis and is often referred to as a PCR primer (Box 3.6). It is the requirement for a region of double-stranded DNA, to initiate DNA synthesis, which has been exploited to make the PCR reaction specific; copies can only be made of DNA regions specified by the primers.

Box 3.6 Oligonucleotides as Probes and Primers

Short synthetic single stranded DNA molecules are often referred to as oligonucleotides, or "oligos" for short, although where they are used to prime DNA synthesis as in the case of PCR, they are usually called primers. As we will see, there are a number of important applications for oligonucleotides in gene cloning. Not only are they used as primers in PCR and DNA sequencing but they can also be used as gene probes. The chemical synthesis of short stretches of single stranded DNA molecules of any sequence up to about 100 nucleotides in length has become a routine and relatively cheap procedure; there are many commercial companies who supply oligonucleotides to order. The process of chemical synthesis is a very different process from that used by DNA polymerase and the sequence of the bases can be specified without the requirement for a template.

Single stranded DNA molecules in solution will anneal to other DNA molecules with a complementary sequence of bases. Hydrogen bonds will form between complementary bases and a double stranded duplex will be formed. The strength of the interaction and hence the stability of the duplex will depend on the number of hydrogen bonds formed between the two molecules. If you mix genomic DNA, which has been denatured to form single strands, with an oligonucleotide, the oligonucleotide will anneal to its complementary sequence wherever that sequence occurs. If the solution is then heated gradually there will come a point where the duplex is no longer stable and the oligonucleotide will dissociate from the genomic DNA. This temperature is called the melting temperature (I_m) and can be calculated for any DNA molecule. The important factors that affect the melting temperature are: the proportion of GC base pairs, and the length of the complementary region. The buffer in which the DNA molecules are dissolved, or more specifically the concentration of monovalent cations (usually Na^+ or K^+) is also important. For some applications of oligonucleotides changing the buffer is used as a way of affecting the melting temperature.

There are many formulae for calculating melting temperatures of oligonucleotides, these need to take into account the proportion of GC to AT bases, the length of the oligo and the salt content of the buffer that the it is dissolved in. These formulae are quite complex, however they can be greatly simplified by assuming that the buffer conditions are standard. The simplest formula (shown below), works well for short oligonucleotides and assumes that the reaction is carried out in the presence of 50mM monovalent cations. This formula simply takes account of how many GC and how many AT base pairs there are.

$$T_m \ ^{\circ}C = 4 \times (G + C) + 2 \times (A + T)$$

where G + C = number of GC base pairs and A + T = number of AT base pairs. For example, to calculate the melting temperature for the oligonucleotide: $^{5'}$GATAAGCCAGGCTACCTTAGC$^{3'}$

$$
\begin{aligned}
T_m \ ^{\circ}C &= 4 \times (5 + 6) + 2 \times (6 + 4) \\
&= 4 \times 11 + 2 \times 10 \\
&= 44 + 20 = 64^{\circ}C
\end{aligned}
$$

It is important when designing primers and gene probes that the sequence is unique so that they will only anneal to one genomic sequence. It is possible to work out how often a sequence is likely to occur in the genome using the same formula which we used to work out how often restriction enzymes will cut (see section 3.7). A 17bp sequence would be expected to occur, by chance, every 4^{17} bp or every 1.7×10^9 bp. The human genome is 2.95×10^9 bp so a 17 bp primer would, at least in theory, be unlikely to anneal to more than one sequence in the human genome. In practice primers of between 17 and 25 bp are commonly used.

Q3.19. Why is the ratio of GC to AT base pairs an important factor in determining the melting temperature of double-stranded DNA molecules?

Q3.20. Can you think of any features of the human genome that would make it likely that a 17 bp primer would anneal to more than one sequence?

Q3.21. List four components necessary to copy DNA *in vitro*.

While using DNA polymerase to make copies of DNA molecules *in vitro* is a routine technique in gene cloning, the insight which Kary Mullis, the inventor of PCR, had was to see that if you use two primers to define each end of the region which you want to copy, and if you perform repeated cycles of priming and copying, you will amplify the region of DNA in a exponential fashion.

3.13 How Does PCR Work?

First, primers are designed which anneal to sequences at either end of the region to be amplified. It is important that the primers are orientated so

Left hand primer

5'**GTCGACGCTGGTTGTTGTG**3'

5'GTCGACGCTGGTTGTTGTGGGGTGGACAGCGCAGATCGATCCAGAGATGATACTACAGGTT 3'

3'CAGCTGCGACCAACAACACCCACCTGTCGCGTCTAGCTAGGTCTCTACTATGATGTCCAA 5'

3'**GGTCTCTACTATGATGTCCAA** 5'

Right hand primer

Figure 3.20 The sequence of a short stretch of double-stranded DNA is shown together with two oligonucleotide sequences (shown in bold) which could be used as PCR primers. The left-hand primer would anneal to the 3' end of the bottom strand and initiate synthesis in the direction of the right hand primer.

that DNA synthesized from them will extend into the product to be amplified, in other words so that 3' ends of the primers point towards each other and one primer anneals to each strand of the DNA (Figure 3.20). For convenience these primers can be regarded as left- and right-handed primers although this designation is arbitrary.

In the first cycle of the PCR reaction, template DNA is heated to 92°C to cause the DNA strands to separate, thus providing single-stranded template for DNA polymerase (Figure 3.21a). The reaction is then cooled to a temperature at which the PCR primers will anneal to the DNA (Figure 3.21b) and DNA polymerase is used to synthesize a complementary DNA strand (Figure 3.21c). In the first cycle the newly formed product will begin at each of the primers and extend beyond the sequence complementary to the other primer. It will terminate when polymerase falls off the template molecule. If we assume, for the sake of simplicity, that there was only one template molecule in the original reaction, then the products of this first round of DNA synthesis will be two partially double-stranded molecules of indeterminate length but beginning at the positions defined by the PCR primers (Figure 3.21c).

In cycle 2 the temperature of the reaction is again raised to 92°C. The newly formed DNA duplexes will dissociate providing four template molecules for DNA synthesis (shown in black in Figure 3.21d). Two will have sequences complementary to primer 1 and the other two will have sequences complementary to primer 2. As the reaction mixture is cooled new primers will anneal and DNA synthesis will begin again. In cycle 2 the primers will anneal to the genomic template DNA (broken lines) as before and synthesis will proceed for an indeterminate distance (Figure 3.21e). However, in cycle 2 primers will also anneal to the molecules produced in the first round of synthesis. In this case synthesis will begin at the position where the primer anneals but will stop when the template runs out, a point defined by the position of the other primer (Figure 3.21e). These new molecules will be exactly the same length as the target region, bounded by the sequences defined by the two primers, and it is these molecules which are amplified in subsequent cycles of PCR.

Cycle 1

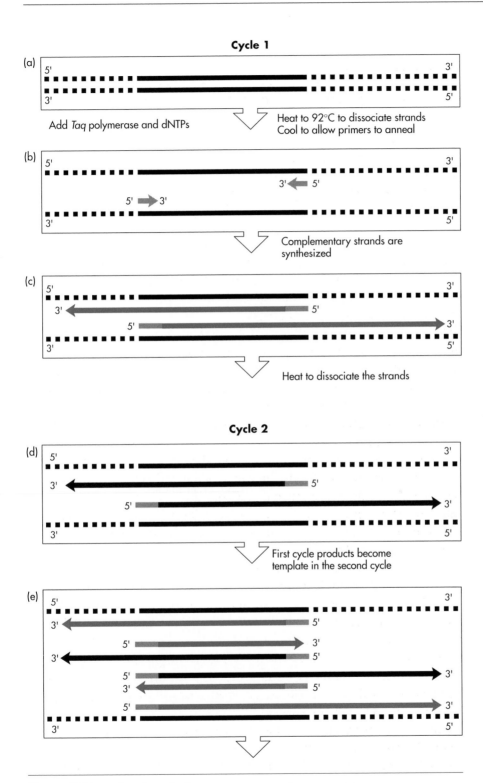

Figure 3.21 *continued overleaf*

Cycle 3

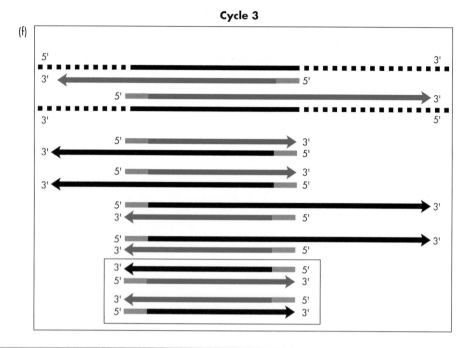

Figure 3.21 Polymerase chain reaction. The target DNA, present at the start of the reaction, is shown in black, and primers in gray; DNA synthesised in each cycle is shown in blue. The region of the target DNA which is to be amplified is shown in unbroken lines; the flanking regions are shown as broken black lines. In the first cycle of the PCR reaction template DNA a) is heated to cause the DNA strands to separate providing single-stranded template for DNA polymerase. The reaction is cooled to a temperature at which the PCR primers will anneal to the DNA b) and DNA polymerase is used to synthesize a complementary DNA strand. c). In the first cycle the newly formed product will begin at each of the primers and extend beyond the sequence complementary to the other primer, it will terminate when polymerase falls off the template molecule. The products of this first round of DNA synthesis will be two partially double-stranded molecules of indeterminate length but beginning at the positions defined by the PCR primers c).

In cycle 2 the temperature of the reaction is again raised to 92°C, the newly formed DNA duplexes will disassociate providing four template molecules for DNA synthesis d). Two will have sequences complementary to primer 1 and the other two will have sequences complementary to primer 2. As the reaction mixture is cooled new primers will anneal and DNA synthesis will begin again. In cycle 2 the primers will anneal to the genomic template DNA as before and synthesis will proceed for an indeterminate distance e). However, in cycle 2, primers will also anneal to the molecules produced in the first round of synthesis. In this case synthesis will begin at the position where the primer anneals but will stop when the template runs out, this will be the position of the other primer e). These new molecules will be exactly the same length as the target region, bounded by the sequences defined by the two primers.

In cycle 3 the process is repeated again f), this time there are eight template molecules. The two molecules from cycle 2, which are the length of the target region, will give rise to a complementary strand, also exactly the length of the target region and defined by the positions of the two primers. These double-stranded molecules are the molecules, which PCR is designed to make (indicated by a box in panel f)

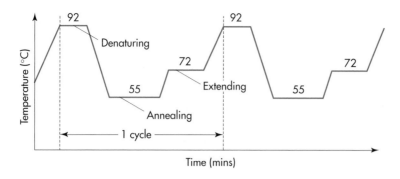

Figure 3.22 A typical PCR reaction involves 30 cycles in which the temperature is raised to 92°C to denature the template, cooled to a temperature suitable for annealing of the primers, 55°C in this example and then raised to 72°C the optimum temperature for Taq polymerase. There are many variations on this basic pattern.

In cycle 3 the process is repeated again (Figure 3.21f). This time there are eight template molecules (shown in black). The two molecules from cycle 2, which are the length of the target region, will give rise to a complementary strand, also exactly the length of the target region and defined by the positions of the two primers (indicated by a box in Figure 3.21f). A graph showing the changes in temperature that take place during a single PCR cycle is shown in Figure 3.22. In all subsequent cycles of PCR there will be three classes of double-stranded molecule produced. The first two are duplexes consisting of one strand of genomic DNA and one newly synthesized strand of indeterminate length (Figure 3.21f), molecules composed of one strand beginning at the position of a primer but with an undefined 3′ end and a second strand in which both ends are defined by the position of the primers. Most important is the third class of product molecule which is double-stranded molecules representing the sequence between the positions of the two primers. While the number of product molecules with one strand of indeterminate length and one strand defined by the positions of the primers will increase in a linear fashion as PCR proceeds, the number of double-stranded target molecules will increase more than two fold in each successive cycle (Table 3.1). It is this exponential increase in the number of target molecules in successive cycles, leading to the amplification of the target sequences, which makes PCR such an efficient technique for copying DNA. Starting with one single template molecule in the first cycle, after 22 cycles you would expect to have over a million target molecules (Table 3.1). In practice you start with many copies of the template, although the process is unlikely to be 100% efficient. Thirty cycles are usually performed to ensure sufficient amplification of the desired region of the template.

PCR requires constant cycling of the reaction mixture between a

Table 3.1. PCR amplification

Cycle Number	Number of Double-stranded Target Molecules
1	0
2	0
3	2
4	8
5	22
6	52
7	114
8	240
9	494
10	1,004
11	2,026
12	4,072
13	8,166
14	16,356
15	32,738
22	1,048,536
30	1,073,741,764

temperature suitable for annealing of the primers and those where the DNA double helix will disassociate. Standard DNA polymerases are denatured by temperatures above 60°C and are not suitable for automated PCR. Kary Mullis's second important insight was to see that use of a thermostable polymerase would remove the need to add fresh enzyme in each cycle and allow for the automation of the process. The enzyme most commonly used in PCR is *Taq* DNA polymerase (Box 3.5).

3.14 Designing PCR Primers
The first step in designing a PCR reaction is to design a pair of oligonucleotide primers. There are a number of important points to bear in mind when designing PCR primers. The major constraints are usually the sequence that you have available and the nature of the target region. The primers should be chosen so that they will anneal in a region of known sequence on either side of the target sequence and so that DNA polymerase will be synthesized from them across the target region. Because DNA

polymerase adds deoxyribonucleoside units to the 3′ hydroxyl terminus of a DNA chain this means that the primer sequences will be derived from two regions of sequence on opposite strands with the 3′ ends of the primers orientated towards each other (Figure 3.20). Primers should be at least 17 bp long to ensure that the sequence is likely to be unique and would normally have a melting temperature between 50 and 65°C (Box 3.6). Primers are usually designed as a pair as the PCR reaction works better if the two PCR primers have similar melting temperatures. It is also important that the primers should not anneal to each other or have significant internal complementarity, which would interfere with the annealing to the target DNA. It has been shown experimentally that primers with a G or C at the 3′ end work well because the stronger hydrogen bonding in a GC base pair ensures that the 3′ bases remain paired; this is often referred to as a G clamp. There are computer programs that aid in the design of PCR primers: given DNA sequence data they will list pairs of primers, which match your criteria. However, many workers still prefer to design primers by close inspection of DNA sequence.

Q3.22. In the DNA sequence in Figure 3.20, the positions of two oligonucleotides are marked. What is the melting temperature, T_m, for each of these oligonucleotides? What length would the PCR product primed by these two oligonucleotides be?

3.15 The PCR Reaction

PCR reactions are carried out in specialized machines called thermal cyclers. These machines are designed to hold small thin walled disposable plastic tubes or special 96-well plates in a solid metal block. The machines are programmable and will cycle through a series of incubations at different temperatures. Modern machines use a piezoelectric heating and cooling system, which allows the temperature to be altered very quickly and achieves very accurate and reproducible temperature shifts. A typical PCR reaction contains a small amount of DNA template, the dNTPs required for DNA synthesis, an excess of PCR primers and *Taq* polymerase.

The PCR program involves cycling through three temperatures. In step 1, the temperature is raised to 92°C which causes the two strands of the template DNA to separate. In step 2, the temperature is then lowered to a temperature suitable for annealing of the primers. This temperature is usually about 5°C lower than the melting temperature of the primers and will be different for every pair of primers. In step 3, the temperature is then raised to 72°C, which is the optimum temperature for activity of *Taq* polymerase, and the reaction is incubated for a short while to allow *Taq* to extend the DNA molecule. These steps are repeated 30 times during which the DNA is amplified many times. Agarose gel electrophoresis is used to analyze the products of PCR reactions.

3.16 Uses for PCR Products

For some applications the analysis by agarose gel electrophoresis of the PCR products may be enough. The presence of a PCR product of the appropriate size indicates the presence in the sample of a particular gene or piece of DNA. Whether or not this product cuts with a particular restriction enzyme may give more information. For instance, the genetic mutation that gives rise to sickle cell anemia results in the loss of the recognition site for the restriction enzyme *Mst*I. It is possible to test for the presence of this mutation by amplifying the appropriate region of DNA from the subject and then treating with *Mst*I. If the PCR product can be cut into two fragments then the normal gene is present. It is also possible to sequence PCR products directly. For most other purposes, such as expressing the protein product of a gene or studying regulation and control, it is necessary to clone the PCR product into a plasmid vector.

3.17 Cloning PCR Products

Sometimes a PCR product may happen to have restriction enzyme sites within it, which are suitable for cloning. If this is the case, the PCR product can be digested with the restriction enzyme or enzymes and the resulting fragment ligated into a suitably cut vector exactly as described above. There are a number of alternative approaches to cloning PCR products, however, and we will look at two in detail.

The first method relies on the fact that although, in theory, PCR should produce blunt-ended double-stranded molecules, in practice *Taq* DNA polymerase tends to add an additional nucleotide, normally adenosine phosphate, to the 3′ end of the molecule. Specially designed cloning vectors are available which have a 3′ T overhang, (Figure 3.23) into which the PCR product can be cloned directly. Because there are no restriction enzymes which give rise to 3′ T overhangs, these vectors are usually supplied by the manufacturers ready for ligation to PCR products.

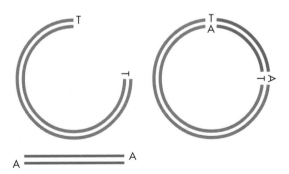

Figure 3.23 PCR products (shown in blue) often have an extra adenosine at the 3′ end; this is exploited in the cloning of PCR products using vectors with a 3′ T overhang.

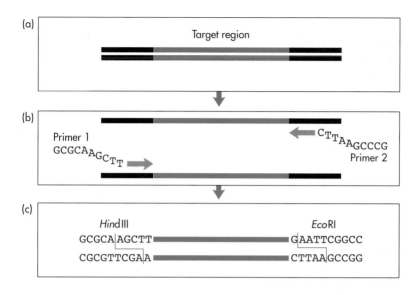

Figure 3.24 PCR reactions can be used to introduce restriction sites into the product, which can be used in cloning. Primers are designed to be complementary to sequences flanking the target region, and a 5′ extension, including a restriction site, is added to each primer. During the first few rounds of PCR the complementary part of the primer anneals to the template, amplifying the target region but also incorporating the restriction site into the product. In later cycles the whole extended primer will be able to anneal and the temperature of the PCR reaction can be raised. The PCR product can be cut with appropriate restriction enzymes and cloned in a vector such as pUC18.

A second technique is to use the PCR reaction to incorporate restriction enzyme sites at the end of the PCR product. PCR primers are designed which recognize sequences flanking the target region. A short extra region of DNA encoding a restriction site is added to the 5′ end of the primers (Figure 3.24b).

Restriction sites right at the ends of DNA fragments do not make good substrates for restriction enzymes so a few extra nucleotides are usually added at the 5′ end; commonly Gs and Cs are used as they ensure good hydrogen bonding. The first few cycles of PCR are carried out at an annealing temperature appropriate to the part of the primers which is complementary to the target DNA. The 5′ end of the primers with the restriction site will not anneal but will not interfere with priming of DNA synthesis. The 5′ ends will be incorporated into the PCR product, however, and when synthesis is primed by the other primer, it will continue to the 5′ end of the molecule, so the restriction site will be incorporated into the newly synthesized molecule. After 10 cycles of PCR the annealing temperature is raised, favoring annealing of the full length extended primer. This will increase the production of double-stranded molecules with the restriction sites incorporated at their ends (see Figure 3.24c). When 30 cycles of PCR are complete the PCR product needs to be cut with appropriate restriction enzymes to produce sticky ends, before cloning into an appropriate vector.

3.18 Real-time PCR for Quantification of DNA

The amount of template DNA in a PCR reaction affects the amount of product produced in a fixed number of cycles, and this can be used as a way of measuring the quantity of target DNA in the original sample. Modern approaches to this use a fluorescent dye, usually SYBR® Green, which only emits significant fluorescence in the presence of double-stranded DNA. As the PCR reaction proceeds, the amount of double-stranded DNA and hence the amount of fluorescence increases; fluorescence is measured in real time. There is a linear relationship between the threshold cycle (C_t) at which fluorescence begins to increase rapidly (Figure 3.25) and the number of template molecules in the starting sample. C_t values can be used calculate the concentration of template in the original sample.

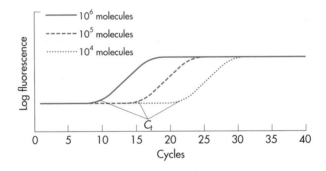

Figure 3.25 PCR can be used to measure the amount of starting material by using fluorescence to monitor product formation in real time. The accumulation of product gives a sigmoidal shaped curve, the curve is flat initially until detectable levels of fluorescence are reached, this is followed by exponential accumulation of product and finally as the substrates become depleted the curve levels off. The threshold cycle C_t is the point where the level of fluorescence begins to increase rapidly. As the amount of template is reduced the C_t value increases in a linear fashion.

3.19 Advantages and Limitations of PCR

PCR is a straightforward technique, which requires only limited technical skill. It allows the amplification of a specific region of DNA without reliance on pre-existing restriction sites. Because of the specificity inherent in the technique there is no requirement for the kind of selection method to allow identification of the clone of interest that is required when trying to clone a novel gene (see Chapter 5).

The most fundamental limitation of PCR is that you need to have some

DNA sequence information before you begin. As the amount of sequence data available in public databases increases this is becoming a less significant constraint. There are many applications of PCR that involve amplifying a region of DNA where the sequence is known; in these cases the experiment is often designed to look for polymorphisms (small changes in DNA sequence in the same gene or region of DNA which are found within the population of a particular species). Examples of this type of application are in genetic finger-printing and detection of genetic diseases (Chapter 13). It is also possible to use PCR to amplify a gene from one organism by designing primers based on the sequence of the same gene in a closely related organism.

The other main limitation to PCR is the length of the region that can be amplified. PCR works well over short stretches of DNA up to about 2 kb. With longer target regions it is necessary to increase the extension time to allow DNA polymerase to complete the synthesis of the longer product. There are PCR kits on the market, which improve the effective range of the PCR reaction. These usually use a second thermostable DNA polymerase with proofreading ability (Box 3.5) in conjunction with *Taq* polymerase and use PCR programs which increase the extension time in the later cycles. These systems make it possible to amplify target regions of up to 20 kb. The use of PCR primers with melting temperatures above 58°C also increases the efficiency of long range PCR as primers with low melting points increase the chance of non-specific annealing leading to preferential amplification of short PCR products.

Taq polymerase does not have proofreading capability (Box 3.5). As a result it has a relatively high error rate: on average it incorporates the wrong nucleotide every 9000 nucleotides. As the products of each cycle are used as template in successive cycles this can lead to a cumulative error rate of 1 in 300 nucleotides over 30 cycles. For many applications this is not a problem. However, the error rate inherent in PCR is especially important when PCR products are to be cloned. In this case a single molecule will be selected by the cloning procedure and copied many times; clearly if a mutation has occurred by mis-incorporation of a base during the PCR process, this will be perpetuated as the cloned DNA is copied in the host. This problem can be avoided by using alternative high fidelity thermostable DNA polymerases with proofreading activity. Generally speaking, even when using high fidelity polymerases it is essential to sequence the PCR product after cloning to ensure that no errors have been incorporated in the product.

Questions and Answers

Q3.1. If you had a culture of *E. coli*, some containing only pBR322 and some only pUC18, how would you select only those with pBR322?

A3.1. *From Figure 3.1, you can see that pBR322 carries genes for resistance to both tetracycline and ampicillin, whereas pUC18 only carries the gene for* *ampicillin resistance. If the culture was plated onto medium containing tetracycline, only those cells containing pBR322 would grow and form colonies.*

Q3.2. Name two features of pBR322 that make it useful as a cloning vector.

A3.2. *Cloning vectors need to be able to be replicated (so they must contain an origin of replication). They must carry selectable markers, such as antibiotic resistance. pBR322 has both of these features.*

Q3.3. If you have a circular plasmid containing a single *Eco*RI site, and you cut it with *Eco*RI, how many pieces of DNA will be formed? What about a circular plasmid containing two *Eco*RI sites? What about a linear piece of DNA, containing one *Eco*RI site?

A3.3. *Cutting a circular piece of DNA (or any other substance!) at a single site will lead to the formation of a single linear piece of DNA. If there are two sites, then two fragments will be formed. Their sizes will depend on the distance between the two sites on the DNA. If a linear piece of DNA containing a single EcoRI site is cut, it will also yield two fragments.*

Q3.4. In a DNA ligation, why is it important to inactivate the restriction enzymes before adding the DNA ligase?

A3.4. *If this is not done, the enzymes will continue to recut the DNA every time the DNA ligase seals the gap.*

Q3.5. List the steps necessary to cut a large DNA molecule up into several restriction fragments with *Eco*RI sticky ends and ligate them into pBR322. What temperature would you use for each step and why?

A3.5. *DNA is first digested with EcoRI at 37°C. This temperature is optimal for most restriction enzyme digestions. The restriction enzyme is then inactivated. After adjusting the relative amounts of fragments and vector, the two are then allowed to anneal at 16°C, and are ligated together at this temperature, which is a compromise between the best temperature for the sticky ends generated by EcoRI digestion to remain annealed to each other, and for DNA ligase to be maximally active.*

Q3.6. Transformation is a very inefficient process, only a very small percentage of bacteria take up plasmid DNA. In a typical gene cloning experiment how do you avoid having to screen each colony on the plate to look for those that have taken up DNA?

A3.6. *Plasmids used in gene cloning contain selectable markers, usually antibiotic resistance genes. By plating onto agar containing the antibiotic only successfully transformed bacteria which have taken up plasmid will be able to grow.*

Q3.7. What restriction enzymes would you use to cut the tetracycline resistance (Tet^r) gene out of pBR322?

A3.7. *EcoRI, ClaI and HindIII are all on one side of the tetracycline resistance gene, and PvuII is at the other end. Therefore any of the first three enzymes in combination with PvuII would do the trick. Note that the fragment would contain the Tet^r gene but also some extra DNA including part of the origin of replication.*

Q3.8. How many restriction fragments would you predict you would get if you digest pBR322 with *Sau*3AI? Why does the predicted number in this question not exactly match with the actual number?

A3.8. *On the basis of size alone, you would predict 4363/256 = 17 Sau3AI sites in pBR322. In fact there are 22. This emphasizes that restriction sites occur at random, and so only an approximate estimate can be made of their frequency of occurrence.*

Q3.9. Suppose you isolate a DNA fragment which has been produced by cutting a DNA molecule with *Sau*3AI, and ligate it with a vector which has been cut with *Bam*HI. Would you always expect to be able to cut the new plasmid with *Bam*HI and recover the *Sau*3AI fragment?

A3.9. *No, because although all BamHI sites contain a Sau3AI site within them, not all Sau3AI sites are flanked by the bases that make up a BamHI site. If a Sau3AI site is ligated into a BamHI site, only in 25% of cases will the base after the Sau3AI site be a C (see Figure 3.10) which will hence restore the BamHI site after ligation. This means that on average only one out of every 16 Sau3AI fragments ligated into a BamHI site will still have a complete BamHI site at each end of the fragment after ligation.*

Q3.10. Why is it important that a cloning vector should have a number of unique restriction enzyme recognition sites?

A3.10. *Unique restriction enzyme recognition sites represent points where the vector can be cut to allow the insertion of foreign DNA. It is more difficult to use restriction enzymes which cut the vector several times as digestion of the vector may result in the loss of an important part of the plasmid.*

Q3.11. You have a restriction fragment generated by digestion with *Pvu*II which you want to clone into the multiple cloning site in pUC18. What restriction enzyme would you cut the vector with?

A3.11. *Refer to Figures 3.1 and 3.9. There is no PvuII site in pUC18 for you to clone into, but PvuII cuts to give blunt or flush ends and pUC18 has a SmaI site in its multiple cloning site, which also cuts to give a blunt end. SmaI would therefore be the enzyme of choice.*

Q3.12. It is very important to inactivate the alkaline phosphatase before you mix the vector and fragment. Can you see why this is?

A3.12. *If you do not inactivate the alkaline phosphatase it will remove the 5′ phosphates from the fragment as well and there will be no ligation.*

Q3.13. The shorthand way of describing the antibiotic resistance phenotype of bacteria which have been transformed with a plasmid carrying an antibiotic resistance gene is to write ap^r (ampicillin resistant) or tet^r (tetracycline resistant), or if they are sensitive to the antibiotics ap^s, tet^s. If bacteria are expressing β-galactosidase they are said to be lac positive, which can be written as lac$^+$; if not, then lac negative or simply lac$^-$. Write down the shorthand for the phenotype of: (a) *E. coli* transformed with pBR322; (b) *E. coli* transformed with a recombinant plasmid derived from pBR322 but with an insert cloned into the tetracycline gene; (c) A strain of *E. coli*, with a deletion of the portion of the *lacZ* gene which encodes the α peptide of β-galactosidase, transformed with pUC18.

A3.13. *(a) The plasmid pBR322 encodes resistance to ampicillin and tetracycline; this can be written as ap^r tet^r. (b) The presence of the insert will inactivate the tetracycline gene, and the resulting bacteria will be ap^r tet^s. (c) The host* E. coli *strain will be lac$^-$ because both the α peptide and the rest of the β-galactosidase enzyme are required to confer the lac$^+$ phenotype. However, transformation with pUC18 will supply the α peptide and also make the transformants ampicillin resistant, in shorthand ap^r lac$^+$.*

Q3.14. What color would the colonies of the ap^r lac$^+$ transformants described in (c) above be if plated on ampicillin and X-gal? What color would they be if a fragment of DNA had been cloned into the multiple cloning site of pUC18?

A3.14. *The ap^r lac$^+$ colonies would be blue because they express a functional β-galactosidase. A DNA fragment cloned into the multiple cloning site would inactivate the lacZ′, resulting colonies would be lac$^-$ and white on X-gal.*

Q3.15. What would be the sizes of the fragments that you would expect to find if you cut this plasmid with (a) *Bam*HI, (b)*Eco*RI, (c) both *Bam*HI and *Eco*RI together?

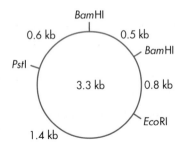

A3.15. (*a*) *If you cut with* BamHI *you would expect two fragments of 0.5 kb and 2.8 kb.* (*b*) *If you cut with* EcoRI *you would expect a single fragment of 3.3 kb.* (*c*) *If you use both restriction enzymes together you would expect three fragments of 0.5, 0.8 and 2 kb.*

Q3.16. The table shows the distance migrated in an agarose gel by each of the eight fragments produced when λ DNA is cut with *Hind*III. This can be used as a size marker.

Distance Migrated (mm)	Size of Fragment (kb)
8.5	23
23.5	9.4
29	6.6
36	4.4
46.5	2.3
49	2.9
69.5	0.56
94	0.13

Draw a graph of the $\log_{10}$ of the size of each fragment against the distance migrated. Work out the size of the DNA fragments that would give rise to a band that migrated 60 mm.

A3.16. *Your graph should look like this:*

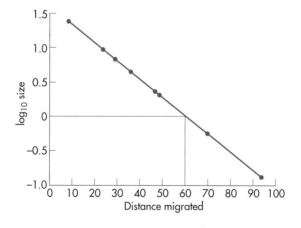

A band which has migrated 60 mm gives a value of 0 on the y axis, remember you have plotted the $\log_{10}$ of the size, the anti-log gives a size of 1 kb.

Q3.17. Imagine you have a 3 kb *Eco*RI fragment of DNA cloned into the *Eco*RI site in the multiple cloning site of pUC18, and that you know that there is a *Bam*HI site in the fragment, which divides it into unequal pieces of 1 and 2 kb. You want to subclone the 1 kb *Eco*RI *Bam*HI fragment into pBR322. Outline the steps you would use to perform this operation. *Hints.* (i) First draw out the plasmid you want to subclone from, and (ii) there are at least two sensible approaches to this problem; in one you will need to use preparative agarose gel electrophoresis and in the other you may use a combination of antibiotic selection and restriction mapping to identify the clone of interest.

A3.17. *Digest the plasmid with EcoRI and BamHI, which will yield fragments of 1 kb, 2 kb and 2.7 kb (the last fragment being the vector). There will also be a very small band of 29 bp consisting of the DNA between the EcoRI and BamHI sites from the multiple cloning site of pUC18, but this will not be visible on the gel as it is too small to see on agarose gels). Then either (method 1) purify the 1 kb fragment from the gel, and ligate it into EcoRI–BamHI cut pBR322, or (method 2) ligate the entire mix of fragments into EcoRI–BamHI cut pBR322. Whichever method you choose you will need to select for ampicillin resistance and you could also choose to screen for tetracycline sensitivity to find recombinants. If you have chosen method 2 you will need to purify plasmid DNA from several clones and use restriction mapping to confirm which ones contain the 1 kb fragment.*

Q3.18. Suppose that in cloning fragments of genomic DNA generated by *Eco*RI digestion into the plasmid pBR322, the sticky ends of two genomic fragments annealed to each other before annealing to the sticky ends of the vector. What would be the consequence of this in terms of what would be recovered after transformation? Would it be a problem? How would you detect this?

A3.18. *The result of this would be a vector containing two EcoRI fragments, derived from totally different parts of the genome. This could be a problem as it might give the impression that these two fragments were actually adjacent on the genome, whereas in fact their association in the plasmid is a matter of chance. This could be detected by cutting the plasmid with EcoRI: if it contains two EcoRI fragments then both will be released and can be seen on a gel.*

Q3.19. Why is the ratio of GC to AT base pairs an important factor in determining the melting temperature of double-stranded DNA molecules?

A3.19. *GC base pairs are more stable than AT base pairs as they have three hydrogen bonds rather than two; this means that a GC-rich DNA molecule will have a higher melting temperature than a molecule of the same length but with a higher proportion of AT base pairs.*

Q3.20. Can you think of any features of the human genome that would make it likely that a 17 bp primer would anneal to more than one sequence?

A3.20. *You will know from Chapter 2 that the genome sequences, including the human, are not truly random. Families of genes exist which have evolved from each other through a series of gene duplication events, the sequences of these genes are similar to each other. If you were to design a 17 bp primer using a sequence of a gene from one of these families there would be a chance that the same sequence would be found in another member of the gene family.*

Q3.21. List four components necessary to copy DNA *in vitro.*

A3.21. *A single-stranded DNA template strand, an oligonucleotide primer complementary to this strand, deoxyribonucleotides, and DNA polymerase. Also necessary is a suitable buffer containing Mg^{2+} ions.*

Q3.22. In the DNA sequence in Figure 3.20, the positions of two oligonucleotides are marked. What is the melting temperature, T_m, for each of these oligonucleotides? What length would the PCR product primed by these two oligonucleotides be?

A3.22. *Using the formula given in Section 3.12 both primers will have a T_m of 60°C when bound to exactly complementary DNA. The amplified fragment will be 60 base pairs long.*

Further Reading

REBASE – restriction enzymes and DNA methyltransferases (2005) Roberts RJ, Vincze T, Posfai J and Macelis D, Nucleic Acids Research, Volume 33 special database issue Pages D230–D232.
REBASE is a collection of information about restriction enzymes and related enzymes. You can access the database from the following url:
http://rebase.neb.com/rebase/rebase.html

Type II restriction endonucleases: structure and mechanism (2005) Pingoud A, Fuxreiter M, Pingoud V and Wendea W, Cellular and Molecular Life Sciences, Volume 62 Pages 685–707.
This is a long and thorough review of the type II restriction enzymes which are the workhorses of gene cloning. You will probably want to dip into this review rather than read it in its entirety

Studies on transformation of *Escherichia coli* with plasmids (1983) Hanahan D, Journal of Molecular Biology, Volume 166 Pages 557–580.
This is a careful analysis of the factors which make E. coli *competent for transformation*

The unusual origin of the polymerase chain-reaction (1990) Mullis KB, Scientific American, Volume 262 Page 56.
This is a short, easy-to-read article describing how PCR came about

4 Gene Identification and DNA Libraries

Learning outcomes:

By the end of this chapter you will have an understanding of:

- the problems involved in identifying a particular gene
- why gene libraries are an important tool in gene cloning
- the differences between genomic and cDNA libraries including the types of sequences they contain, their construction, and the circumstances where each is the most appropriate tool
- the differing roles of plasmids, bacteriophage λ, cosmids, YACs and BACs as vectors in the construction of libraries
- simple calculations which allow you to calculate how many clones are required in a library

4.1 The Problem

In the previous two chapters we have discussed the basic gene cloning techniques used to manipulate DNA. We have also discussed the nature of DNA in a range of different organisms. In this chapter, we will begin to address one of the commonest questions raised by students when they start to understand molecular biology, which is: how do you actually go about cloning a particular gene?

The problem is as follows. If we wish to clone a particular gene, either for study or to make a protein product, then we have to use methods that enable us to isolate this gene from all the other genes in the organism of interest. To show the scale of this problem, imagine that you want to clone a human gene – perhaps one that is responsible for an inherited genetic disease. Remember from Chapter 2 that the human genome is 3,289,000,000 bp long, encoding about 25,000 genes. If you are going to make a detailed study of the gene responsible for the genetic defect, you will need to get your hands on a manageable-sized piece of DNA, encoding just the gene of interest and perhaps its associated regulatory sequences. If you extract DNA from human cells you will then have to locate the part containing the gene that is of interest to you, a task equivalent to the proverbial search for a needle in a haystack.

The task of cloning human genes has been made much easier by the progress of the human genome project. In 1998 a map of the human genome was available giving the location of hundreds of genes and 30,000 genetic markers. The announcement of the completion of the first draft of the human genome was made in January 2000. If the location of the gene you are interested in is known you can simply look it up on a suitable website such as the Ensemble genome browser (http://www.ensembl.org/). This will give you the sequence of the gene (including the many introns that it is likely to contain) and of the DNA upstream and downstream of it. You can then design PCR primers which will amplify exactly the region you want to work with (Section 3.12).

Q4.1. Once you have amplified the region containing your gene by PCR you will need to clone it into a vector. How would you do this?

The PCR amplification approach will only work if you know something in advance about the gene you are interested in, such as its DNA sequence, the sequence of a gene which is closely related to it from a different organism, or the sequence of the protein which it encodes. (Often, however, the only information that you may have about the gene will be that mutations in it cause a particular phenotype. We will look at routes for cloning genes about which there is little available information in Chapter 6.)

The other approach to the problem of how to clone a particular gene, which was used before whole genome sequences became available and is still useful in many cases, is to construct what is called a DNA library. A DNA library is made by taking DNA from the organism of interest, breaking it into small pieces, and cloning these individually into a suitable vector. Thus, a DNA library consists of a large collection of clones, each containing a part of the genome of the organism. The collection needs to be large enough so that you can be fairly sure that one of the clones will contain the gene of interest. This collection of clones is called a library because each clone is like a book in that library containing part of the information contained in the genome. Unfortunately, the books in a DNA library do not have classification marks printed on their spines, nor is there an index that would allow you to locate the book you want. The process of selecting the clone which contains the gene you are interested in from the library is called screening and is the subject of the next chapter. In this chapter, we will discuss the nature and construction of the two basic types of DNA library: genomic and cDNA libraries. Each of these has advantages and disadvantages in different situations.

Q4.2. What are the basic steps you would need to construct a library containing genomic DNA from a bacterium? *Hint.* This is just a variation of the techniques outlined in Chapter 3. Instead of cloning a particular DNA fragment, you want all the DNA from an organism.

4.2 Genomic Library

A genomic library is a collection of recombinant clones representing the entire genome of an organism. Each clone is like a book in that library containing part of the information contained in the genome. Thus a genomic library does not only contain the coding regions of the genes but also the regions between genes which may include important sequences involved in gene regulation. If the library is made from a eukaryotic organism, a genomic library would also contain introns, repetitive and junk DNA (Section 2.3).

To see how to construct a genomic library we will use a relatively straightforward example. Soil organisms such as species of *Pseudomonas* possess enzymes that enable them to use a wide range of compounds as sources of carbon and energy. One such enzyme, catechol 2,3-oxygenase, carries out one of the steps involved in the breakdown of xylene; it is encoded by the *xylE* gene. The presence of a functioning *xylE* gene can be detected by spraying colonies that express the *xylE* gene with catechol: as the catechol is broken down the colonies turn yellow. We can use this as a simple and direct way of selecting our clones from the library. Alternative approaches to screening DNA libraries are discussed in Chapter 5.

4.3 Constructing a Genomic Library

Genomic DNA extracted from *P. aeruginosa* (Box 4.1) is cut with a restriction enzyme such as *Bam*HI. The *P. aeruginosa* genome is 6.3 Mb, larger than that of *E. coli*, and *Bam*HI will cut it into about 1500 fragments with an average size of 4000 bp. The restriction fragments are then ligated into the *Bam*HI site of a suitable vector, such as pUC18, introduced into *E. coli* by transformation (Figure 4.1), and plated onto agar containing ampicillin.

The colonies that grow on the agar plate make up the library. Remember that each individual colony is made up of identical bacteria containing the same recombinant plasmid. Different colonies will contain different recombinants each consisting of pUC18 with a different piece of *Pseudomonas* DNA in it. As the average size of a bacterial gene is about 1 kb and the average size of the inserts in this example is about 4 kb it is likely that each clone in the genomic library will contain the DNA for more than one gene.

Having constructed our library the next step is to find out which of the clones in our library encode the *xylE* gene. If we spray the plate with catechol, the colonies that are expressing catechol 2,3-oxygenase will turn yellow. These are the colonies we are interested in. Plasmid DNA can now

Box 4.1 Preparation of Chromosomal DNA

The first steps in the preparation of chromosomal DNA are very similar to those in the purification of plasmids (Section 3.6). To prepare chromosomal DNA from bacteria you would typically start with an overnight culture. Cells are lysed and most of the non-nucleic acid material is removed. The main problem in the purification of chromosomal DNA, particularly for applications like library construction, which require high molecular weight starting material, is to keep the large DNA molecules intact. Shaking or vigorous pipetting is likely to fragment the molecules. The size of the chromosomal DNA molecules can be exploited in their purification, in some protocols ethanol is layered on top of the lysate and the chromosomal DNA is spooled out by winding it around a glass rod. Alternatively, the lysate can be mixed with the ethanol and the DNA, which is precipitated, can be collected by centrifugation. The DNA preparation can be further purified by centrifugation on a cesium chloride (CsCl) gradient; a specialized form of density gradient centrifugation. When mixed with ethidium bromide and placed in a solution of CsCl, this is centrifuged at high speeds, which causes a gradient to form, the chromosomal DNA will form a discrete band separate from both RNA and any remaining protein contamination.

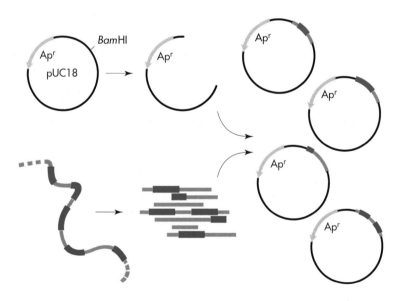

Figure 4.1 Construction of a genomic library. The genomic DNA (shown as a gray line with genes represented by blue boxes) is cut into random fragments each one of which is cloned separately into a vector.

be extracted from these colonies, and used in further experiments. Unfortunately, the process of selecting the clones you are interested in from a library is seldom as straightforward as this example where you are able to look for expression of a gene directly. Different techniques for selecting the correct colony (or "screening the library") are the subject of the next chapter.

Q.4.3. Five different colonies in your *Pseudomonas* genomic library turn yellow when sprayed with catechol. How would you extract plasmid DNA from them? How would you find out whether all of the five colonies carried the same piece of genomic DNA?

4.4 How Many Clones?

An important consideration to be taken into account when making a library is to decide how many individual colonies it should contain to ensure that you will be able to find the gene that you are interested in. Because you do not start with a single genome and cut it up into pieces, and because you have no way of ensuring that each and every part of the original genome will be equally represented in your library, you have to use probabilities to work out how many colonies you need. The formula you use is shown below. It takes into account the size of the genome (g), and the average size of the fragments (f) in your clones:

$$N = \frac{\ln (1 - P)}{\ln (1 - f\!/\!g)}$$

where N is the number of clones and P is the probability that any given gene will be present. Using the formula you can calculate for any given number of clones the probability of finding the gene you require (Table 4.1). You will notice that (as you might expect) the bigger the library the better the chance of finding your gene. For the *Pseudomonas* genome

---

Table 4.1. Relationship between library size and probability

Probability	Number of Clones
0.5	1,091
0.9	3,625
0.99	7,251
0.999	10,876

The table shows the relationship between library size and probability that a particular gene will be represented for a genomic library with an average fragment size of 4 kb made from an organism with a 6.3 Mb genome.

which is 6.3 Mb, to have a 99% chance of finding any particular gene, you would need to have 7251 clones. Notice that because the calculation is based on probabilities and you have no way of ensuring that each clone in the library is unique, you need considerably more clones than the 1500 restriction fragments *Bam*HI would cut the *Pseudomonas* genome into. With 10,876 clones the probability goes up to 99.9% and it would probably not be worth increasing the library size any further than this.

> **Q4.4.** How many clones would you need in a genomic library made from (a) *E. coli* or (b) *Saccharomyces cerevisiae* to have a 99% chance of cloning a particular gene? Assume an average fragment size of 4000 kb as before.
>
> **Q4.5.** How many clones of *E. coli* would you need to have a 99.9% chance of cloning a particular gene?

4.5 Some DNA Fragments are Under-represented in Genomic Libraries

Some stretches of the genome are harder to clone than others because of the burden placed on the host cell by the genes encoded on them. A good example of this is genes for proteins which are located in membranes. If such proteins are over-expressed they can often cause damage to the membranes of the host bacterium, and reduce its viability. Cloned genes may be expressed at higher levels than in their normal setting because they have been cloned into a multiple copy number plasmid and also possibly removed from the normal regulation mechanisms. This may also place an unacceptable burden on the host bacterium.

Larger DNA fragments are also harder to clone than small ones, mostly because as they are big there is a greater chance that they will contain sequences which make them unstable or which place a burden on the host cell. Because restriction enzyme recognition sites occur randomly throughout the genome a particular gene may be located on a very large fragment, which may be under-represented in the genomic library. Some stretches of DNA are intrinsically less stable when cloned than others, often because they contain repetitive sequences which may cause mistakes during DNA replication. These regions are also likely to be under-represented in DNA libraries.

The probability calculations are based on the assumption that any particular piece of the genome is just as likely as any other piece to be successfully cloned. This is clearly not true; in practice most workers will use a library with at least two or three times as many clones as are theoretically needed, to give a high probability of finding a particular gene.

4.6 Using Partial Digests to Make a Genomic Library

One other reason why you may not be able to find your gene of interest in your library is that it contains within its sequence one or more sites for the

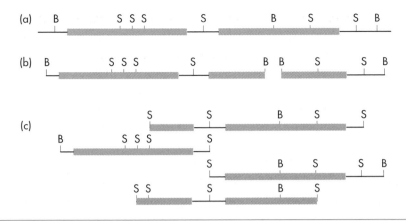

Figure 4.2 Using partial digests in the construction of a genomic library. a) A piece of genomic DNA encoding two genes; the diagram shows the genes as blue boxes and the positions of the BamHI (B) and the Sau3AI (S) sites. b) A library made from BamHI digested DNA would contain clones in which the first gene, but not the second, was intact. Cutting to completion with Sau3AI would result in many small fragments, but a partial digest with Sau3AI would yield many different fragments of which a few are shown in c). A library made from a partial digest with Sau3AI would contain some clones in which the first gene was intact and some in which the second was intact.

restriction enzyme used in making the library. In our example we have constructed a library of BamHI fragments, if the xylE gene has a BamHI site in it then the gene will be split between two clones in the library (Figure 4.2). No single clone will be able to make catechol 2,3-oxygenase so nothing will be detected when the plate is sprayed with catechol.

For this reason genomic libraries are often made using partial digests. In a partial digest the reaction conditions are designed to be sub-optimal so that the restriction enzyme will not cut at all of the possible recognition sites. The usual ways of achieving this are to reduce the enzyme concentration or the incubation time, or both.

In the construction of genomic libraries, partial digests using a restriction enzyme with a 4 bp recognition site such as Sau3AI are usually used (Figure 4.2). Because the sites which are cut are essentially random, there will be a range of sizes of fragment from a few base pairs to many kilobases. As many copies of each genome are present in the original reaction, even if in some cases the gene of interest is cut by Sau3AI, there will be some large fragments containing the full-length gene where the Sau3AI sites have not been cut.

For a number of reasons you may wish to control the range of sizes of fragments in a genomic library. If you are trying to clone a very large gene or a group of genes which act together you will need fairly large fragments.

Also, if all the fragments are of similar size there is less bias in the way individual sections of the genome are represented. To achieve this, a subset of the total digested DNA can be prepared by a process called "size fractionation". This is usually achieved by running the digested DNA on an agarose gel and cutting out that portion of the gel where the fragments are of the desired size. These can then be extracted from the gel and used to construct the library (Section 3.11).

Q4.6. Using a partial *Sau*3AI digest you have isolated DNA fragments from *Pseudomonas* with an average size of 10 kb and constructed a genomic library. How many clones would you need to have a probability of 0.99 of finding the gene of interest?

Q4.7. Using a *Sau*3AI partial digest presents a problem when you come to cutting your cloning vector. Can you see what the problem is? Can you think of a cloning strategy which would overcome this problem? *Hints.* How many times would you expect *Sau*3AI to cut pUC18? Is this a problem? Is there a restriction enzyme with sticky ends, which are compatible with those produced by *Sau*3AI, with a unique site in pUC18?

4.7 Storage of Genomic Libraries

One of the advantages of making a library is that once made it can be stored and re-plated many times to allow screening of other genes. This is usually done by scraping the colonies off the plates and making a suspension in a medium containing glycerol. This suspension can then be stored for many months at –20°C without loss of viability. It is important that each time you want to screen a library you re-plate from this original stock culture because each time you grow the library you will introduce bias selecting in favor of clones that grow well and are stable.

Q4.8. What factors may cause some clones to be less well represented in the library than others?

4.8 Advantages and Disadvantages of Genomic Libraries

To recap, the main reason that we have discussed above why you might make a genomic library would be because you wanted to identify a clone from the library which encodes a particular gene or genes of interest. Genomic libraries are particularly useful when you are working with prokaryotic organisms, which have relatively small genomes.

On the face of it, genome libraries might be expected to be less practical when you are working with eukaryotes, which have very large genomes containing a lot of DNA which does not code for proteins. A library representative of a eukaryotic organism would contain a very large number of clones, many of which would contain non-coding DNA such as repetitive DNA and regulatory regions (Chapter 2). Also, eukaryotic genes often contain introns, which are untranslated regions interrupting the coding sequence (Section 2.3). These regions are normally copied into mRNA in the nucleus but spliced out before the mature mRNA is exported to the cytoplasm for translation into protein. Prokaryotic organisms are unable to do this processing so the mature mRNA cannot be made in *E. coli* and the protein will not be expressed. If your screening method requires that the gene be expressed it will not work with a genomic library from a eukaryotic organism. However, genomic libraries from eukaryotic organisms are very important since they allow you to study the genome sequence of a particular gene, including its regulatory sequences and its pattern of introns and exons. Ways in which you can screen a library for the presence of a gene which is not expressed will be discussed in the following chapter.

4.9 Cloning Vectors for Gene Libraries

The number of individual clones which can be obtained by cloning into plasmids is limited by the relative inefficiency of transformation of *E. coli* and the small number of individual colonies which can be distinguished on an agar plate.

> **Q4.9.** Assuming that you can work at a density of 500 colonies per plate, how many plates would you need to make a 99% representative (a) genomic library containing 10 kb fragments of *Pseudomonas* DNA, (b) genomic library containing 20 kb fragments of human DNA?

An alternative type of vector, which gets round some of the problems with plasmids, is bacteriophage λ (see Section 2.7). Bacteriophage λ actively infects host bacteria and so its DNA is taken into the cell with a much higher efficiency than occurs in chemical transformation. Infection can be seen by plating infected bacteria on an indicator lawn giving rise to plaques resulting from lysis of a susceptible bacterial strain (Figure 4.3). The plaques that form on the lawn of *E. coli* are also smaller than colonies so you can work with a much greater density.

4.10 Vectors Derived from Bacteriophage λ

Before bacteriophage λ can be used as a cloning vector it needs to be modified to allow it to carry inserted DNA. The process by which the λ genome is inserted into the phage particles is called packaging (Figure 4.4) and relies on specific cos sequences that delineate individual copies of the genome (Section 2.7). The size of the DNA fragment that can be packaged into λ is restricted because the phage can only accommodate about 5%

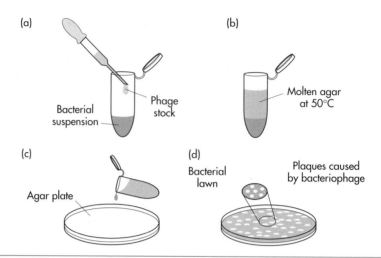

Figure 4.3 Plating bacteriophage λ on an indicator lawn. a) A small amount of phage stock is mixed with an actively growing culture of a strain of susceptible *E. coli*. b) This is then mixed with molten agar cooled to 50°C and c) poured onto the surface of an agar plate. d) The bacteria grow in this "top agar" producing an even lawn of growth rather than the colonies seen when they are spread on the surface of the agar. Bacteria infected with λ phage are lysed, releasing phage particles to infect neighboring cells. As this process of infection and bacterial lysis goes through several rounds it gives rise to a visible cleared area, called a plaque, within a few hours.

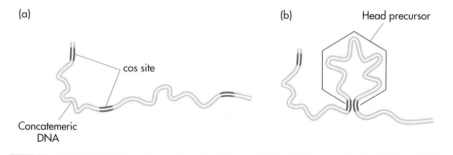

Figure 4.4 Packaging of bacteriophage DNA. a) In the later stages of infection by bacteriophage viral DNA is copied in continuous concatemers in which viral genomes are arranged end to end and separated by cos sites. b) A single genome's length of DNA is contained in the immature head precursor and the concatemeric DNA is cleaved at the cos site to generate single-stranded cohesive ends. Further head proteins and the separately assembled tail structure are added to produce the mature virus. One significant consequence of the way in which concatemeric DNA is packaged into phage head particles in genome sized chunks is that it only works if the cos sites are between 37 and 52 kb apart (75–105% of the size of the λ genome). This feature has been exploited in the design of many λ vectors where large portions of the genome have been deleted. These vectors are too small to be successfully packaged unless extra DNA has been added; hence there is a natural selection for recombinants containing a DNA insert.

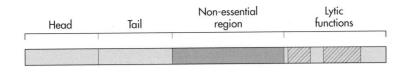

Figure 4.5 Physical map of bacteriophage λ**.** The regions responsible for head and tail assembly and also the lytic functions are shown. The regions that can be deleted without affecting the normal functions (gray stripes) and the region that can be replaced by cloned DNA (dark blue) are shown.

more DNA than is present in its normal genome. However, parts of the λ genome can be deleted without affecting the ability of the bacteriophage to infect host cells and assemble new virus particles (Figure 4.5). Cloning vectors which have had these regions removed are called insertion vectors, and they can accommodate about 12 kb of foreign DNA (Figure 4.6a). In addition these vectors are generally engineered to have a unique restriction site for a commonly used restriction enzyme such as *Eco*RI.

A second class of λ phage vectors called replacement vectors (Figure 4.6b) is based on exchange of the non-essential region (Figure 4.5) for the DNA to be cloned. These vectors can accept larger inserts because they have had more of the λ DNA removed. Because of the size constraints placed on successful packaging (Figure 4.4), these vectors contain a stuffer fragment making them large enough for replication and packaging. This stuffer fragment is removed during cloning and replaced with the insert DNA. Using λ replacement vectors it is possible to clone fragments between 9 and 23 kb.

Bacteriophage λ DNA can be manipulated *in vitro* in the same way as plasmid DNA. It can be cut with restriction enzymes and ligated with genomic DNA similarly treated to give recombinants, which can be used in the construction of a library. It is possible to introduce these recombinant molecules into *E. coli* by transformation where they will direct the host to produce phage particles. This approach, however, fails to take account of one of the most useful features of phage λ, namely its ability to introduce DNA into *E. coli* with a high frequency by its normal infection process.

4.11 Packaging Bacteriophage λ *In Vitro*

Because bacteriophage will essentially self-assemble, it is possible to package DNA into virus particles in the test tube and then to use these to infect *E. coli*. This process, referred to as *in vitro* packaging, can yield about 10^9 plaques per mg of DNA. Mixing concatemeric recombinant DNA with viral head precursors, phage tails and other proteins required for packaging results in mature viral particles containing recombinant DNA, which can be used to infect *E. coli* cells and which will produce plaques on a bacterial

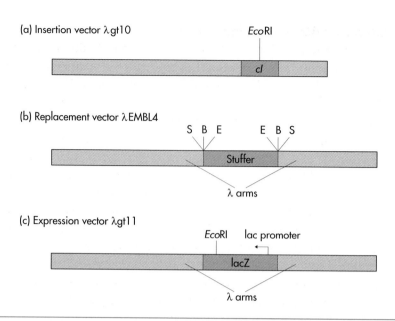

Figure 4.6 Examples of λ cloning vectors. a) λgt10 is an insertion vector commonly used in cloning cDNA; it can accept up to 7.6 kb of DNA inserted into the EcoRI site in the cl gene. The cl gene encodes the λ repressor protein, which switches the life cycle to the lysogenic rather than the lytic phase (Section 2.7). Insertion at this site results in the loss of the λ repressor protein and all phage enter the lytic phase resulting in clear plaques with a high titer of phage. b) λEMBL4 is a replacement vector (restriction enzyme recognition sites are indicated: S SmaI, B BamHI, E EcoRI). These vectors have had more of the λ DNA removed allowing for cloning of larger inserts. The result of this is that when the two arms, either side of the cloning site, are religated the resulting molecule is too small to be efficiently packaged into λ heads. During propagation of the vector it is necessary to have a replaceable fragment of DNA in the insertion site; this is often referred to as the stuffer fragment and is removed and replaced with the DNA to be cloned. Using λ replacement vectors it is possible to clone fragments between 9 and 23 kb. c) λgt11 is an example of a vector suitable for construction of a cDNA expression library. It contains the E. coli lacZ under the control of the lac promoter. Insertion of cDNA into the unique EcoRI restriction site results in insertional inactivation of the lacZ gene which can be detected by plating on media containing X-gal and IPTG. Proteins encoded by the cDNA clones are expressed as fusions to the β-galactosidase protein under the control of the lac promoter. The cDNA sequences need to be inserted in the correct orientation and reading frame to be expressed. The overall size of the library is usually increased to have a good probability that fusions in all six possible reading frames will be present.

lawn in the usual way. The process of introducing *in vitro* packaged λ DNA into *E. coli* is often described as transfection to distinguish it from transformation (introduction of DNA by chemical treatment) and infection (the process by which the phage introduces its own DNA into bacteria).

4.12 Cloning with Bacteriophage λ

If you want to use a λ replacement vector to make a DNA library, you firstly need to cut out the stuffer fragment. Replacement vectors are usually designed with multiple cloning sites at the margins of the stuffer fragment to make this easy (Figure 4.6b). Digesting with a second restriction enzyme, which only cuts the stuffer fragment, will reduce the stuffer fragment to a number of small fragments. The sticky ends of these smaller fragments are not compatible with those of the λ cloning arms so the stuffer fragment is unlikely to be involved in the recombinants. Fragments of genomic DNA in the region of up to 23 kb with compatible sticky ends are ligated with the λ arms to form concatemers which can then be packaged *in vivo* into bacteriophage particles and used to transfect *E. coli*. You will see from Figure 4.7 that the concatemeric DNA which is produced in the ligation reaction can contain λ arms joined together without insert molecules in place of the stuffer fragment. These non-recombinant molecules are not wanted in the DNA library, since they increase the number of plaques that need to be screened to identify the plaque containing the clone of interest. However, as the cos sites in these molecules will be less than 37 kb apart they will not be packaged successfully into phage heads and will not be present in the library. Many λ vectors use this type of size selection for recombinant molecules.

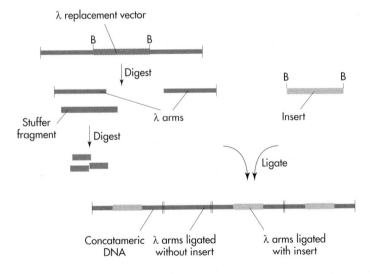

Figure 4.7 Cloning with a bacteriophage λ replacement vector. Linear bacteriophage λ is cut with a restriction enzyme to separate the λ arms and a stuffer fragment. Digestion with a second restriction enzyme which only cuts the stuffer fragment reduces the chances of it religating to the arms in the ligation step. The λ arms are mixed with insert DNA and ligated. Concatemers are formed consisting of λ arms and insert DNA. Because there is a minimum distance between cos sites for efficient packaging of λ DNA any arms which ligate to each other will not package successfully.

The strictly defined range of fragment sizes that λ can accommodate can also be used to ensure that the library contains DNA fragments of uniform size. Although packaging into λ will effectively select only those DNA fragments which are approximately 23 kb it is usual when making a library to pre-select fragments of approximately the correct size. This prevents packaging of several small fragments which have become ligated together, which in turn could give the mistaken impression that fragments from different parts of the genome are actually adjacent to each other.

> **Q4.10.** How would you prepare insert DNA suitable for cloning into a bacteriophage λ replacement vector? *Hint.* You need large fragments of DNA so complete digestion with a six cutter will not work. You will also need to select fragments of the correct size.

4.13 Calculating the Titer of your Library

Your DNA library now consists of a suspension of bacteriophage λ particles containing recombinant DNA. You need to know how many plaques will be produced if you plate this stock onto *E. coli*. This figure is a measure of the concentration of the suspension and is known as the titer of the library. To calculate the titer you take a small sample of your library stock and make serial dilutions. Each dilution is mixed with *E. coli* and plated out in top agar (Figure 4.3). You count the plaques and multiply by the dilution factor to give you the titer of your stock in plaque forming units per milliliter (pfu mL^{-1}).

> **Q4.11.** You have performed an experiment to measure the titer of λ library stock. You make serial 10-fold dilutions of your stock. You take a 10 μL sample from each dilution and mix it with 100 mL of *E. coli* and 2 mL of molten agar and pour onto an agar plate. After incubation at 37°C you count the plaques. You find that the 10^{-3} dilution gives 512 plaques per plate, the 10^{-4} dilution gives 54 and the 10^{-5} dilution gives four plaques. Calculate the titer of the library stock.
>
> **Q4.12.** How much of what dilution of this stock would you use to produce plates for screening with about 200 plaques each?

4.14 Cosmid Libraries

Even with λ replacement vectors it is not possible to clone inserts larger than about 25 kb. Plasmid vectors can accommodate much larger inserts but the transformation frequency, especially with large plasmids, is low. What is needed is a type of vector that combines the high efficiency of λ transfection with the ability to clone large pieces of DNA into plasmid

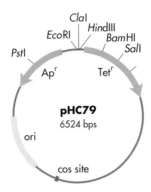

Figure 4.8 Map of the cosmid vector pHC79. This vector has the ampicillin and tetracycline resistance genes and origin of replication from pBR322. In addition it has the bacteriophage λ cos site allowing it to be packaged *in vitro*.

vectors. Such vectors which have characteristics of both phage and plasmids are called cosmids. As we have seen, concatemeric DNA can be packaged *in vitro* as long as there are cos sites approximately 50 kb apart. This fact has been exploited in the construction of cosmids. These are plasmids which have had a small piece of bacteriophage λ DNA, including the cos site, cloned into them (Figure 4.8). The presence of the cos site means that they can be packaged *in vitro* allowing us to exploit the higher transfection frequencies of bacteriophage λ. Because of the large insert size and the high transformation efficiencies possible with these vectors, cosmids have been used extensively in the construction of DNA libraries (Figure 4.9).

4.15 Making a Cosmid Library

As with bacteriophage λ it is necessary to do a partial digest and size fractionate the genomic DNA before using it to make a cosmid library. The cosmid is linearized with a compatible restriction enzyme and treated with alkaline phosphatase to prevent vector molecules ligating to each other (see Section 3.8). The vector and genomic fragments are mixed at a ratio of at least 3 moles of vector to each mole of genomic fragments and at a high overall concentration. These conditions are designed to ensure that concatemers are formed. In fact, if alkaline phosphatase treatment is successful, with these high vector ratios the most likely product of the ligation reaction is a single fragment of genomic DNA bounded by a cosmid molecule at each end. This will package successfully into bacteriophage λ particles *in vitro*.

Q4.13. What range of sizes of fragments of genomic DNA would be suitable for constructing a cosmid library using pHC79?

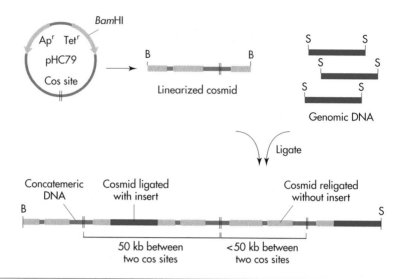

Figure 4.9 Construction of a genomic library in the cosmid pHC79. The cosmid is linearized by cutting at a unique restriction site, *Bam*HI. Fragments of size fractionated genomic DNA with compatible sticky ends are ligated to the linearized vector to form concatemers. Whenever a genomic fragment is cloned between two cosmid molecules, the cos sites will be 50 kb apart and the cosmid can be packaged by bacteriophage λ.

Q4.14. Why is it important, when cloning into a cosmid vector, to size fractionate the genomic DNA to ensure that the fragments are in the correct size range?

4.16 YAC and BAC Vectors

There are two further types of vector which are used in library construction and are particularly useful for cloning very large fragments. These are yeast artificial chromosomes (YACs) and bacterial artificial chromosomes (BACs). YACs contain all of the sequences required for replication of a chromosome in yeast: a centromere, telomeres as well as a bacterial origin of replication (Figure 4.10a) and are hence referred to as artificial chromosomes. For cloning the YAC is cut with restriction enzymes to give two arms, each with a selectable marker (Figure 4.10b). Insert DNA is cloned between these two arms to give a linear artificial chromosome, with a centromere and telomeres at either end (Figure 4.10c). Because this is essentially a chromosome it can maintain DNA inserts of up to 1 million base pairs, although for routine library construction insert sizes are usually in the 250 to 400 kb range.

Unlike YACs, BACs are not true artificial chromosomes, they are simply bacterial plasmids designed to carry very large inserts. BACs are based on

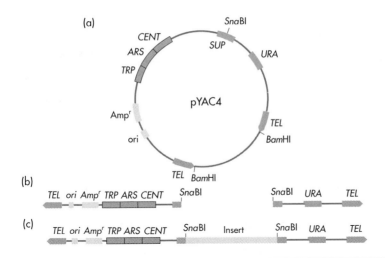

Figure 4.10 The main features of a yeast artificial chromosome (YAC).
a) These are artificial constructs which contain all of the necessary elements for replication inside yeast as well as a bacterial origin of replication (ori). The centromere (*CENT*) is important to ensure that the YAC is distributed to daughter cells and *TEL* is a sequence on which the host yeast will be able to build telomeres to protect the ends of the YAC. It also contains two selectable markers, *URA* and *TRP*, for selecting transformants. Blunt-ended fragments are cloned into the *Sna*BI inactivating the third marker *SUP*; this can be detected by looking for a color change. b) For cloning, the YAC is cut with *Bam*HI in addition to *Sna*BI to give two arms. c) When the insert DNA is ligated into the *Sna*BI site the construct adopts the structure of a simplified artificial chromosome and is capable of carrying large inserts in a stably inherited fashion.

the large, stable, bacterial F or fertility plasmid. With replication and partitioning functions from the F plasmid BACs (Figure 4.11), are capable of maintaining large inserts in a stably inheritable fashion. These vectors are relatively easy to prepare and to manipulate and can accommodate inserts of at least 300 kb. For many routine purposes BACs have superseded YACs as they are more stable and easier to manipulate.

Whether you choose to make the library in a plasmid, bacteriophage λ, a cosmid, a BAC or a YAC depends on a number of factors including the size of the genome and the likely size of the gene or group of genes you are interested in. When working with eukaryotic organisms the large capacity of BACs and YACs can significantly reduce the number of clones that you will need in your library (Table 4.2). It might also depend on the method you intend to use to screen the library for your gene. These methods are discussed in the next chapter.

4.17 cDNA Libraries

Genomic libraries are very useful when you are working with prokaryotic organisms with relatively small compact genomes. They do present a

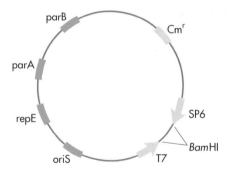

Figure 4.11 The main features of a typical bacterial artificial chromosome (BAC). The oriS and *repE* genes derived from the *E. coli* fertility or F plasmid, are required for replication. The partitioning genes *parA* and *parB* are required to ensure that each daughter cell receives a copy of the plasmid after cell division. BACs usually contain a selectable marker, in this case Chloramphenicol resistance (Cm^r). T7 and SP6 are universal promoters to ensure gene expression from the cloned fragment. In the example shown large DNA fragments can be cloned between the two *Bam*HI sites.

number of problems, however, if you are working with eukaryotes (Section 4.8). For example, for a eukaryotic organism, particularly one with a large genome, you may need a prohibitively large number of clones to make a representative library (Table 4.2). Genomic libraries also contain all the untranslated sequences including introns; if you need your protein to be expressed in *E. coli* you will need clones without introns.

An alternative approach, when you want to find a clone carrying your gene of interest, instead of making a library from the entire genome of an organism, is to make one from the mRNA. This will include only those parts of the genome that are expressed in the particular cell type from which the mRNA is prepared, and will contain the coding regions for eukaryotic

Table 4.2. Number of clones required for a representative library in different types of vector

Vector Insert Size	Plasmid (20 kb)	Bacteriophage λ (25 kb)	Cosmid (45 kb)	BAC (300 kb)	YAC (1 Mb)
E. coli	1,066	852	473	69	19
Yeast	2,761	2,208	1,226	182	53
Human	757,318	605,854	336,584	50,486	15,144
Rice	96,706	77,365	42,979	6,445	1,932

The table shows the number of clones required to make a library, with a 99% chance of obtaining any particular fragment, from a range of organisms of differing genome size; the values have been calculated using the formula given in section 4.4.

genes with the introns removed. A library made from mRNA has another advantage. To illustrate this, imagine that you are trying to clone the gene for α-globin, one of the protein subunits of hemoglobin. This protein is expressed at high levels in reticulocytes. In these immature red blood cells, over 95% of the protein synthesized is hemoglobin and most of the mRNA encodes either α-globin or β-globin. If you make a library from mRNA isolated from reticulocytes, a high proportion of the clones will encode α-globin, making screening much easier.

It is not possible to clone mRNA directly, because it is single-stranded and not a substrate for DNA ligase. Instead, you need to make a DNA copy of the mRNA. This is called complementary or cDNA, the library is constructed from this. Making cDNA from mRNA is a two-step process. First, the mRNA is copied into DNA using the enzyme reverse transcriptase (Box 4.2). Then the mRNA strand is removed from the mRNA–cDNA duplex, and DNA polymerase is used to make the second DNA strand by copying the first strand of DNA. To make a cDNA library the cDNA is then cloned into a vector and used to transform *E. coli* as for a genomic library.

Box 4.2 Reverse Transcriptase

In nature the flow of genetic information is from DNA to RNA to protein. Enzymes like DNA and RNA polymerases, which make either DNA or RNA by reading a DNA template, are well known. It was long held that enzymes capable of doing the reverse, making a DNA copy of an RNA template would not exist in nature. In 1970 two groups independently discovered that some viruses have an RNA-dependent DNA polymerase. Certain viruses, now called retroviruses, have RNA genomes; their replication involves reverse transcription of the infecting RNA genome into complementary DNA. The enzyme responsible for this, reverse transcriptase, has become a key reagent in gene cloning.

Q4.15. What are the most important considerations to take into account when choosing the starting material from which to extract mRNA for the construction of a cDNA library?

Q4.16. What tissue would you choose as your starting material to isolate mRNA from to clone genes for (a) myosin, (b) a protein from the light-harvesting complex of plants?

4.18 Making a cDNA Library

While it is theoretically possible to make a cDNA library from a prokaryotic organism, there would be little to gain from attempting it; cDNA libraries are used to make libraries from eukaryotic organisms.

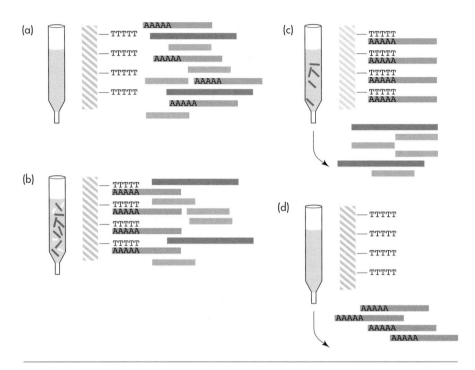

Figure 4.12 Purification of eukaryotic mRNA. a) Total cellular RNA is passed down a column on which polyT oligonucleotides are immobilized on an inert column matrix. b) The polyA tails of the mRNA base pair with the polyT oligonucleotides and are immobilized on the column. c) The other species of RNA can be washed off the column and d) the purified mRNA is finally eluted with a low salt buffer.

The first thing you need to do in making a cDNA library is to isolate mRNA from your chosen starting material. This usually involves disrupting the cells, removing major impurities and then isolating total cellular RNA. As the bulk of the RNA in cells is rRNA and tRNA you will need to separate the mRNA from these. The most straightforward way of doing this is to exploit the fact that, in eukaryotic organisms, a polyA tail is added to the 3′ end of each mRNA transcript. By passing total cellular RNA down a column on which polyT oligonucleotides are immobilized you can purify mRNA (Figure 4.12).

Once you have purified the mRNA you can make a complementary copy of it using reverse transcriptase (Figure 4.13). This enzyme, like other DNA polymerases, requires a short double-stranded region and a free 3′ hydroxyl group to initiate synthesis. Since we want to make a cDNA copy of all the mRNAs from the cell, we need a universal primer to provide the double-stranded region. We can, again, use the fact that eukaryotic mRNA has a polyA tail and use a polyT oligonucleotide primer, usually of about 20 nucleotides in length. This will anneal to all of the polyA tails of the

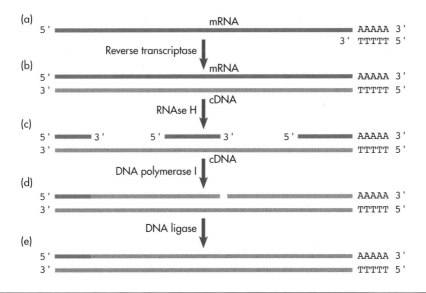

Figure 4.13 Synthesis of cDNA. a) After addition of a polyT oligonucleotide primer, which will anneal to the polyA tail of the mRNA, reverse transcriptase can synthesize complementary DNA using the mRNA as template b). RNAse H is added which will partially remove the RNA strand c), short regions of RNA remain annealed which can act as primers for DNA synthesis by DNA polymerase. The strand displacing activity of DNA polymerase I will ensure that the RNA primers are removed d). Finally DNA ligase is used to mend any nicks in the new DNA strand e).

different mRNA molecules allowing reverse transcriptase to synthesize a strand of DNA complementary to each of the mRNAs in the cell. You now need to replace the mRNA template with DNA; this can be achieved using the enzyme RNAse H which selectively breaks down RNA in an RNA–DNA duplex. A second DNA strand is synthesized by DNA polymerase using short pieces of RNA which remain as primers.

Q4.17. List, in order, the enzymes you would need to use to make cDNA from an mRNA template. In each case say what the enzyme is used for.

4.19 Cloning the cDNA Product

You now have your cDNA. To make a cDNA library, this needs to be cloned into a vector and introduced into *E. coli*. Bacteriophage λ vectors are commonly used for the construction of cDNA libraries because of the high efficiencies associated with these vectors. As you will be cloning relatively small, gene-sized pieces of DNA you can use an insertion vector (Figure 4.6a), but if you are constructing an expression library you will need to use

Box 4.3 Converting a Blunt End to a Sticky End

Although T4 DNA ligase is capable of joining blunt-ended DNA molecules this is less efficient than ligation of molecules with sticky ends. It is possible to introduce a sticky end onto a blunt-ended DNA molecule by the use of linkers or adaptors. These are short synthetic oligonucleotides, which encode a restriction enzyme recognition site. They can be ligated to the end of blunt-ended DNA molecules to introduce a restriction site that is required for cloning.

Linkers

(a)

P CGGGATCCCG OH

HO GCCCTAGGGC P

(b)

P CGGGATCCCG OH
GCCCTAGGGC
HO P

*Bam*HI site

(c) Ligate linkers

(d) Cut with *Bam*HI

Linkers are made from two complementary oligonucleotides (a). They will anneal to each other to form a blunt-ended double-stranded molecule containing a restriction enzyme recognition site (b). When mixed at high concentrations with blunt-ended DNA fragments multiple linkers will be ligated onto each end of the molecule (c). The resulting molecule is still blunt-ended, but cutting with the restriction enzyme results in removal of the extra linkers and generates sticky ends (d). The main problem with this approach is that if your DNA fragment contains the same recognition site as the linker it will also be cut.

Adaptors

These are also short double-stranded oligonucleotides but in this case they have one blunt and one sticky end. Adaptors are often generated from two identical oligonucleotides (a) which anneal in such a way as to give the recognition site for a restriction enzyme, which results in blunt ends,

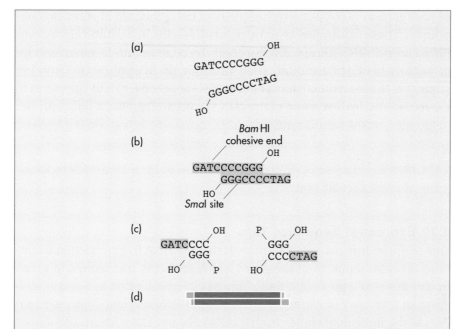

flanked by sticky ends compatible with a second restriction enzyme. The oligonucleotides used to make adaptors are synthesised without the 5′ phosphate, this ensures that they cannot ligate to each other (b). The molecule is then cut to give two identical adaptor molecules that are asymmetric, having one blunt and one sticky end (c). In this case a single adaptor will be ligated to each end of the blunt-ended DNA fragment (d), they already have sticky ends so there is no need to cut with a restriction enzyme. Once any unligated adaptor molecules have been removed the 5′ phosphate is added to the sticky end, using DNA polynucleotide kinase, so that it can be used with a dephosphorylated vector.

a vector with a suitable promoter to ensure the cloned genes are expressed (Figure 4.6c). As cDNA does not have sticky ends it may be quite difficult to clone directly. One common solution to this problem is to add linkers or adaptors (Box 4.3) to the ends of the cDNA molecules. These provide sticky ends and make cloning more efficient.

Q4.18. What type of vector would be suitable for cloning cDNA products to make a library?

There are many variations on this basic technique for making cDNA, all designed to improve the quality of the product. In some techniques a second primer is used to prime second-strand synthesis. Obviously, you cannot use a specific primer unless you know what the sequence of the 3′ end of your gene is. A common technique is to use a mixture of short random primers. These will anneal at many points along the first-strand cDNA template and prime synthesis of the second strand. Random hexanucleotides or nonanucleotides are often used in this type of application. Another strategy is to use terminal deoxynucleotidyl transferase to add T residues to the terminal 3′ hydroxyl end of the cDNA, you can then use a polyA primer to initiate second-strand synthesis.

Q4.19. How many hexanucleotides would you need to synthesize to have one of every possible sequence?

4.20 Expressed Sequence Tags

One of the features of a cDNA library is that the DNA fragments that it contains are representative of the genes which are being transcribed in a particular tissue at a particular time. This has been exploited as a way of creating an inventory of expressed genes by partially sequencing large numbers of the individual clones from a cDNA library. These short, sequenced regions act as tags identifying individual transcribed regions and are dubbed expressed sequence tags (ESTs). High-throughput sequencing methods (Section 7.7) are used to generate raw partial sequences from one or both ends of individual cDNA clones rapidly and cheaply. They are not intended as definitive sequence but they are of sufficiently high quality to identify similar sequences in sequence databases; this will often make it possible to identify the likely product of the gene they are derived from using homology searches (Section 8.8). As ESTs represent regions of the genome which are transcribed into mRNA, they are useful in helping to identify protein coding regions (Section 8.2) and new genes in DNA sequencing projects. ESTs have also proved invaluable as genetic markers in mapping projects (Section 6.6) and, if ESTs have been generated from both ends of the cDNA clones, they can be used to design PCR primers. Also, because the ESTs from a given cDNA library represent a snapshot of the genes expressed in the tissue the library was derived from, they can be used as a way of comparing gene expression in different tissues at different times. This basic idea has been developed into the micro-array for global studies of gene transcription (Section 11.6).

4.21 What are the Disadvantages of a cDNA Library?

cDNA libraries contain only the parts of genes which are found in mature mRNA. We saw earlier that this can be an advantage because introns will have been removed. However, if you are interested in the sequences before and after the gene, for example those which are involved in the regulation of expression of the gene, you will not find them in a cDNA library. If your cloning strategy is dependent on expression of a eukaryotic gene, you may well have chosen to make a cDNA library so that the cloned gene will have had the introns spliced out and can be expressed in *E. coli*. In this case you will have to clone it into a vector that contains a suitable promoter. See Section 9.3 for more discussion of expression vectors. If the gene you are trying to clone is only ever expressed at low levels, then there will be very little mRNA for it in any cell type and it may be completely swamped by the more abundant species. In this case construction of a cDNA library would not be the best approach for isolating the gene, unless it can be coupled with an approach for screening very large numbers of colonies for the one containing the desired gene.

Q4.20. What are the two main differences between a genomic and a cDNA library?

Q4.21. In each of the following examples decide whether a cDNA or a genomic library would be the most appropriate and give your reasons.

(a) You are trying to clone the genes from a bacterium involved in the synthesis of a novel antibiotic, and your preliminary research suggests that several genes, probably all part of the same operon, are involved.

(b) You are trying to clone the genes for the enzyme rubisco from spinach. This protein, which is responsible for fixing CO_2, is a multimeric protein composed of two subunits. It is the most abundant protein on Earth, and makes up 15% of all chloroplast proteins.

(c) You are trying to clone the dystrophin gene. Mutations in this gene give rise to the human sex-linked disorder muscular dystrophy.

Questions and Answers

Q4.1. Once you have amplified the region containing your gene by PCR you will need to clone it into a vector. How would you do this?

A4.1. *In section 3.17 we described two methods by which PCR products can be cloned. You could take advantage of the fact that* Taq *polymerase tends to add a 3′ adenosine residue to the PCR product and clone into a T-tailed vector. Alternatively, you could incorporate restriction enzyme recognition sites into the PCR primers, which can then be used for cloning.*

Q4.2. What are the basic steps you would need to construct a library containing genomic DNA from a bacterium? *Hint.* This is just a variation of the techniques outlined in Chapter 3. Instead of cloning a particular DNA fragment, you want all the DNA from an organism.

A4.2. *Extract chromosomal DNA. Cut both DNA and vector with the same restriction enzyme, mix and ligate. Transform into competent* E. coli *and plate on selective agar. Remember from Chapter 3 that each colony, which grows on an agar plate, is a collection of individual bacteria descended from a single organism which arose from a single transformation event. You started with a heterogeneous mixture of DNA fragments representing the whole genome of your bacteria, therefore each colony will contain vector carrying a different DNA fragment. If you have enough colonies, you should find any particular part of the genome in one or other of the colonies.*

Q4.3 Five different colonies in your *Pseudomonas* genomic library turn yellow when sprayed with catechol. How would you extract plasmid DNA from them? How would you find out whether all of the five colonies carried the same piece of genomic DNA?

A4.3. *You would need to subculture each of the colonies. This would involve growing them in separate broth cultures until a high density of bacterial cells had been achieved. You could then extract the plasmid DNA using a method such as the miniprep method outlined in Section 3.6. To find out whether the clones carry the same genomic DNA insert you would need to restriction map them (as described in Section 3.10). This means digesting the extracted plasmid DNA with a range of different restriction enzymes and comparing the patterns generated.*

Q4.4. How many clones would you need in a genomic library made from (a) *E. coli* or (b) *Saccharomyces cerevisiae* to have a 99% chance of cloning a particular gene? Assume an average fragment size of 4000 kb as before.

A4.4. (*a*) *For* E. coli *with a genome size of 4.64 Mb*

$$N = \frac{\ln (1 - 0.99)}{\ln (1 - 4/4640)} = \frac{\ln 0.01}{\ln 0.99914}$$

$$= \frac{-4.61}{-0.00086} = 5340.$$

Because the E. coli *genome is smaller than the* Pseudomonas *genome you would only need 5340 clones in your library to have a 99% chance of a particular gene being present.*

(*b*) *For* Saccharomyces cerevisiae *with a genome size of 12.1 Mb,*

$$N = \frac{\ln (1 - 0.99)}{\ln (1 - 4/12,100)} = \frac{\ln 0.01}{\ln 0.99967}$$

$$= \frac{-4.61}{-0.00033} = 13,970.$$

Because the S. cerevisiae *genome is larger, than the bacterial genomes we have considered so far, you would need 13,970 clones in your library to have a 99% chance of a particular gene being present.*

Q4.5. How many clones of *E. coli* would you need to have a 99.9% chance of cloning a particular gene?

A4.5. *You would need 8010 clones to have a 99.9% chance of cloning your gene. This is nearly twice as many clones and you have only increased your chances by 0.9%. This is sometimes referred to as the law of diminishing returns; you have to decide whether increasing the number of clones is worthwhile.*

Q4.6. Using a partial *Sau*3AI digest you have isolated DNA fragments from *Pseudomonas* with an average size of 10 kb and constructed a genomic library. How many clones would you need to have a probability of 0.99 of finding the gene of interest?

A4.6. *Using the formula with* f = 10, g = 6300 *and* P = 0.99,

$$N = \frac{\ln\,(1 - 0.99)}{\ln\,(1 - 10/6300)} = \frac{\ln\,0.01}{\ln\,0.9984}$$

$$= \frac{-4.61}{-1.5885 \times 10^{-3}} = 2899.$$

With these larger sized fragments you would only need 2900 clones compared with over 7000 for a library made with 4 kb fragments.

Q4.7. Using a *Sau*3AI partial digest presents a problem when you come to cutting your cloning vector. Can you see what the problem is? Can you think of a cloning strategy which would overcome this problem? *Hints.* How many times would you expect *Sau*3AI to cut pUC18? Is this a problem? Is there a restriction enzyme with sticky ends, which are compatible with those produced by *Sau*3AI, with a unique site in pUC18?

A4.7. *Four base cutters like* Sau3AI *cut on average every 256 bp so it is likely that any cloning vector will be cut into many pieces rather than being opened up at a single site. Refer to Figure 3.10; you can cut the vector with* BamHI *or* BglII *and ligate the* Sau3AI *fragments into the site. Many commonly used cloning vectors have a unique restriction site for at least one of these enzymes, both of which produce sticky ends which are compatible with those from* Sau3AI.

Q4.8. What factors may cause some clones to be less well represented in the library than others?

A4.8. *See Section 4.5. Factors such as the relative burdens placed on the host bacterium by different clones, the size of the insert and the characteristics of the DNA sequence all play a role in the likelihood that a clone will continue in the library on repeated subculture.*

Q4.9. Assuming that you can work at a density of 500 colonies per plate how many plates would you need to make a 99% representative (a) genomic library containing 10 kb fragments of *Pseudomonas* DNA, and (b) genomic library containing 20 kb fragments of human DNA?

A4.9. *(a) Using a P value of 0.99 you would need 2899 colonies in a representative* Pseudomonas *library. You would not be able to distinguish the individual colonies if you plated this number of clones on a single agar plate. Normally you would plate out a small sample of your transformation mix to work out the titer (i.e. how many colonies you would expect to get from 1 mL) you can then plate your library out at a density of 500 colonies per plate. At this density you would need six plates; that is a reasonable number of plates to screen.*

(b) For the whole human genome of 3.29×10^6 kb you would need 757,000 colonies, or about 1500 plates!

Q4.10. How would you prepare insert DNA suitable for cloning into a bacteriophage λ replacement vector? *Hint.* You need large fragments of DNA so complete digestion with a six cutter will not work. You will also need to select fragments of the correct size.

A4.10. *A partial digest with a restriction enzyme with a 4 bp recognition site, such as Sau3AI, can be controlled to give fragments in the correct size range. You will need to run the digested DNA on a low percentage agarose gel. You can then cut out the region where the 20–25 kb fragments are and extract them from the gel. The reason that you cannot use a six cutter like BamHI is that, although if you cut to completion with BamHI you will also have some fragments of the correct length for packaging, they will not be representative of the whole genome. Areas of the genome where BamHI sites are less than 20 or more than 25 kb apart will not be represented.*

Q4.11. You have performed an experiment to measure the titer of λ library stock. You make serial 10-fold dilutions of your stock. You take a 10 μL sample from each dilution and mix it with 100 μL of *E. coli* and 2 mL of molten agar and pour onto an agar plate. After incubation at 37°C you count the plaques. You find that the 10^{-3} dilution gives 512 plaques per plate, the 10^{-4} dilution gives 54 and the 10^{-5} dilution gives four plaques. Calculate the titer of the library stock.

A4.11. *Most of the plates will have so many plaques that you cannot count them accurately. You need to choose the plate with enough plaques to be sta-*

tistically significant but be sure that you can easily distinguish each plaque. In this case it is probably the 10^{-4} dilution. Then the calculation needs to take into account the volume of stock plated out and the dilution. Multiply the number of plaques by the dilution factor:

$$54 \times 1/10^{-4} = 5.4 \times 10^5$$

this gives you the number of plaques in 10 µL, and

$$5.4 \times 10^5 \times 100 = 5.4 \times 10^7 \; pfu \; mL^{-1},$$

multiply by 100 to find the concentration per milliliter.

Q4.12. How much of what dilution of this stock would you use to produce plates for screening with about 200 plaques each?

A4.12. *Ten milliliters of a 1/2700 dilution would give you 200 plaques per plate. This dilution would normally be achieved by diluting 1/2.7 and then doing three 10-fold dilutions. These operations often involve an awkward dilution like 1/2.7; this is usually achieved by mixing 100 µL of the λ library stock with 170 µL of buffer.*

Q4.13. What range of sizes of fragments of genomic DNA would be suitable for constructing a cosmid library using pHC79?

A4.13. *Since it is possible to package molecules with cos sites between 37 and 52 kb apart (75–105% of the size of the λ genome, see Figure 4.4) and pHC79 is only 6.5 kb, fragments ranging in size from 30.5 to 45.5 kb would be suitable.*

Q4.14. Why is it important, when cloning into a cosmid vector, to size fractionate the genomic DNA to ensure that the fragments are in the correct size range?

A4.14. *Concatemers with very large inserts would not package at all, and it is advisable to remove these from the preparation because they would tie up vector molecules in unpackagable concatemers. The main problem is with smaller fragments, which could ligate together to make molecules large enough to package. This could give a mistaken impression of the relative positions of these fragments in the genome by bringing together, into the same cosmid, fragments from different parts of the genome. Once again it is advisable to size fractionate the genomic DNA (see Section 4.6) after restriction enzyme digestion.*

Q4.15. What are the most important considerations to take into account when choosing the starting material from which to extract mRNA for the construction of a cDNA library?

A4.15. *Most importantly the starting material must contain cells in which the gene of interest is expressed, preferably at high levels. A second important consideration is that you can get reasonable quantities of the starting material easily.*

Q4.16. What tissue would you choose as your starting material to isolate mRNA from to clone genes for (a) myosin, and (b) a protein from the light-harvesting complex of plants?

A4.16. *Muscle is highly enriched in myosin and so would be a suitable choice for (a). Light harvesting complex is present at high levels in photosynthetic tissue, so green leaves would be the ideal source of RNA to try to isolate the genes for components of this complex (b).*

Q4.17. List, in order, the enzymes you would need to use to make cDNA from an mRNA template, in each case say what the enzyme is used for.

A4.17. *Reverse transcriptase makes a DNA copy from an mRNA template, RNAse H removes the RNA strand from the RNA–DNA duplex, DNA polymerase synthesizes the second DNA strand using the first as a template, DNA ligase mends any single-stranded nicks in the second strand.*

Q4.18. What type of vector would be suitable for cloning cDNA products to make a library?

A4.18. *It is quite common to use λ insertion vectors in the construction of cDNA libraries because of the high transformation frequencies which can be achieved. λ insertion vectors are used in preference to λ replacement vectors as cDNAs are generally not very large (Figure 4.6).*

Q4.19. How many hexanucleotides would you need to synthesize to have one of every possible sequence?

A4.19. *This is basically the same question as: how often would a restriction enzyme with a six base pair recognition site cut? There are four possible bases at each of the six positions. Therefore you would need 4^6 or 4096 oligonucleotides. In fact it is unlikely that these would be synthesized individually, it is more usual to synthesize a mixture of random oligonucleotides, each six nucleotides in length.*

Q4.20. What are the two main differences between a genomic and a cDNA library?

A4.20. *In a genomic library, at least in theory, all the sequences from the genome are equally represented. However, a cDNA library only contains sequences for genes expressed in the starting material. This means that some genes may be highly represented while others are completely absent from the library. All of the sequences in the genome, including promoters and terminators, introns and junk DNA will be included in a genomic library whereas a cDNA library will only contain coding regions.*

Q4.21. In each of the following examples decide whether a cDNA or a genomic library would be the most appropriate and give your reasons.

(a) You are trying to clone the genes from a bacterium involved in the synthesis of a novel antibiotic, and your preliminary research suggests that several genes, probably all part of the same operon, are involved.

(b) You are trying to clone the genes for the enzyme rubisco from spinach. This protein, which is responsible for fixing CO_2, is a multimeric protein composed of two subunits. It is the most abundant protein on Earth, and makes up 15% of all chloroplast proteins.

(c) You are trying to clone the dystrophin gene. Mutations in this gene give rise to the human sex-linked disorder muscular dystrophy.

A4.21. *(a) A genomic library would be the most suitable. Bacterial genomes are usually small so the library would be of a manageable size. If you want to express the antibiotic, which would be the easiest way to isolate the clone from the library, you will need all of the genes, together with promoters and other regulatory sequences to be cloned on the same fragment. This is only possible with a genomic library. You would probably want your library to consist of quite large DNA fragments so a cosmid would be a good choice of vector. Note that this approach will only work if all the genes are grouped together on the chromosome as in an operon. Preliminary genetic mapping may be required to test whether or not this is true.*

(b) In this case a cDNA library would probably be the most suitable. You should have no difficulty in getting hold of large quantities of leaf from which to isolate the mRNA and you would expect to find that a large proportion of the clones in your library contained the gene you are looking for. You might consider making a chloroplast cDNA library in the hope of making a library in which a higher proportion of the clones would encode rubisco, but in this case you would only find the genes for the small subunit as the large subunit is encoded on a chromosome.

(c) In this case the answer is less clear cut. The dystrophin gene is very large and split into many exons, although the protein itself is fairly small. This suggests that a cDNA library would be a good approach as it is the only way you are likely to get a single clone encoding the whole gene. However, the

dystrophin gene is not highly expressed even in muscle tissue and may be hard to find in a cDNA library, you may have to use a genomic library. Because muscular dystrophy is a sex-linked disease you can deduce that dystrophin is encoded on the X chromosome so you need only include the X chromosome in your library. In fact the cloning of the dystrophin gene was a complex process in which the researchers made use of both genomic and cDNA libraries and a great deal of genetic mapping data.

Further Reading

Library construction is described in laboratory manuals and, of course, text books like this one. The recommended reading for this chapter refers to each of the different types of specialist vectors used in library construction

A small cosmid for efficient cloning of large DNA fragments (1987) Hohn B and Collins J, Gene, Volume 11 Pages 291–298.
This paper describes an early cosmid vector

λ-replacement vectors carrying polylinker sequences (1983) Frischauf AM, Lehrach H, Poustka A and Murray N, Journal of Molecular Biology, Volume 170 Pages 827–842.
A description of a family of λ-replacement vectors for genomic library construction

Cloning and stable maintenance of 300-kilobase-pair fragments of human DNA in *Escherichia coli* using an f-factor-based vector (1992) Shizuya H, Birren B, Kim UJ, Mancino V, Slepak T, Tachiiri Y, *et al.*, Proceedings of the National Academy of Sciences of the United States of America, Volume 89 Pages 8794–8797.
This paper describes construction of a human genomic library in a bacterial artificial chromosome

Cloning of large segments of exogenous DNA into yeast by means of artificial chromosome vectors (1987) Burke DT, Carle GF and Olson MV, Science, Volume 236 Pages 806–812.
This paper describes the cloning of large segments of human and yeast DNA into the yeast artificial chromosome pYAC2

5 Screening DNA Libraries

Learning outcomes:

By the end of this chapter you will have an understanding of:

- *the various approaches that can be used to identify the clone of interest from a library*
- *which approach would be most suitable in a given set of circumstances and the limitations of the techniques available*
- *DNA–DNA hybridization*
- *a range of approaches to labeling DNA*
- *how to clone a gene by homology with a known gene*
- *how PCR can be used to screen a DNA library*

5.1 The Problem

In the previous chapter we discussed the problem of obtaining a clone containing the part of the genome which encodes the gene or genes we are interested in. In many cases there may be enough genome sequence data available for you to design PCR primers flanking the region of interest; you can then simply amplify the part of the genome you are interested in by PCR. In situations where genome sequence data is not available you may need to make a DNA library. Having constructed the DNA library you now have a collection of probably many thousand clones in the form of either plaques or colonies on a plate. The process of determining which of those clones carry the gene or genes you are interested in is called screening; this is the subject of this chapter.

In Section 4.4 we calculated that if we made a genomic library containing fairly small (4 kb) fragments of bacterial genomic DNA we would need more than 5000 individual clones in the library to have a 99% chance of a particular gene being present. For a eukaryotic organism such as the yeast *Saccharomyces cerevisiae* this number increases to about 14,000 clones and for a human being or a higher plant it would be higher still. Using bacteriophage and cosmid vectors, which make it possible to clone larger DNA fragments, the number of clones in a genomic library can be reduced, but it is still a formidable task to detect the one or two clones containing your gene of interest.

> **Q5.1.** Using a cosmid vector such as pHC79 how many clones would you need to have a 99% chance of finding a particular gene in a genomic library of (a) *Saccharomyces cerevisiae* and (b) human? *Hint.* You will need to know the size of insert which can be cloned in pHC79 and the size of the *S. cerevisiae*, and human genomes. You will also need to use the formula in Section 4.4.

In the case of a cDNA library the problem may be less daunting. If you have been able to make your cDNA library from a starting material that is rich in the mRNA transcribed from the gene of interest, a high proportion of clones in the library may encode the gene of interest. However, few genes are expressed at levels as high as in that of the example, α globin in reticulocytes, used in Section 4.4 and you will still need a method of detecting the clones containing the gene you are trying to clone.

The method you use to detect the clone of interest depends on what information you already have about the gene you are trying to clone. The sorts of information you might have, which would be useful in designing a screening strategy, would be information about the DNA or protein sequence, or information about the phenotype that the cloned DNA should confer on a positive clone. Using the information you have, you need to design a screen that will allow you to test many thousand clones as quickly and easily as possible. Screening methods use two main approaches: those that rely on expression of the gene and those where you look for a particular DNA sequence.

5.2 Screening Methods Based on Gene Expression

In order to use a screening method based on expression of the gene of interest it is necessary to construct an expression library in a vector designed to ensure that the cloned genes are expressed (see Figures 4.6 and 9.1). You will then need a way of detecting expression of your particular gene. In the example used in Chapter 4 of cloning the *xylE* gene from *Pseudomonas aeruginosa* we were lucky because the enzyme, catechol 2,3-oxygenase, which is the product of the *xylE* gene, can be detected by a simple colorimetric test (Section 4.2). In this case we could plate out the library, spray each plate with catechol and pick out the yellow colonies (Figure 5.1). A range of bacterial genes can be screened for in this way including those involved in the utilization of specific sugars and the genes involved in antibiotic production and resistance to antibiotics.

> **Q5.2.** Devise a screening method that would allow you to detect clones encoding the genes for resistance to the antibiotic tetracycline from a genomic library constructed from a resistant strain of *Escherichia coli*.

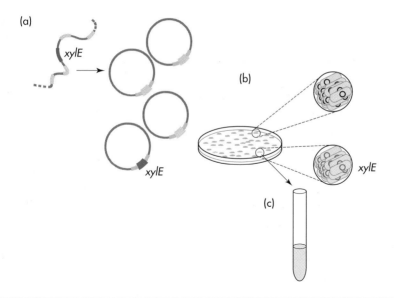

Figure 5.1 Screening a DNA library by detection of a phenotype. It is possible to detect the activity of the *xylE* gene of *Pseudomonas aeruginosa* using a simple colorimetric test. a) A genomic library is made by cloning fragments of *P. aeruginosa* DNA into a vector and transforming into *E. coli*. b) The library is plated out and the colonies are sprayed with catechol. Clones expressing *xylE* will turn yellow and can be isolated and c) grown up in liquid culture.

5.3 Complementation

E. coli has a very versatile metabolism and can grow on a minimal medium containing only simple sugars and minerals. Unlike human cells *E. coli* cells are able to synthesize organic molecules like amino acids. Now imagine that you want to clone the genes responsible for the synthesis of, for example, the amino acid lysine. You might be able to devise a simple test for the presence of lysine but, as it is unlikely to be secreted from the bacteria, any test is likely to involve some complex manipulations. One approach to this problem is to use complementation. It is relatively easy to make auxotrophic mutants of *E. coli*, that are unable to grow on minimal media without supplementation with specific nutrients, by treating the cells with UV light or a chemical mutagen. If such a mutant is unable to grow in the absence of lysine (Lys⁻), it is a reasonably safe assumption that the mutations fall in one or other of the enzymes involved in the synthesis of lysine. To screen a genomic library, constructed from a wild-type strain of *E. coli*, for clones expressing genes involved in the synthesis of lysine, you need to introduce your library into the mutant Lys⁻ strain and plate onto minimal medium without lysine. The colonies which grow must contain a clone which is able to complement the mutation in the Lys⁻ strain. These colonies can be isolated and cultured for further investigation.

One obvious problem with this approach is that you will only identify the gene in which the original mutation lies. In the case of lysine synthesis, as with most biochemical pathways, there are a whole series of enzymes which must act in sequence. In bacteria the genes encoding these enzymes are often arranged next to each other on the chromosome, sometimes forming a functional unit called an operon. If your library was constructed from relatively large fragments of DNA you may well have already cloned many of the other genes in the pathway. If there are still missing components of the pathway you may need to screen a whole bank of mutations by complementation.

Screening by complementation, or marker rescue as it is sometimes known, has been very useful in the elucidation of biochemical pathways both in bacteria and also in eukaryotic organisms such as the yeast *Saccharomyces cerevisiae*. It relies on being able to make a mutation in a suitable host strain such as *E. coli* that can be complemented by the gene you are trying to clone. Auxotrophic mutants of *E. coli* have been used successfully in screening for genes involved in metabolism in a wide range of bacteria, as the genes are usually similar enough to complement the mutations in *E. coli*.

5.4 Immunological Screening of Expression Libraries

The approaches to screening DNA libraries by looking for protein expression which we have discussed so far rely on techniques that are specific to the protein being studied. Ideally, we would like a technique that would be adaptable to detect a wide range of proteins. Antibodies have been used in a wide range of applications, such as western blotting (Box 5.1), to detect specific proteins immobilized on a sheet of nitrocellulose (Figure 5.2). This

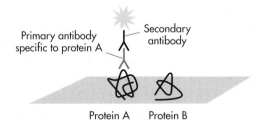

Primary antibody specific to protein A

Secondary antibody

Protein A   Protein B

Figure 5.2 Using specific antibodies to detect proteins. Proteins are usually immobilized on a suitable inert support. The primary antibody is applied and will bind only to the specific protein it was raised to detect; excess antibody is washed off. A secondary antibody is used to detect bound primary antibody. Antibody molecules of a particular class (i.e. IgG) raised in a particular species have a common epitope; this means that a single antibody will be capable of detecting all rabbit IgG antibodies for instance. However, if your primary antibody was raised in a rat you will need a different secondary antibody. Secondary antibodies are commercially available and are conjugated to detection systems, commonly alkaline phosphatase or horseradish peroxidase, which are enzymes that can be detected by simple colorimetric tests.

Box 5.1 Gel-based Methods for Analyzing Proteins

SDS polyacrylamide electrophoresis (SDS PAGE)
Proteins can be separated according to their size on a polyacrylamide gel in much the same way as DNA molecules (Section 3.11), but unlike DNA molecules, proteins do not have a uniform charge. The anionic detergent SDS is included in the gel and the loading buffer, this has the effect of disrupting the secondary structure of the protein and imparting a uniform negative charge allowing proteins to be separated solely according to size. Proteins are denatured by heating in a buffer containing SDS and mercaptoethanol before loading onto the gel and SDS is included in the gel to ensure that the proteins remain denatured. In SDS PAGE the gel has two components: the stacking gel which separates the proteins according to size and the resolving gel which spreads them out so that they can be distinguished more easily.

Western blotting
Once separated on a gel, proteins can be transferred to a nitrocellulose membrane and probed with antibodies. This process is analogous to Southern blotting and northern blotting, but in the case of proteins an electrical current has to be applied to move the proteins from the gel to the membrane. One of the remarkable features of western blotting is that the proteins re-fold on the nitrocellulose membrane; this is important as antibody recognition sites may be made up of amino acids which are not adjacent to each other unless the protein is correctly folded.

Isoelectric focusing (IEF)
In this method proteins are separated according to charge rather than size, by electrophoresis on a gel with a pH gradient. During electrophoresis the proteins will migrate through the gel to the point where they have no net charge and are hence not affected by the electrical current. This point, the pH where the protein has no net charge, is called the isoelectric point or pI. IEF is often used in conjunction with SDS PAGE in two-dimensional electrophoresis where proteins are separated according to their isoelectric point in the first dimension and then according to their molecular weight in the second giving rise to spots rather than bands in the finished analysis.

technology has been adapted for screening DNA libraries. To use an antibody to screen a library, the library is plated out and the bacteria lysed to release intracellular proteins. Proteins are then transferred to a solid support and probed with the antibody. This approach has a very wide range of applications and can be used to screen for both prokaryotic and eukaryotic genes.

The main difficulty with antibody-based technology is that it is necessary to raise a specific antibody for each protein that you wish to detect. This usually means being able to purify the specific protein in sufficient quantity to be able to inject it into a small mammal, usually a mouse, rat or rabbit, in order to cause an immune response. Serum is then harvested from the animal and the antibody may be further purified before use. This is a lengthy and costly procedure and can only be carried out successfully with proteins which can be produced in reasonably large amounts.

It is common to use vectors derived from bacteriophage λ (Section 4.10) in the construction of libraries for antibody screening. This is because when you plate a bacteriophage λ library, plaques are produced where the bacteria have been lysed (Figure 4.3). This means that proteins are released from the bacterial cells and are readily available to the antibodies and there is no need for subsequent lysis of the bacterial colonies.

> **Q5.3.** You want to construct a DNA library, which you intend to screen with an antibody, for expression of a eukaryotic gene. (a) What kind of DNA library would be most suitable? Give your reasons. (b) What type of vector would you construct your library in? Explain your choice.

Preparation of filters

To use immunological screening we need to immobilize the proteins, produced by each of the phage in our library, on a solid support, in such a way that we know which original plaque they were produced by. This technique, often called a "plaque lift", is outlined in Figure 5.3. For immunological screening nitrocellulose filters are used, as proteins will bind to them. The filters are then bathed in a solution containing the primary antibody, which will bind specifically to the protein it was raised against. Excess antibody is washed off and the bound primary antibody is detected using a secondary antibody conjugate (Figure 5.2). Using this technique it is possible to pinpoint which of the original plaques was produced by a phage expressing the protein of interest. It is often difficult to be sure exactly which plaque corresponds to the signal on the filter, especially if the library was plated at high density. In this case a plug of agar is removed from the area of the positive signal, phage are grown up from it and replated at a lower density and the screening process is repeated. Once a single plaque on an agar plate has been identified, which is expressing the protein of interest, the phage can be isolated for further study.

> **Q5.4.** Describe the approach you would use to clone the gene for (a) β-galactosidase (see Section 3.9) from the Gram negative bacterium *Citrobacter*, and (b) human insulin. You should consider what type of DNA library would be the most suitable and what form of screening for expression you could use.

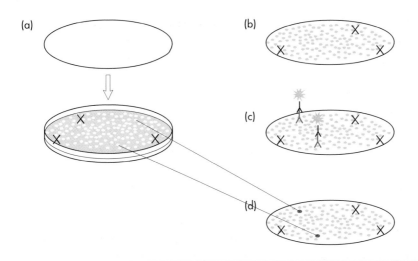

(a)

(b)

(c)

(d)

Figure 5.3 Immunological screening of a phage library. a) A nitrocellulose disk is placed onto the surface of an agar plate on which the phage library has been plated (see Figure 4.3). Marks are made on the agar plate and the nitrocellulose disk so that they can be realigned later. b) When the nitrocellulose disk is lifted off again, proteins released from the bacteria by phage lysis are bound to the nitrocellulose support. c) This can be probed with specific antibody (Figure 5.2). d) Plaques formed by bacteriophage that are expressing the protein detected by the antibody will "light up" (shown in blue). The agar plate and the nitrocellulose disk can be realigned and the plaques which relate to the positive clones can be identified.

5.5. Screening Methods Based on Detecting a DNA Sequence

This is perhaps the most commonly used approach to library screening. It relies on the fact that a single-stranded DNA molecule will hybridize to its complementary sequence (Figure 5.4). In this approach the single-stranded DNA molecule is used as a probe in a hybridization experiment to bind to and thus identify specific sequences. DNA probes used in this way are usually either synthetic oligonucleotides or fragments of cloned DNA. DNA probes can be used to probe DNA released from bacterial colonies directly onto the nylon membrane (colony blot, Section 5.8) or, as described in Box 5.2, purified DNA samples which are either applied directly to the membrane (dot blot) or size separated on an agarose gel prior to analysis (Southern blot).

In order to use a DNA probe to screen a library you will need some DNA sequence or cloned DNA from which to derive the probe. In some experimental situations you may already know all or part of the sequence of the gene you are trying to clone. An example of this would be when you have a clone from a cDNA library but you want a genomic clone of the same gene. This might be because you are interested in sequences outside the coding

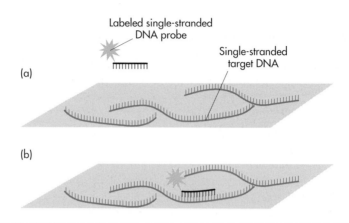

Figure 5.4 Hybridization of DNA probes. a) The DNA to be probed is denatured to make it single-stranded and immobilized on a solid support; a labeled DNA probe is added. b) The probe will anneal to molecules with a complementary sequence of bases, hydrogen bonds will form between complementary bases and a double-stranded duplex will be formed. This process is often referred to as hybridization. In this way molecules complementary to probe will be specifically labelled and hence can be identified.

region, either regulatory sequences or adjacent genes. Alternatively, you might know the sequence of the gene in a related organism; DNA hybridization using a probe derived from one species can be used to detect homologous genes either in the same organism or in other species. If you are trying to clone a gene for which there is no sequence data it may still be possible to use DNA hybridization, as long as you can purify the protein product of the gene of interest. If you can determine the sequence of part, usually the N-terminus, of the protein you can then derive the DNA sequence from this.

5.6 Oligonucleotide Probes

Short synthetic DNA molecules similar to those used as primers in PCR and sequencing can be used as probes in hybridization procedures. Oligonucleotide probes can be designed from known DNA sequence but more usually are derived from amino acid sequence. Relatively small samples, containing as little as 5 pmol of purified protein, can be sequenced by the Edman degradation. This is a sequential procedure in which the N-terminal amino acid is chemically cleaved from the protein and identified by liquid chromatography; the procedure is repeated with subsequent amino acids. This is now an automated process; it can reliably identify the first 20 amino acids of a protein, although some proteins are blocked making sequencing by this approach very difficult.

Q5.5. Why is it not possible to derive an unambiguous DNA sequence from the amino acid sequence of a protein?

Box 5.2 Dot Blots and Southern Hybridization

These are both techniques which are used to probe purified DNA samples immobilized on a nylon membrane. In a dot blot the DNA is applied directly to the nylon membrane whereas in Southern hybridization DNA fragments are first separated by agarose gel electrophoresis and then transferred to a nylon membrane. The process of probing the filters is the same in both procedures.

In Southern hybridization DNA is purified and cut with a restriction enzyme to produce a series of fragments; these are separated according to size by agarose gel electrophoresis (a). The DNA fragments are then transferred to the nylon membrane (b). This can be done very simply by placing the gel onto a filter paper wick dipped in buffer, the nylon membrane is placed on top and absorbent pads are used to draw the buffer through the gel and membrane. As the buffer passes through the gel the DNA fragments are drawn up onto the membrane where they are deposited. The DNA can be fixed to the membrane and probed with a labeled oligonucleotide or DNA fragment as described for colony hybridization in Section 5.8 (c). This will reveal only those fragments or spots which have DNA sequence complementary to the DNA probe.

The real strength of Southern hybridization is that it allows you to focus in on the specific part of the DNA which is of interest. This is particularly useful when examining genomic DNA where a restriction digest will give rise to so many fragments that they will appear as a smear on a gel. Southern blotting allows examination of individual fragments such as RFLPs (Box 6.4). Unlike colony and plaque hybridization and dot blots, Southern hybridization allows you not only to detect the presence of the sequences but to calculate the size of the restriction fragment that they are present on.

An important application of both dot blots and Southern hybridization is to look for homologous sequences in the genomic DNA of a range of related organisms using a probe derived from the gene of interest in one of the organisms; this is described as a zooblot.

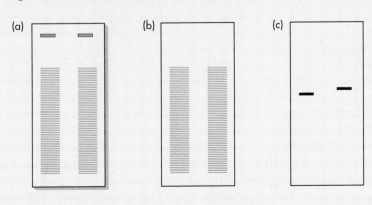

(a)   (b)   (c)

N terminus

Phe	Val	Asp	Glu	His	Leu	Cys

Direction of chemical synthesis

TTT	GTT	GAT	GAA	CAT	CTT	TGT
TTC	GTC	GAC	GAG	CAC	CTC	TGC
	GTA				CTA	
	GTG				CTG	
					TTA	
					TTG	

Figure 5.5 Deriving a nucleotide sequence from an amino acid sequence. The first seven amino acids of the sequence of the B chain of human insulin are shown written in the three-letter code. Shown underneath each amino acid are all the possible triplet codons which encode each amino acid. (Refer to the genetic code in Table 8.1). DNA synthesizers add new residues to the 5′ phosphate terminus unlike biosynthesis of DNA where additions are to the 3′ hydroxyl.

Designing an oligonucleotide probe

Imagine that we want to use an oligonucleotide to screen a human DNA library for a clone of the insulin gene. Have a look at Figure 5.5 which shows the N-terminal sequence of the B chain of human insulin. We would need $2 \times 4 \times 2 \times 2 \times 2 \times 4 \times 2$ or 1024 different oligonucleotides of 21 bp in length to be sure of having all the possible combinations. By thinking carefully about how long the probe really needs to be we can reduce the number of combinations required. If a 20 bp oligo would be long enough then there would only be 512 combinations because you do not need to take into account the last thymine, cytosine alternative. If 17 bp would do, the number is dramatically reduced to 128.

Q5.6. What factors would be important in deciding how long an oligo probe needs to be?

You can see that when using DNA probes derived from an amino acid sequence you will need to use a mixture of oligonucleotides to represent all of the possible DNA sequences, which could encode the amino acid sequence. However, the individual sequences do not have to be synthesized separately. The automated DNA synthesizer can be programmed to use a mixture of nucleotide monomers at specific positions in the molecule, resulting in a mixture of oligonucleotides representing all the possible sequences. In the case of the 20 bp oligo derived from the insulin sequence shown in Figure 5.5 the synthesizer would be programmed to start with a guanine (G) residue. It would then add a thymine (T). The oligos produced by this stage will all be identical. For the next residue it would use a mixture of all four residues; these would be incorporated randomly into the products giving four different classes of product. Then a thymine would be added to the end of each of the different products, and so on. In this way a mixture of all 512 oligonucleotides would be made in one run on the synthesizer. This mixture is referred to as a degenerate oligonucleotide.

Q5.7. Would you expect more than one of the oligonucleotides in your mixture to anneal to the human insulin gene?

5.7 Cloned DNA Fragments as Probes

Cloned DNA fragments can be used as probes in hybridization reactions. As we have mentioned before this is useful if you have a cDNA clone but you now want to identify a clone from a genomic library because for example you want to study the regulatory sequences which will not be part of the cDNA clone. The other very powerful way in which cloned DNA fragments can be used as probes is in the cloning of homologous genes. Proteins often form families, the members of which have evolved from a common ancestor, and which have similar structures and functions. These homologous proteins will be encoded by similar DNA sequences, the degree of sequence similarity reflecting their evolutionary relationship. It is possible to use a gene cloned from, for instance, a mouse to probe a human DNA library.

Labeling DNA probes

If you want to use a DNA probe to identify DNA molecules with a complementary sequence you will need a way of tagging or labeling the probe molecules. By detecting this label you will be able to tell where the probe has bound. Incorporating a radioactive or other marker into the molecule is a common method of labeling probes (Boxes 5.3 and 5.4). One of the other advantages of this type of labeling is that the methods of detecting the labeled probe often involve an amplification process so that you can detect the presence of small quantities of bound probe.

5.8 Colony and Plaque Hybridization
Preparing the filters

Once you have labeled your DNA probe, you need to transfer the DNA from individual library members onto a solid support, in such a way that they can be related back to the original library member, before screening can begin. This is done using a method analogous to the plaque lift described in Section 5.4, the library is plated out and a small amount of bacteria from each colony is transferred to a sterile filter (Figure 5.6). The filters are then

Box 5.3 Types of DNA Labeling

In Chapter 3 we saw that DNA can be visualized by staining with ethidium bromide (Section 3.11). However, many techniques require either the specificity or sensitivity that can be obtained by attaching a label to DNA molecules. Once a DNA molecule is labeled it can be detected in a mixture of unlabeled DNA molecules, for instance when it is being used as a probe (Section 5.5). Additionally, as most methods of detecting the labels involve amplification of the signal it is possible to detect very small amounts of labeled DNA for example in DNA sequencing (Section 7.3). Commonly used DNA labels include radioactive isotopes, haptens, which can be detected immunologically, and fluorescent dyes.

Radioactive labels

Radioactive isotopes such as phosphate can be incorporated into nucleosides (^{32}P or ^{33}P) (a) or can replace elements in the nucleoside (^{35}S) (b). When DNA is being labeled for analytical purposes the end result of the experiment is often an electrophoresis gel where the labeled DNA is detected by autoradiography. The gel is placed in direct contact with a sheet of photographic film for several hours, the energy released by the radioactive isotopes causes the formation of dark silver grains and when the film is developed these appear as bands or spots.

Labels which are detected immunologically

For reasons of safety and convenience non-radioactive labels are increasingly replacing radioactive ones. Haptens such as biotin and digoxigenin are small molecules which can be coupled to nucleotides and can then be incorporated into nucleic acid molecules by DNA polymerases. These molecules are immunogenic and can be detected by specific antibodies conjugated with a moiety which can be visualized by fluorescence, chemiluminescence or a color reaction (c).

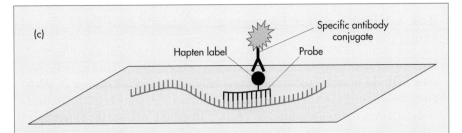

Fluorescent labels

Fluorescent dyes or fluorophores now dominate as labels in a range of applications including DNA sequencing, forensic analysis and DNA-based diagnostic tests. Fluorophores are coupled to DNA molecules and can be detected by measuring the fluorescence produced when they are excited by light. An additional advantage of fluorescent dyes is that they are available in a range of different colors and so can be combined to detect several different DNA molecules in a single reaction.

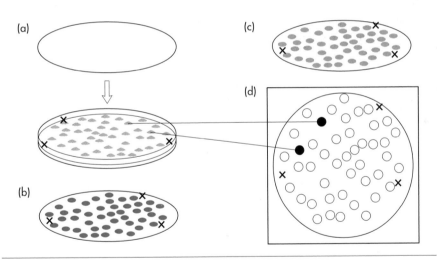

Figure 5.6 Screening a DNA library by hybridization with a DNA probe. a) A nylon filter disk is placed onto the surface of an agar plate on which the library has been plated. Marks are made on the agar plate and the nylon disk so that they can be realigned later. b) When the nylon filter is lifted off again, part of the colonies will adhere to the filter in a mirror image copy of the original plate, which is retained as the master plate. The filter can be placed, colony side up, on a fresh agar plate and grown overnight if desired. c) The colonies are then lysed to release the DNA, treated with alkali to separate the strands and then fixed to the nylon filter (Figure 5.7). The filter can be probed with a labeled DNA probe which can be detected by autoradiography. d) The autoradiograph is aligned with the master plate and the colonies containing library members with a cloned gene complementary to the probe are identified.

Box 5.4 Methods for Incorporating Labels into DNA Molecules

Random prime labeling

This method involves incorporating a labeled nucleotide during *in vitro* DNA synthesis and is used to make DNA probes. Remember from Chapter 3 that DNA polymerase requires a free 3′ hydroxyl; usually provided by a primer. One of the most straightforward ways to provide this is to use random hexanucleotides in a process called random prime labeling.

(a)

(b)

(c)

The target DNA is denatured by heating. a) Random hexanucleotide primers are added and the mixture is cooled to allow the primers to anneal. b) DNA synthesis, in the presence of four dNTPs, is initiated by the primers and labeled nucleotides are incorporated. c) The DNA is heated to separate the strands before being used as a probe. Because the hexanucleotides will anneal to random positions on the target DNA a whole series of labeled overlapping fragments will be produced. The advantage of this technique is that you do not have to make specific primers for each fragment to be labeled; the random hexanucleotide mixtures are commercially available.

Nick translation

Another approach to labeling DNA fragments is nick translation. In this case DNA synthesis is initiated at a single-strand break in the DNA duplex; the remains of the other strand is displaced as synthesis proceeds.

DNase I introduces single-stranded nicks into the DNA, these provide a free 3′ hydroxyl for DNA synthesis. In this case DNA polymerase I is used in preference to the Klenow fragment as the 5′ and the 3′ exonuclease activity will extend the regions which can be labeled.

One of the advantages of techniques where a labeled nucleotide is incorporated during *in vitro* DNA synthesis is that many labeled nucleotides are incorporated into each probe molecule giving a stronger signal than with an end-labeled probe.

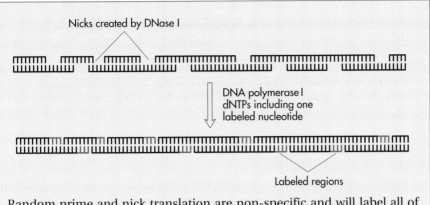

Nicks created by DNase I

DNA polymerase I
dNTPs including one
labeled nucleotide

Labeled regions

Random prime and nick translation are non-specific and will label all of the DNA included in the reaction. If there is significant homology between the cloning vectors used in constructing the probe clone and the library you will need to isolate the cloned fragment before labeling.

taken through the series of treatments outlined in Figure 5.7 to produce filters with DNA from the bacterial colonies bound to them. Firstly, the bacteria are lysed by treatment with the detergent SDS and sodium hydroxide, to release the DNA, which will include chromosomal DNA as well as library DNA encoded on the plasmid. The alkaline conditions will also denature the DNA making it single-stranded and available to hybridize with the probe molecules. Next the alkali has to be neutralized, otherwise it will interfere with annealing of the probe. DNA is covalently cross-linked to the filters before hybridization. It is bound to the filter by its sugar–phosphate backbone leaving the bases free to hybridize with the complementary bases of

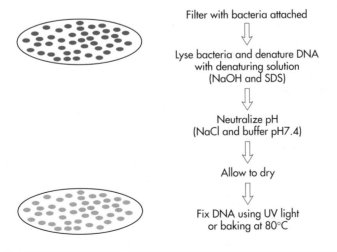

Filter with bacteria attached

Lyse bacteria and denature DNA
with denaturing solution
(NaOH and SDS)

Neutralize pH
(NaCl and buffer pH7.4)

Allow to dry

Fix DNA using UV light
or baking at 80°C

Figure 5.7 An outline of the steps involved in preparing filters for colony hybridization.

the DNA probe. Both nylon and nitrocellulose filters can be used in this procedure as DNA can be bound to both substrates. However, nylon filters are usually preferred as they are more robust than nitrocellulose and the DNA can be fixed to them either by heating or exposure to UV light. The same procedure can be used to screen a phage library, in which case the lysis step is not necessary as the phage plaques represent lysed bacteria from which the DNA has already been released.

Probing the filters

In DNA hybridization techniques it is important to use conditions where the probe will only bind to the specific DNA sequence it is designed to probe for. In the case of colony hybridization this usually means using conditions of low stringency initially, in these conditions the probe will bind to many sequences even where there is only a partial match. Higher specificity is then achieved by a series of washes of increased stringency that remove probe that is bound non-specifically. The usual ways of increasing stringency are to raise the temperature and either reduce the sodium chloride concentration, or increase the SDS concentration in the buffer. With oligonucleotide probes you can use higher stringency conditions with probes with higher melting temperatures as they bind more stably to their complementary sequences. If you are using degenerate oligonucleotides or probes derived from a homologous sequence you would decrease the stringency of the conditions compared with those you would use for a probe that was an exact complement to the sequence you are screening for.

> **Q5.8.** Why would you need to use lower stringency conditions with (a) a degenerate oligonucleotide rather than an exact match and (b) a probe derived from a homologous sequence rather than one derived from the identical sequence?

5.9 Differential Screening

DNA hybridization can be used to answer more complex questions than simply to identify a gene in a library. Imagine that you are trying to clone a gene from a plant and that you know it is expressed in the petals but not in any other part of the plant. You need a screening method that is able to detect the difference between the genes expressed in the different parts of a plant; this is called differential screening. In this case you would make a cDNA library from the petals, this will include clones representing the gene of interest. You would also make another cDNA library from part of the plant where the gene is not expressed, for instance the leaves. By hybridizing one library against the other you should be able to identify the clones of genes expressed only in the petals. To do this you need to label DNA from all of the individual clones in the leaf cDNA library and use this as the probe in a hybridization experiment to screen the petal library. Those clones detected by the leaf cDNA probes will represent genes found in both tissues. The few that are not detected will be genes expressed in the petal and not in the leaf; your gene of interest should be one of these. This is a very powerful tool and has been used to examine which genes are expressed in different tissues or at different stages of development. This approach has been combined with modern micro-array technology in the development of a new area of study called transcriptomics, which is discussed in detail in Section 11.6.

A further refinement of differential screening that can been used to identify genes expressed under specific conditions, but not in control conditions, is the construction of a subtractive library. In a subtractive library, cDNAs common to both the test and control conditions are eliminated, leaving only those cDNAs which are specific to the test conditions. This is usually achieved by hybridization of cDNA from the two populations and elimination of the double-stranded molecules, which represent sequences present in both samples. Because a subtractive library is enriched for the target sequences it can be particularly effective in identifying genes which are expressed at low levels.

5.10 Using PCR to Screen a Library

If you were to extract DNA from a clone from a library and use it as the template in a PCR reaction using specific primers, and you were to get a product of the size you expected, that would be a way of demonstrating that the clone most probably contained the specific DNA sequence of interest. This principle has been developed into a method for library screening. The method involves performing PCR reactions on pools of clones to reduce the total number of PCR reactions which would be needed to screen every clone in a library.

In order to adapt PCR to screening a library, without having to perform a separate PCR reaction for each individual clone, the PCR reaction is carried out on pools containing DNA extracted from many individual clones from the library. In the same way in which it is possible to perform PCR on a sample of human genomic DNA and to produce a single PCR product that is specific to the set of primers used, it is also possible to perform a PCR reaction on DNA extracted from a pool containing many individual clones. In this case you will only amplify a PCR product if the pool from which you made your DNA template contained the clone recognized by the primers. You would need to perform a second round of PCR screening of each of the individual clones in the pool to identify the clone precisely.

Various techniques have been developed to reduce the number of clones which need to be screened. One of these is to create pools in such a way that each clone is present in more than one sample. This effectively gives each clone an address in the library. The principle behind this strategy is easier to

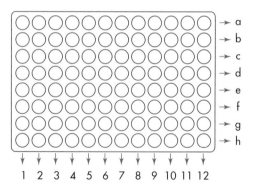

Figure 5.8 PCR screening of pools from a library. Pools are made by taking a small sample from each well in, for instance, row a and mixing them; this is pool a. The process is repeated for rows b–h and for columns 1–12 giving a total of 20 pools. DNA extracted from each pool is used as the template in a PCR reaction and the reactions are analyzed by agarose gel electrophoresis. If a band is seen for DNA extracted from pools 4 and c, a single well in the multi-well plate can be identified containing the clone that gave rise to the band.

understand if we start by considering a small library consisting of 96 clones, each of which has been inoculated into a separate well in a multi-well plate (Figure 5.8). A positive signal in the pool created by mixing all of the clones in row c and in the pool created from all of the clones in column 4 gives a unique address to the well where row c and column 4 intersect. In this way any individual clone can be identified from 20 PCR reactions rather than performing one for each of the 96 wells. This principle can be extended so that each clone has an address in three or even four dimensions rather than only in two, allowing 9600 clones to be screened in only 40 PCR reactions.

To prepare a library for screening by this method it needs to be gridded out in multi-well plates. This can be done in one of two ways. The library can be plated as plaques or colonies on agar plates and individually inoculated into the wells of the multi-well plate. This is a labor intensive process and, because it involves the experimenter inoculating wells with individual clones, it can lead to bias in favor of larger colonies or plaques. The alternative method involves diluting the library. A small part of the original library, the packaging mix in the case of a phage library, or transformation in the case of a plasmid library, is plated out and the titer of the library calculated (see Section 4.13). A larger sample is then diluted to give a titer of 100 colonies per mL. Dispensing 100 µL into each well will theoretically give a 10 clones in each well. These are then pooled as described above and PCR reactions performed to identify the wells containing the clone of interest. This technique is often used to screen commercially available libraries which can be supplied ready gridded.

In this chapter we have considered methods for screening DNA libraries by looking either for expression of a particular protein or by detecting the presence of a specific DNA sequence. However, in many cases there is not enough information to use these strategies. In many cases, for example in the cloning of human disease genes, you may know little more than the phenotype resulting from mutations in the gene; techniques for cloning genes in these circumstances are discussed in the next chapter.

Q5.9. In each of the following cases decide which approach you would use to screen the library.

(a) Remember the question at the end of Chapter 4 where a cDNA library seemed to be the most appropriate choice when trying to clone the genes for the enzyme rubisco from spinach. What approach would you use to identify the genes for rubisco from a spinach leaf cDNA library?

(b) You have a clone from a wheat grain cDNA library which encodes a gene involved in starch synthesis. You want to study the upstream regulatory DNA sequences. How would you obtain a clone containing these sequences?

(c) β-lactamase is a bacterial enzyme, which is capable of inactivating β-lactam antibiotics including penicillin and hence gives rise to antibiotic resistance. You have cloned the chromosomal *ampC* β-lactamase gene from *Pseudomonas aeruginosa* but now you want to clone β-lactamase genes from a range of other bacteria.

(d) You are interested in the human glutathione *S*-transferase (GST) which encodes an enzyme important in detoxification in the liver. You can look up the sequence of the gene on the human genome project database but you need a cDNA clone for your studies. Human cDNA expression libraries derived from a number of different tissues are commercially available, typically containing as many as 96,000 individual clones.

(e) You have a genomic DNA clone representing a small part of the dystrophin gene. Remember from Chapter 4 that mutations in this gene give rise to the human sex-linked disorder muscular dystrophy. You want to find the rest of the gene and you have both cDNA and genomic libraries.

Questions and Answers

Q5.1. Using a cosmid vector such as pHC79 how many clones would you need to have a 99% chance of finding a particular gene in a genomic library of (a) *Saccharomyces cerevisiae* and (b) human? *Hint.* You will need to know the size of insert which can be cloned in pHC79 and the size of the

S. cerevisiae, and human genomes. You will also need to use the formula in Section 4.4.

A5.1. (*a*) *The cos sites in a cosmid need to be about 50 kb apart for successful packaging by bacteriophage* λ. *pHC79 is 6.5 kb in length so you can clone about 43.5 kb of insert DNA into it. The yeast genome is 12.1 Mb long. Using the formula with* f = 43.5, g = 12,100 *and* P = 0.99, *then*

$$N = \frac{\ln (1 - 0.99)}{\ln (1 - 43.5/12,100)}$$

you would need 1297 clones to have a 99% chance of finding a particular gene in a genomic library of S. cerevisiae.

(*b*) *the human genome is* 2.95 × 10⁶ *kb long. Using the formula with* f = 43.5, g = 2.95 × 10⁶ *and* P = 0.99, *then*

$$N = \frac{\ln (1 - 0.99)}{\ln (1 - 43.5/2.95 \times 10^6)}$$

and you would need 3.12 × 10⁶ *clones to have a 99% chance of finding a particular gene in a human genomic library.*

Q5.2. Devise a screening method that would allow you to detect clones encoding the genes for resistance to the antibiotic tetracycline from a genomic library constructed from a resistant strain of *E. coli*.

A5.2. *This should be a fairly straightforward procedure. If you plate out your library onto agar containing tetracycline the ones which grow must be resistant to tetracycline and are likely to encode the genes you are trying to clone. There are a couple of important considerations to take into account in designing your library in the first place. You must use an* E. coli *host that is not resistant to tetracycline and you would want to make your library in a vector like pUC18, which does not encode tetracycline resistance, rather than in one like pBR322.*

Q5.3. You want to construct a DNA library, which you intend to screen with an antibody, for expression of a eukaryotic gene. (a) What kind of DNA library would be most suitable? Give your reasons. (b) What type of vector would you construct your library in? Explain your choice.

A5.3. (*a*) *The important consideration in answering both parts of this question is that, for antibody screening to work, it is essential that the gene you are interested in is expressed. A cDNA library would be a good choice as it is made from mature mRNA with the introns removed, eukaryotic genomic DNA cannot be processed correctly by bacteria and would be unlikely to be expressed.*

(*b*) *We have already mentioned the advantages of phage libraries in immunological screening but you will also probably want to use a special-*

ized expression vector (Section 9.3) to ensure that the protein is expressed at high enough levels to be detected.

Q5.4. Describe the approach you would use to clone the gene for (a) β-galactosidase (see Section 3.9) from the Gram negative bacterium *Citrobacter*, and (b) human insulin. You should consider what type of DNA library would be the most suitable and what form of screening for expression you could use.

A5.4. *(a) With a bacterial gene, like the gene for β-galactosidase, a genomic library would be your best option. It is probable that the gene from* Citrobacter *would be expressed in* E. coli *but you may choose to use a plasmid expression vector to make sure that the gene is expressed at significant levels in* E. coli. *Complementation of a* lacZ *mutant of* E. coli *should work here as* Citrobacter *and* E. coli *are closely related, but you could also look for direct expression of β-galactosidase using* X-gal *and IPTG. In any case you will still need to use a* lac⁻ *strain of* E. coli *to eliminate background expression of β-galactosidase.*

(b) With a human gene you will probably want to make a cDNA library, the pancreas would be a good source of cells with insulin mRNA. You will have to use immunological screening so you will probably use a phage vector as phage lysis will release proteins from the cells making screening easier. You will need to raise an antibody to insulin, which means that you will need a pure sample of the protein; again this can be prepared from pancreas tissue. As you are using immunological screening you will need to be sure that the insulin is expressed in the E. coli *host, you will need to make your cDNA library in a specialized expression vector.*

Q5.5. Why is it not possible to derive an unambiguous DNA sequence from the amino acid sequence of a protein?

A5.5. *Because of the degeneracy of the genetic code any given amino acid may be encoded by up to six different triplet codons. For example, leucine is encoded by CUU, CUC, CUA, CUG, UUA and UUG (Box 8.1).*

Q5.6. What factors would be important in deciding how long an oligo probe needs to be?

A5.6. *The considerations are the same as with designing oligonucleotide primers for PCR (Box 3.6). Probes need to be at least 17 bp long to ensure that the sequence will not occur by chance. They are not usually more than about 25 nucleotides long, partly because of expense but also because of the problems created by the degeneracy of the genetic code. As with PCR primers the melting temperature is an important consideration (Section 5.8).*

Q5.7. Would you expect more than one of the oligonucleotides in your mixture to anneal to the human insulin gene?

A5.7. *Yes. A one or two base pair mismatch should not prevent the oligonucleotide annealing so more than 1/512 of the oligos in your mixture should be useful probes. You can also alter the hybridization conditions to favor annealing of imperfect matches, as described in Section 5.8.*

The other way to reduce the number of alternative oligos that you need to synthesize is to try and predict which of the alternatives the organism you are trying to clone from is likely to use. Because many organisms have a distinct bias in their use of alternative codons for each amino acid (Box 8.1), it is often possible to make an educated guess about which alternative is the most likely. This is discussed in more detail in Chapter 8.

Q5.8. Why would you need to use lower stringency conditions with (a) a degenerate oligonucleotide rather than an exact match and (b) a probe derived from a homologous sequence rather than one derived from the identical sequence?

A5.8. *(a) While your degenerate oligonucleotide probe will contain some molecules which are an exact match to the target sequence, it will contain many more which are partial matches. These molecules will effectively have a lower T_m because hydrogen bonds will not form between the mismatched bases. You will need to use lower stringency conditions to ensure that these molecules remain bound to the target sequence.*
(b) Similarly a probe derived from a homologous sequence will be similar to but not identical with the target sequence, so lower stringency conditions are necessary.

Q5.9. In each of the following cases decide which approach you would use to screen the library.

(a) Remember the question at the end of Chapter 4 where a cDNA library seemed to be the most appropriate choice when trying to clone the genes for the enzyme rubisco from spinach. What approach would you use to identify the genes for rubisco from a spinach leaf cDNA library?

(b) You have a clone from a wheat grain cDNA library which encodes a gene involved in starch synthesis. You want to study the upstream regulatory DNA sequences. How would you obtain a clone containing these sequences?

(c) β-lactamase is a bacterial enzyme, which is capable of inactivating β-lactam antibiotics including penicillin and hence gives rise to antibiotic resistance. You have cloned the chromosomal *ampC* β-lactamase gene from *Pseudomonas aeruginosa* but now you want to clone β-lactamase genes from a range of other bacteria.

(d) You are interested in the human glutathione *S*-transferase (GST) which encodes an enzyme important in detoxification in the liver. You can look up the sequence of the gene on the human genome project database but you need a cDNA clone for your studies. Human cDNA expression libraries derived from a number of different tissues are commercially available, typically containing as many as 96,000 individual clones.

(e) You have a genomic DNA clone representing a small part of the dystrophin gene. Remember from Chapter 4 that mutations in this gene give rise to the human sex-linked disorder muscular dystrophy. You want to find the rest of the gene and you have both cDNA and genomic libraries.

A5.9. (*a*) *Because this is a very abundant protein it will probably be fairly easy to purify, so a sensible approach to screening for this gene would be to design an oligonucleotide probe based on the amino acid sequence of the first few N-terminal amino acids from each of the protein subunits. The amino acid sequences can be back-translated into a degenerate nucleotide sequence representing all the possible ways of encoding the amino acids. Remember that you can either make a degenerate oligonucleotide with all the possibilities present or you can use codon preference data to predict the most likely sequence for a gene from spinach.*

(b) *As the cDNA clone will not contain any of the upstream regulatory sequences you will need to screen a genomic library. If you have sequenced your cDNA clone you could design an oligonucleotide probe, although it would probably be easier to use the cDNA clone as your probe. It could be labeled by random primer labeling or nick translation.*

(c) *You could label the cloned* P. aeruginosa ampC *gene and use this to probe genomic libraries from the other bacteria. You should detect clones encoding genes homologous to the* ampC *β-lactamase. You would probably use a low stringency wash initially because the probe and the target sequence are unlikely to be exactly complementary, this would be a particularly important consideration if you are screening a library from a distantly related organism.*

(d) *Using the sequence from the human genome project you could design PCR primers which would amplify the genomic DNA encoding the gene for GST, but being the genomic sequence this would include the introns. You could label this PCR product, by including a radioactive nucleotide in the last few cycles of the PCR reaction, and use this to probe a liver cDNA library. Alternatively you could use the PCR primers directly in a PCR screen of pools of clones from the cDNA library.*

(e) *This is a fairly complex example, there are several appropriate screening strategies that you could try, and in practice you would probably need to employ several of them. Assuming that the genomic library has been made using a partial digest and contains overlapping clones you could therefore use your genomic clone to reprobe the same library. This will allow you to detect clones which overlap with your probe clone and represent sequences on either side. You could repeat this process several times extending out from the original clone, this process is referred to as chromosome walking. You could also use your genomic clone to probe the cDNA library. In this case if you find any positive clones they should represent the whole of the coding sequence of the dystrophin gene. Because a large proportion of the dystrophin gene comprises non-coding sequences, it is likely that any single genomic clone will contain part of one of the large introns. In this case you will not detect the dystrophin cDNA clone as it does not contain the intron sequences. However, if you use a number of overlapping clones from the genomic library you have a better chance of detecting the cDNA.*

Further Reading

Colony hybridization – method for isolation of cloned DNAs that contain a specific gene (1975) Grunstein M and Hogness DS, Proceedings of the National Academy of Sciences of the United States of America, Volume 72 Pages 3961–3965.
This paper is the first description of the colony hybridization method, the technique used today is somewhat simplified but based on the principles described here

Screening λ gt recombinant clones by hybridization to single plaques *in situ* (1977) Benton WD and Davis RW, Science, Volume 196 Pages 180–182.
This very short paper describes plaque hybridization with a nucleic acid probe

Functional expression of cloned yeast DNA in *Escherichia coli* (1977) Ratzkin B and Carbon J, Proceedings of the National Academy of Sciences of the United States of America, Volume 74 Pages 487–491.
This early example describes the cloning of a eukaryotic gene based on screening for complementation of an auxotrophic mutant of E. coli

6 Further Routes to Gene Identification

Learning outcomes:

By the end of this chapter you will have an understanding of:

- transposon tagging as a method for cloning genes in both eukaryotes and prokaryotes
- signature-tagged mutagenesis, a refinement of transposon tagging, as a method for cloning genes with phenotypes which are hard to select
- map-based cloning

6.1 How Do We Get From Phenotype to Gene: A Fundamental Problem in Gene Cloning

In Chapter 4 we looked at a variety of techniques which would enable you to construct a gene library, either directly from genomic DNA or indirectly by reverse transcription of mRNA and cloning of the products. In Chapter 5 we then looked at how such a library can be screened to identify a clone or clones that carry the particular gene you are interested in. Screening can be by function, by screening for the protein product, or by using a screening method based on DNA–DNA hybridization. These methods all have associated advantages and drawbacks, with which you should now be feeling quite familiar.

One major drawback of all these methods is that they require some prior knowledge of the nature of the gene or its protein product. While it is often the case that this information is available, it is also often the case that it is not. Indeed, many of the most interesting challenges in molecular biology have been where nothing at all was known about the gene or genes of interest except the phenotype that resulted when they are lost or become defective. Since the objective of molecular biology is to explain properties of organisms at the molecular level, it is clearly essential that we should be able to identify the genes, and hence the proteins, which produce these properties. Such identification should greatly increase our level of understanding about the particular aspect of the phenotype in which we are interested. This interest is often not just academic, but includes serious matters such as genetic diseases, where the absence of a functional gene, or the presence of a defective gene, can lead to disability or death. In such cases, the identification of the gene concerned can help in early diagnosis, including early screening in pregnancy and screening of embryos created by *in vitro* fertilization (IVF), and has great potential in devising treatments for the condition.

We can illustrate this point with one of the best known examples: the disease cystic fibrosis (CF). This disease has long been known to be caused by mutations in a single gene. These mutations are recessive, but if an individual carries two copies of the gene both of which are defective, then they will develop CF. The defective gene is inherited in a classical Mendelian fashion for a single gene, so if both parents are carriers (i.e. heterozygous for the defective gene) then one in four of their children, on average, will develop the disease, and half of their children, on average, will be carriers. The symptoms of the disease were well known for many years, but the underlying cause of the disease – the nature of the gene, its role in its normal non-defective state, and how mutation of the gene causes the disease – were mysteries until the gene was successfully cloned. We will return to this classic story in Section 6.10.

Thus, methods have had to be developed for cloning genes when nothing is known in advance about the sequence of the gene or the nature of the protein which it encodes. These methods can be divided into two groups, which are very different in their approach and applicability. The first of these, gene tagging, involves generating mutations in such a way that the gene is marked so that it can be recovered in subsequent cloning experiments. The second, map-based or positional cloning, requires the development of powerful maps of the genome which can be used to focus the search for a particular gene on smaller and smaller regions. We will consider each of these methods in turn in the next two sections.

6.2 Gene Tagging: A Method That Both Mutates and Marks Genes

Mutations in genes often produce changes in phenotype. To identify a gene which has been mutated and where the phenotype has consequently changed, it would be useful to be able to identify the gene purely on the basis that it carries a mutation. This is the idea behind gene tagging, which involves generating mutations using something which not only randomly inactivates genes, but at the same time marks them so that they can be easily detected, for example by screening a gene library. It involves essentially four steps.

In the first step, a large number of mutations are generated, using as a mutagen a piece of mobile DNA that can insert at random in the genome, disrupting and blocking the function of any gene into which it inserts. This results in a population of organisms where the only difference between them is the position of the mutation. The mobile DNA used for the mutagenesis is usually a transposon; that is, a piece of DNA which is able to jump to different positions in the genome. Ideally, the transposon is one which is not present in the genome of the organism being studied, but nevertheless is capable of transposition when introduced into that organism (Box 6.1 and Figure 6.1). Usually, the transposon will carry a selectable

Box 6.1 Transposons

Transposons are pieces of DNA that can "jump" into novel positions in the genome. There are numerous different types of transposon which differ from each other in many ways. They have been exploited very extensively in molecular biology research. They can insert themselves into DNA without requiring any similar sequence to be present in themselves and their "target". This makes them powerful mutagens, since their insertion into a gene will generally prevent the gene from producing a functional protein. As mutagens, their effect is not always completely random, as some do have preferences for certain sequences or types of chromatin, and others will transpose preferentially to adjacent regions rather than randomly throughout the genome.

Generally, a minimum of two things are required for transposition. The first is a gene which encodes a transposase enzyme, and which can recognize the ends of the transposon, excise the transposon (or a copy of it) from its starting point, cut DNA randomly or semi-randomly elsewhere, and catalyze the movement of the transposon from its starting point to its new target. The transposase can be encoded either by the transposon itself (in which case the transposon is often referred to as being "autonomous") or elsewhere – on another transposon, for example, or on a plasmid. Transposons which rely on a transposase which they do not themselves carry are called "non-autonomous". They occur naturally, and any autonomous transposon can be converted to a non-autonomous one by removing the transposase gene from it. The second requirement for transposition are the sequences at the ends of the transposon, which are recognised by the transposase. This minimal requirement for an autonomous transposon is shown below. There are many variants on this theme!

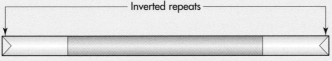

Transposons use many different mechanisms to jump around the genome, but they all boil down to two basic types: replicative and non-replicative (shown in Figure 6.1). Replicative transposons copy themselves when they transpose, and leave one copy at the original site while inserting the second copy elsewhere. Non-replicative transposons are excised from their starting position and inserted elsewhere in the genome.

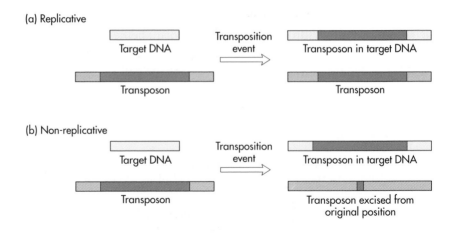

Figure 6.1 Replicative and non-replicative transposition. In replicative transposition a), the transposon makes a copy of itself which is inserted randomly in the target DNA. In non-replicative transposition b), the transposon excises from its original position (usually leaving a small change in the sequence at the excision point) and reinserts at random in the target DNA.

marker, such as an antibiotic resistance marker, which means that only those individuals where the transposon has inserted into the genome are selected for further screening. Next, individuals in the population which result from this process are examined to detect those which display the mutant phenotype of interest. In the third step, DNA from these individual mutants is then isolated and used to make a gene library. Finally, the library is screened to identify clones which contain the transposon. Various methods for doing this are discussed below. Among the clones isolated using this approach, some or all will have the mobile element present in the gene whose loss of function causes the phenotype. Figure 6.2 illustrates the basics of this approach using a very simple hypothetical example.

Q6.1. Why is it better for transposon tagging to use a transposon which is not normally present in the target genome?

For this approach to work, a number of conditions need to be satisfied or at least optimized. First, the transposon used for the tagging must transpose at a sufficiently high frequency to generate enough mutants to screen, but not so high that the mutants are unstable or the transposon rapidly spreads to new sites in the genome. Many systems have been devised where the transposon is initially introduced on plasmid or viral vectors which carry the genes needed for transposition but which cannot themselves replicate in the target organism. This makes a single event transposition more likely,

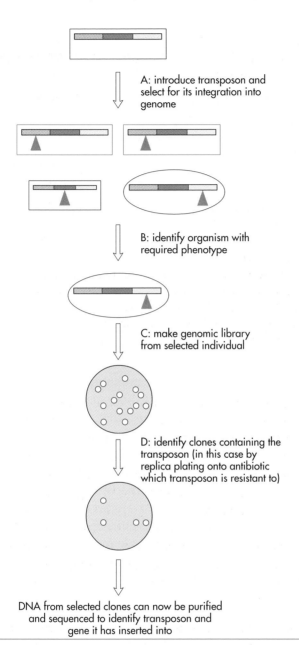

A: introduce transposon and select for its integration into genome

B: identify organism with required phenotype

C: make genomic library from selected individual

D: identify clones containing the transposon (in this case by replica plating onto antibiotic which transposon is resistant to)

DNA from selected clones can now be purified and sequenced to identify transposon and gene it has inserted into

Figure 6.2 The basics of transposon tagging. To illustrate how this principle works, imagine a very simple organism (shown as a box) with three genes (shown shaded), which control respectively size, color and shape. We wish to identify the gene that controls shape. In the first step a) we generate and select for a pool of individuals with random insertions of a transposon in the different genes, shown as triangles at the point of insertion. In step b), we identify an individual where the transposon has inserted into the shape gene: we can do this because the organism has changed its shape as a consequence of this mutation. In step c) we prepare a genome library from this organism, and in step d) we identify clones that carry the transposon, and hence also will carry the shape gene into which the transposon has inserted.

where the transposon jumps from the vector to some position on the genome, but where no further transposition occurs because the genes for transposition are lost when the cell divides.

> **Q6.2.** Why is it important to limit the number of transpositions made by the transposon when attempting to clone genes by transposon tagging?

Second, some method must be available for detecting the novel phenotype. Since the number of mutants generated overall may be very large, and the proportion of these which are actually in a gene of interest will be very small, a key factor in determining the success of this strategy is to have a very robust method for distinguishing between the mutants of interest and all the others. Ideally, a selective approach can be used, where only the mutants of interest survive and all the others die. But other approaches are also possible, as we will see in the examples below.

Third, although clones in a library with the transposon present can be identified by DNA hybridization, it is common to use a transposon which has been modified to make its detection in the library easier. For example, the selective marker (almost always an antibiotic resistance marker) on the transposon can also be used to select those clones from the library that contain the transposon. This requires that the selectable marker is expressed both in the target organism where the original mutagenesis is done, and also in the organism (almost always *E. coli*) in which the library is screened. Alternatively, some transposons used for mutagenesis have been engineered to have two selectable markers – one expressed in the target organism and one in the *E. coli* host – to get round this potential problem. An even more useful approach is to put a bacterial origin of replication onto the transposon as well as a selectable marker. The way in which the region into which a transposon has jumped can be cloned is illustrated in Figure 6.3.

> **Q6.3.** Before looking at Figure 6.3, can you see how including a bacterial origin of replication on a transposon which is being used for gene tagging is helpful in recovering the region of DNA containing the transposon after tagging has taken place?

Having used this approach to identify candidate genes, a number of careful experiments need to be done to confirm that the gene which has been targeted is indeed the one responsible for the phenotype of interest. Some indication of this may come from the gene sequence, and the protein sequence that arises from its translation. Thus if, for example, you are hunting for a gene producing a protein involved in some aspect of

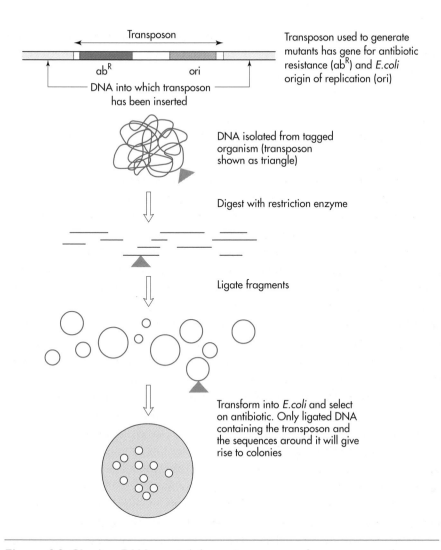

Transposon

ab^R   ori

DNA into which transposon
has been inserted

Transposon used to generate
mutants has gene for antibiotic
resistance (ab^R) and *E.coli*
origin of replication (ori)

DNA isolated from tagged
organism (transposon
shown as triangle)

Digest with restriction enzyme

Ligate fragments

Transform into *E.coli* and select
on antibiotic. Only ligated DNA
containing the transposon and
the sequences around it will give
rise to colonies

**Figure 6.3 Cloning DNA containing a transposon from an organism
where a transposon-tagged mutant has been made.**

signaling across the cell membrane, and a candidate gene proves to encode
a trans-membrane protein, this is good evidence to support the hypothesis
that the correct gene has been cloned. The best experiment, however, is to
clone the wild-type version of the gene and introduce it into the organism
carrying the mutated gene: if this restores the wild-type phenotype, this is
very strong evidence that the correct gene has been identified.

Having obtained a mutant generated by a transposon, how could you
obtain a clone of the wild-type gene? The classic way in which this can be
done would be as follows. A library would be constructed from the mutant.
Clones containing the transposon plus flanking DNA would be selected by
screening for the presence of the selectable marker on the transposon. The

DNA fragment in these clones would then be sequenced. As the sequence of the transposon will be known, it is straightforward to distinguish the regions of sequence derived from the transposon from those of the wild-type gene. A complete clone of the wild-type gene could now be obtained by using part of this sequence to probe either a cDNA or a genomic library from the non-mutated parent of the mutant, to identify the clone containing the wild-type gene. In practice, this approach could be short circuited by using inverse PCR (Box 6.2) to clone the regions flanking the inserted transposon, using primers internal to the transposon. Although this approach has been used successfully, it may run into problems if the gene which has been mutated is very large, as may well be the case if it contains large introns, for example.

The great advantage of gene tagging is that nothing at all needs to be known about the gene or genes which are being looked for: all that is needed is a detectable phenotype, and a transposon that will work in the organism being studied. There are downsides to this approach, however. In particular, it cannot be used in humans, where much of the interest of molecular biologists lies. It also ideally requires transposition to be random,

Box 6.2 Inverse PCR

Inverse PCR is a technique which can be used to amplify DNA flanking a known sequence, usually a transposon insertion in genomic DNA. The genomic DNA is cut with a restriction enzyme which does not cut inside the transposon. This will generate many genomic fragments, one of which will include the transposon with the sequences on either side of the insertion point (a). These fragments are then ligated in dilute conditions which favor ligation of the two ends of the same molecule (b). Now that the fragment containing the transposon has been circularized, if a PCR reaction is performed using primers complementary to the ends of the transposon, the flanking sequences will be amplified (c).

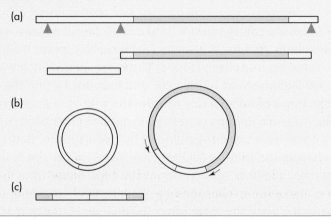

whereas most transposons do show some tendency to insert preferentially into certain positions in the genome. Finally, it is hard to identify genes which are essential using this method, since individuals with mutations in such genes will not survive. Care must also be taken in the interpretation of phenotypes' which may arise directly from the presence of the insertion in a particular gene or indirectly as a consequence of other effects such as alteration of expression of the genes downstream of the transposon.

Gene tagging has been a very successful tool for cloning genes from a range of different organisms. In the examples that follow, some different refinements to the basic technique are illustrated; many others can be found in the research literature.

6.3 A Simple Example of Transposon Tagging in Bacteria: Cloning Adherence Genes from *Pseudomonas*

Among the many examples where transposons have been used to identify bacterial genes is a study on adherence genes in *Pseudomonas aeruginosa*, a bacterium that can cause infection under some circumstances. This study is summarized in Figure 6.4. Adherence (i.e. the tendency of bacteria to stick to certain types of surface) can be a key factor in determining whether or not bacteria can cause infections, and for this reason much attention has been paid to defining the proteins required for adherence. In one study designed to look for the genes encoding these, a transposon was used to generate mutants which could no longer adhere to cells. This was done by transforming a strain of *P. aeruginosa* with a plasmid that could not replicate in the strain, but that carried a transposon which was derived from a transposon called Tn5. Such a plasmid is referred to as a suicide plasmid, because of its inability to replicate and hence to survive in certain strains or bacterial species. The transposon was modified to express a gene encoding gentamicin resistance, and was hence called Tn5G. After transformation with the suicide plasmid carrying Tn5G, cells were spread onto plates containing gentamicin. Only those cells that had received a copy of the Tn5G transposon by transposition from the suicide plasmid should be capable of growing on these plates. Tens of thousands of colonies arise from such transpositions, each one representing a unique transposition event to a unique site on the genome. To identify mutants from this large pool that could no longer bind to cells, colonies from the transformation were pooled and grown in broth, and then exposed to cultured epithelial cells growing on plastic plates. After an hour, the culture media was removed and added to a new layer of epithelial cells. This process was repeated a total of 12 times. The rationale for this was that any mutants that had arisen which could not adhere to cells would always be in the culture media fraction, together with bacteria that have simply failed to adhere during the reaction for other reasons. By repeatedly exposing the super-natant to cell layers, the non-adhering mutants became progressively

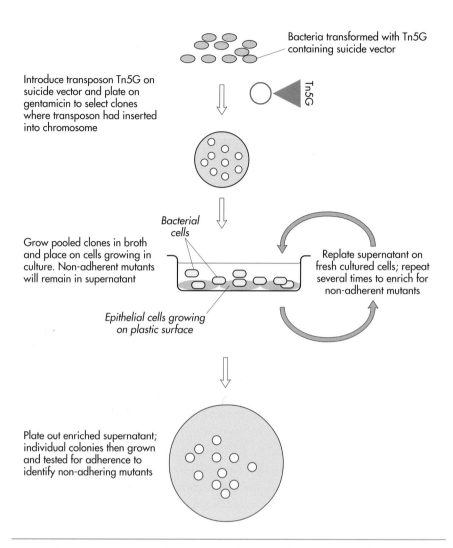

Introduce transposon Tn5G on suicide vector and plate on gentamicin to select clones where transposon had inserted into chromosome

Bacteria transformed with Tn5G containing suicide vector

Tn5G

Bacterial cells

Grow pooled clones in broth and place on cells growing in culture. Non-adherent mutants will remain in supernatant

Replate supernatant on fresh cultured cells; repeat several times to enrich for non-adherent mutants

Epithelial cells growing on plastic surface

Plate out enriched supernatant; individual colonies then grown and tested for adherence to identify non-adhering mutants

Figure 6.4 Generation and isolation of non-adherent bacterial mutants by transposon tagging.

enriched until all, or nearly all, of the bacteria left in the supernatant were incapable of adhering to cells. At this point the cultures were diluted and plated out, and individual colonies tested to see whether they could adhere to cells or not.

Having isolated several transposon-induced, non-adherent mutants, the next step was to identify the gene (or genes) which had become mutated. There are various different ways of cloning the gene into which the transposon has inserted. In this study, one gene was identified as follows. A cosmid library was constructed from wild-type *P. aeruginosa*, and the library was transformed into the mutant non-adherent strain. Using the same tissue culture model of adherence described above, a cosmid was identified which

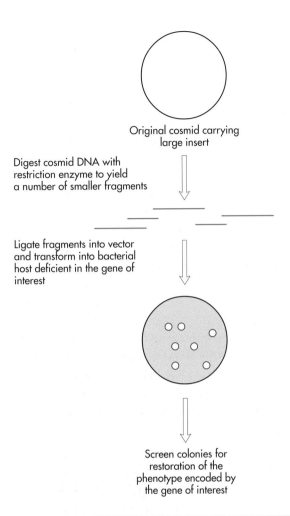

Original cosmid carrying
large insert

Digest cosmid DNA with
restriction enzyme to yield
a number of smaller fragments

Ligate fragments into vector
and transform into bacterial
host deficient in the gene of
interest

Screen colonies for
restoration of the
phenotype encoded by
the gene of interest

Figure 6.5 Identification by subcloning and complementation of the region of a cosmid clone carrying a bacterial gene.

was able to restore the property of adherence to the mutant. The DNA on this cosmid was likely to carry the unmutated copy of the gene and hence to be able to complement the mutation caused by the transposon insertion. The cosmid was quite large and contained several genes: the question then became which of these different genes was the one which was defective in the mutant? The simplest way to determine this is to subclone into plasmids smaller fragments of DNA from the large single fragment of the chromosome which was initially cloned in the cosmid, and test these plasmids individually for their ability to complement the defective gene. This approach is shown in Figure 6.5. By a series of subcloning experiments, a small (less than 1 kb) fragment of DNA was eventually identified from the complementing cosmid which was capable of restoring adherence to the non-

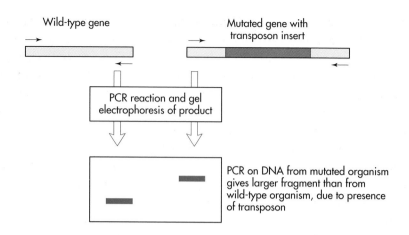

Figure 6.6 Using PCR to confirm the presence of a transposon in a particular gene.

adherent mutant. DNA sequence analysis of this fragment revealed it carried one complete gene, which was thus a very strong candidate for the gene which had been inactivated by the Tn5G insertion.

To finally prove that the gene on the small fragment is indeed the wild-type copy of the gene mutated by Tn5G, we would design two primers which flank the gene and use PCR to amplify up the corresponding DNA from the mutant: this would be expected to be larger because it would also contain the Tn5G sequence. This approach is shown in Figure 6.6. Alternatively, primers inside Tn5G could be designed and used to amplify up the DNA flanking it using inverse PCR. This DNA could then be sequenced to confirm it was the same gene as was contained on the complementing plasmid. Before PCR was developed as a technique, a piece of chromosomal DNA carrying the Tn5G insertion could be cloned by making a library from the mutant strain and selecting for gentamicin resistance, and sequencing the DNA to confirm that this sequence is of the same gene as that of the wild-type copy, as shown in Figure 6.3.

Q6.4. Why is it important to demonstrate that the complementing gene is indeed the one into which Tn5 had inserted in generating the original mutant?

6.4 Signature-tagged Mutagenesis: Cloning Bacterial Genes with "Difficult" Phenotypes

The use of transposons in research on bacteria has been developed over the years to a high degree. One of the most ingenious methods is proving very useful for identifying genes that are involved in complex events such as infection. Since infection by definition takes place inside an organism, it is hard to identify the genes that it requires. One approach would of course be to take a bacterial species which can cause infection (using an example where a good animal model is available, that is, where the organism can infect an animal under laboratory conditions), produce a large number of random transposon mutants of this bacterium, and use them one at a time to infect laboratory animals. Those that failed to infect would be candidates for having a transposon inserted into a gene which was required for the infection to occur, and by the methods described above this gene could be identified. The problem with this approach is that it would require thousands of experiments, since each transposon-generated mutant would have to be individually tested. How much better it would be if all the mutants, or at least a large pool of them, could be tested at the same time in a single experiment, and those mutants that do not cause an infection identified from this pool.

The method known as signature-tagged mutagenesis (often abbreviated to STM) effectively allows this precise experiment to be done. The method relies on the production of a large number of transposons which are non-identical in sequence, using the following ingenious method, which was originally devised to identify genes involved in the ability of the bacterium *Salmonella typhimurium* to cause an infection. An oligonucleotide synthesizer is used to produce an oligonucleotide that is 80 bases in length, where the 20 bases at the 5′ end and the 20 bases at the 3′ end have a fixed sequence, but where the sequence of the 40 central bases is semi-random. (The production of such mixes of oligonucleotides has been discussed in Section 5.6.) In the original paper where this method was first described, this 40 base sequence was referred to as $(NK)_{20}$, where N was A, C, G or T, and K was G or T. Examples of different sequences which all have the form $(NK)_{20}$ are shown in Figure 6.7, and it can be shown that, if such a sequence is made, two identical sequences will be found only once in 2×10^{17} molecules. The varying region is referred to as the signature tag. Having produced this oligonucleotide it was then converted into a double-stranded

```
Sequence 1  AGTGTTATCTATAGTGTTATAGGTCGATATATCTAGATCT
Sequence 2  TGATATCTATGGTTGTAGTGTGGGGTAGTGTGATCGTTCT
Sequence 3  CGTGTTGTTGAGAGTGCTAGCTTGTTTGATGTCTATAGAT
Sequence 4  CGTTGTGGCGCGGGATCTTGCGTGCTGGCGATGTCGGTCT
Sequence 5  CTGGAGTGCGATGTGGCTTTGGAGGGTTTGTGCTCGTGCT
Sequence 6  TTAGGGGGTGTTAGCTATCTAGATATTGAGAGAGAGAGCT
Sequence 7  TTGTTGAGTTATCGGTTGAGCTATGGTTGGCTTTAGTGGT
Sequence 8  CTGTTTGTCGAGTTGTATGGCGATATTTTTCGCTTTCTCT
```

Figure 6.7 The types of sequences used as tags in sequence-tagged mutagenesis. Eight examples of sequences of the form $(NK)_{20}$, where N is A, T, C, or G, and K is G or T are shown.

molecule using PCR with primers homologous to the 20 bases at either end of the oligonucleotide, and ligated into a restriction site in a transposon called miniTn5Km2. The transposon was carried on a plasmid that can replicate in *E. coli* but not in the target bacterium, *S. typhimurium*. The transposon carries the ends of the transposon Tn5 and a selectable marker (kanamycin resistance), but the transposase gene has been removed from the transposon and inserted onto the plasmid vector.

Large numbers of plasmids carrying these varying transposons were then introduced into *S. typhimurium*, and colonies that arose from transposition events of the miniTn5Km2 transposon from the suicide plasmid onto the chromosome were selected by growth on plates containing kanamycin. Each colony represented a unique transposition event, with the transposon inserted randomly at an unknown location on the chromosome, and where the sequence of the semi-random 40 base-pair region was different for each transposon. As the plasmid carrying the transposase gene could not replicate it was now lost from the cell, so no further transposition events occurred.

The next step was to inoculate individual colonies in small amounts of broth in the individual wells of 96-well plates, and to allow them to grow. Once the cultures had grown, samples of the cultures were printed from the multi-well plates onto replicate nylon filters and allowed to grow by placing the filters on agar plates. The filters were then treated to release the DNA from the colonies and fix it in position to the nylon filters. This is a development of the colony hybridization procedure described in Section 5.8. The investigators now had large numbers of multi-well plates with 96 different cultures, one in each well, each containing a mutant with a transposon inserted at a different place in the chromosome. Because each transposon is unique (due to the presence of the signature tag), each mutation now had a distinct tag. The DNA from these sets of 96 cultures had also been fixed, in the same relative positions as the cultures, to a nylon membrane. This grid thus records the mutations and their individual tags in the position that they appear in the multi-well plate. Each set of 96 cultures was then pooled and used to infect experimental mice, and total DNA was prepared from each pool. After 3 days, the mice were killed, and viable bacteria recovered from the spleens of the infected animals.

Consider what will have happened during this infection. Within the pool of 96 different transposon generated mutants, certain individual mutants will have arisen where the transposon has jumped into a gene which is required for the successful growth of the bacteria in the infected animal. In these cases, the bacteria will fail to grow and hence will not be present among the bacteria that are recovered from the spleen of the infected animal. DNA is now prepared from the bacteria recovered from the spleen and the signature tags are amplified from this DNA. The signature tags that are amplified will be those from mutations that did not affect the ability of the

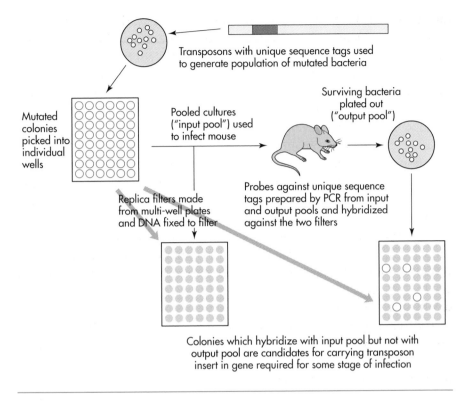

Transposons with unique sequence tags used to generate population of mutated bacteria

Mutated colonies picked into individual wells

Pooled cultures ("input pool") used to infect mouse

Surviving bacteria plated out ("output pool")

Replica filters made from multi-well plates and DNA fixed to filter

Probes against unique sequence tags prepared by PCR from input and output pools and hybridized against the two filters

Colonies which hybridize with input pool but not with output pool are candidates for carrying transposon insert in gene required for some stage of infection

Figure 6.8 Signature-tagged mutagenesis.

clone to infect the mouse. Those signature tags that were present on the transposons that jumped into a gene required for infection will no longer be present, as the bacteria carrying these mutations could not be recovered from the spleen. By comparing the signature tags present before and after infection, these genes required for growth in the infected animal can be identified. This is illustrated in Figure 6.8, and in the question below.

Q6.5. The pool of 96 colonies before infection is referred to as the input pool. The pool of the colonies grown from the animal after infection is referred to as the output pool. If the signature tags from the transposons in the input and the output pools are amplified by PCR and labeled with radioactivity, and are hybridized against the replicate filters, can you predict what you would see with (a) the filter hybridized with amplified signature tags from the input pool and (b) the filter hybridized with amplified signature tags from the output pool?

This method allows colonies to be identified which have arisen from individual bacteria where the transposon had inserted into a gene required for infection. The genes containing the transposon can then be cloned by digesting chromosomal DNA from each colony and ligating it into the standard vector pUC18, using selection for kanamycin resistance to identify those plasmids which had received a piece of DNA containing the transposon. The DNA around the transposon site can then be sequenced, and used to search all the known *S. typhimurium* genes to see which gene the transposon inserted into. In the first report of this method, 28 mutants that failed to infect were identified, 13 of which were in genes which had previously been identified in other experiments as being genes involved in infection, thus acting as a useful positive control to establish that the method was indeed working. A further six were found in genes which were known but which had not previously been shown to have a role in infection, and the remaining nine were in novel genes which had not previously been discovered. This method thus provides a powerful route to identifying novel genes which have a role to play in processes such as infection which are hard to study experimentally at the genetic level.

6.5 Gene Tagging in Higher Eukaryotes: Resistance Genes in Plants

The use of gene tagging as a technique is definitely not limited to microorganisms. Much what we now know about the genetics of development, for example, comes about from studying simple model organisms such as the fruit fly *Drosophila melanogaster* and the nematode *Caenorhabditis elegans*, and in both of these organisms transposon-tagging has been widely used to isolate genes that produce novel phenotypes. Gene cloning from plants has also used transposon tagging. Plants are particularly amenable to this method of gene cloning, as they have large genomes with large amounts of repetitive DNA, which can make other cloning methods difficult and laborious. Transposon tagging in plants is helped by the fact that there are a considerable number of well studied transposons found in plants (which indeed is where they were first discovered), and it is often found that transposons that work in one plant species can be induced to do the same in other species. It is also relatively easy with plants to generate the large numbers of offspring that are needed to search for those ones which have a particular phenotype of interest.

An example which illustrates the use of transposon tagging in plants is the first cloning of a plant gene involved in resistance to infection. Plants show a characteristic response when they are infected with agents such as bacteria, viruses or fungi, with the cells close to the site of the infection dying, which helps to limit the spread of infection through the plants. For this to happen, the plant has to detect the presence of the infecting organism, and in most cases this requires the plant to express a protein which in turn recognizes, in some way, the presence of a particular protein from the infecting organism. The gene encoding the plant protein is referred to as a resistance gene, and the gene in the infecting organism which codes for the protein which is recognized is called an avirulence (*avr*) gene. If the plant lacks the resistance gene, it cannot defend itself against the infecting organism. Obviously, it would be useful to have these genes cloned as they could then be introduced into plant varieties that did not posses them, thus making them resistant to the infecting organism. This process would otherwise have to be done by conventional plant breeding, which takes much longer.

Tomato is infected by a fungus called *Cladosporium fulvum*, and upon infection the tomatoes carrying the resistance gene (referred to as *cf-9*) show cell death around the site of infection. The gene was mapped approximately to the short arm of chromosome 1 by conventional genetics using crosses between different strains, but the nature of the gene was unknown. To identify the actual gene, researchers took advantage of the fact that the avirulence gene from *C. fulvum* had already been cloned, and it had been shown that if this gene was expressed in tomato plants which also were carrying the *cf-9* gene, the whole plant would become brown and die, effectively as if it had suddenly suffered from a massive infection by the fungus in every cell.

To tag the *cf-9* gene, researchers crossed two lines of tomato. One had been genetically engineered to carry a copy of the *avr* gene from *C. fulvum*, and then bred to produce a line that was homozygous for the *avr* gene. This line did not carry a functional *cf-9* gene, and so the presence of this gene was not deleterious to the plant (remember that the cell death reaction requires the presence of both the *avr* gene and the *cf-9* gene). The second line was homozygous for the *cf-9* gene, and also carried (on the same chromosome) a single copy of a transposon called a Ds element which had been introduced from maize. Ds elements were among the first transposons to be discovered, and are non-autonomous elements which require transposase to be provided from another element, called an Ac element. Although initially discovered in maize, they have been introduced into a variety of other plant species, where they are still capable of transposing. The Ds element used in this experiment had been engineered in advance to contain an antibiotic resistance gene and an *E. coli* origin of replication. This line also carried an Ac element introduced from maize, carried on a different chromosome.

When these two lines (shown in Figure 6.9) are crossed, the resulting seeds will carry a copy of the *avr* gene (from the *avr* homozygous parent) and a copy of the *cf-9* gene (from the *cf-9* homozygous parent). They will therefore rapidly die after germination, due to the presence of both genes. In a few cases, however, the Ds element will be activated to transpose by the Ac element, and in a small fraction of these cases, it will transpose into the *cf-9* gene. This will disrupt the *cf-9* gene and so these plants will germinate and grow normally. They are thus very easy to pick out against a large background of dead seedlings. In practice, some 63 examples were found

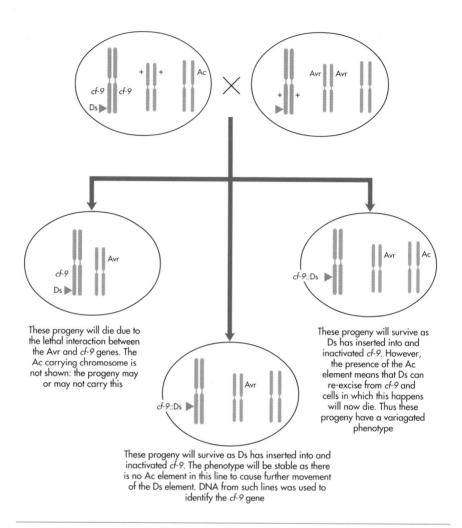

These progeny will die due to the lethal interaction between the Avr and cf-9 genes. The Ac carrying chromosome is not shown: the progeny may or may not carry this

These progeny will survive as Ds has inserted into and inactivated cf-9. However, the presence of the Ac element means that Ds can re-excise from cf-9 and cells in which this happens will now die. Thus these progeny have a variagated phenotype

These progeny will survive as Ds has inserted into and inactivated cf-9. The phenotype will be stable as there is no Ac element in this line to cause further movement of the Ds element. DNA from such lines was used to identify the cf-9 gene

Figure 6.9 The method used to generate transposon-tagged mutants of the tomato *cf-9* gene. Only relevant chromosomes are shown in the cross.

which grew on germination, out of 160,000 seeds which were germinated. Of these, some were completely normal, whereas others showed patches of cell death.

Q6.6. Can you think of a reason why some plants were completely normal and others showed patches of cell death in this experiment?

By digesting total chromosomal DNA from some of these mutants, allowing it to self-ligate, transforming it into *E. coli* and selecting for the antibiotic resistant colonies, plasmids containing part of the *cf-9* gene with the Ds element inserted into it could be recovered. Fragments of these plasmids containing the DNA sequences which flanked the Ds insertion sites were then used to screen a cDNA library for the presence of clones containing the intact *cf-9* gene, and the sequence of the entire gene could then be determined.

6.6 Positional Cloning: Using Maps to Track Down Genes

Although transposon tagging is a powerful technique, it has the major disadvantage that it cannot be used to clone human genes. This is for both ethical and technical reasons: ethically, it would of course be unacceptable to deliberately generate mutations in human populations, and technically, the genetic tool-kit of transposons and vectors that would be required to do so has therefore never been developed. Some other method is therefore needed to clone human genes, and genetic mapping has provided the means to do this.

Genetic maps are exactly what their name implies – a map of the genetic features of an organism, shown in relation to each other. In the same way as we can use a map to navigate from a known position to an unknown one, so a genetic map can be used to navigate to the position of a gene about which nothing is known other than the consequences of mutating it. But genetic maps are unlike the normal maps of everyday life in one very important respect, which is that the features on the map are sites where variation occurs between different individuals. In order to understand how positional cloning is done, it is important to have a clear concept first of exactly what a genetic map is.

The earliest forms of genetic maps were produced long before either the chemical nature of DNA or the role of chromosomes was known. The earliest of true genetic experiments (those done by the monk Gregor Mendel) showed that some traits assorted randomly at meiosis. In a celebrated experiment, he crossed a pure breeding line of peas that had round, yellow peas with one that had wrinkled, green peas. The progeny had only round and yellow seeds. When these progeny were self-fertilized, the members of the next generation showed a 3:1 segregation for each of the two traits (three times as many peas were yellow as green, and three times as many peas were round as wrinkled). The two traits were independent of each other, and hence had the familiar 9:3:3:1 ratio when considered together. Mendel was lucky: the genes he chose to study were all either on different chromosomes or else were a long way apart on the same chromosome. Subsequent studies showed that not all sets of traits showed this independent assortment, and it came to be realized that not only were some traits linked, but some were more tightly linked than others. We now know that the degree of linkage between two traits (i.e. two phenotypes arising from differences in two different genes) is roughly equivalent to the physical distance between the genes on a chromosome. This in turn is because recombination (the phenomenon in meiosis that separates two genes which start off on the same chromosome) is less likely to occur between two genes which are very close together than two which are very far apart. By doing large numbers of crosses with different combinations of mutants, geneticists were able to build up quite detailed genetic maps which showed the orders of genes along chromosomes, long before anyone dreamed that it might one day be possible to isolate and study these genes.

For a genetic map to be constructed, the genes under study must display some variation. If all the individuals in a population had exactly the same phenotype, and there were no variants available, then it would be impossible to map the gene since there would be no way of telling how it was being inherited. The variation needed for mapping will arise initially from mutation, but in many cases mutations which occurred a long time ago will have risen to a relatively high frequency in the population. These are referred to as polymorphisms, defined as any region of the genome which shows variation in greater than 1% of the individuals in a population. In order to use variation to construct a genetic map, data must be available from several generations so that the patterns of inheritance can be followed.

How are genes cloned on the basis of a genetic map? The idea behind this type of cloning, referred to as map-based or positional cloning, is quite simple, but the practice can be very difficult. Suppose we wish to identify a gene for which we have a phenotype – perhaps a disease that results when the gene is mutated in some way. To clone the gene, we need to identify other genes which have already been cloned and which are near to the gene of interest and hence show linkage to it. Once such genes have been identified, we should in principle be able to move along the chromosome by identifying a series of overlapping pieces of DNA, cloned for example in a genomic library, until we arrive at the gene of interest. A useful analogy, which illustrates both the principle and the difficulties inherent in this approach, is to imagine that you are trying to find a particular book in a pitch black room. The first thing you might reasonably try to identify is the bookshelf, since this is where the book will be. This is analogous to identifying the stretch of the chromosome upon which the gene is to be found. You can then work along the bookshelf by feel, analogous to working your way along the chromosome. However, the big question is: how do you know when you have got to the right book? As we will see, various different methods exist to turn the light on at this point, so that we can identify the book in question, but it can be a very dim and uncertain light!

The steps in positional cloning are thus:

• Identification of a marker which is linked to the gene of interest
• Moving from the marker towards the gene of interest
• Identifying the correct gene when it has been reached

We will look now at each of these steps in turn before illustrating them with the particular example of the cloning of the CF gene.

6.7 Identification of a Linked Marker

Although we have been talking up to now in terms of beginning a positional cloning project by finding a known gene which is linked to the gene of interest, in practice the known gene does not have to be a gene in the classical sense of the word; that is, a stretch of DNA which codes for a protein. Cloning experiments in fact rely far more on polymorphisms which are not in genes at all, but rather are found in the non-coding DNA which makes up the bulk of the human genome (Section 2.3). Initially, the most widely used were restriction fragment length polymorphisms (RFLPs) (Box 6.3). RFLPs are not genes, but simply points in the DNA where there are sequence differences between individuals. However, they are of course inherited exactly like traditional genes which code for proteins, and hence can be used in linkage analysis to build up detailed maps of the genome. What RFLPs do is to give us a series of markers which are specific for particular positions on the genome. When we wish to clone a particular gene for which we have no information, one of the first steps we can therefore take is to try to identify RFLPs which are close to this gene and are hence tightly linked to it. The more tightly linked a given RFLP is with a particular phenotype the more useful it is, since we can now attempt to move from this RFLP to the gene we are after, as discussed in the next section. The principle of linking an RFLP to a particular gene of interest is shown in very simplified form in Figure 6.10.

RFLPs have now been largely superseded in positional cloning by other polymorphisms which are more frequent and more informative. The most useful of these are SNPs or single nucleotide polymorphisms (see Section 2.3). However, the broad principles of how these polymorphisms can be used in these sorts of experiments are the same, and these other polymorphisms will not be discussed further.

6.8 Moving From the Marker Towards the Gene of Interest

Finding a marker linked to the gene that we want is not the same as finding the gene. Indeed, even for very tightly linked markers, the actual physical distance from the marker to the gene of interest may still be many thousands of bases. Having identified a polymorphism that is close to our gene of interest, where do we go next? The aim of subsequent experiments is to construct a detailed map of the region adjacent to the RFLP, and to clone the DNA corresponding to the adjacent region, both of which will assist in finding the gene of interest. To do this, a number of methods are used, among them chromosome walking and chromosome jumping. In chromosome walking, an initial cloned fragment, shown by mapping to be linked to the gene of interest, is hybridized against a genomic library. Clones that hybridize to the initial cloned fragment will represent stretches of DNA which are adjacent to the initial fragment. By taking these clones, and re-screening the genomic library with them, the next fragment along can be

Box 6.3 Restriction Fragment Length Polymorphisms (RFLPs)

An RFLP is a region of DNA which differs between the individuals in a population in such a way that, in some individuals, this stretch of DNA contains a particular restriction site, whereas in other individuals it does not. Very large numbers of such polymorphisms were identified in the 1980s in human DNA (and in the DNA of other organisms too). To find RFLPs, random pieces of DNA were cloned and then used as probes in Southern blots (Box 5.4) against total DNA digested with a range of different enzymes and taken from different individuals. If the DNA sequence of a given fragment were the same in all individuals, the pattern of bands seen in the Southern blot would be the same in all individuals, which from a mapping point of view would be totally uninformative. However, if in some cases a certain part of the population possessed an extra site in this sequence, then the pattern revealed after digestion with this enzyme would be different in these individuals from all the others.

a) To visualize an RFLP, total DNA from several individuals is digested with a restriction enzyme (in this case *Eco*RI) and run through an agarose gel and transferred to a nylon membrane.

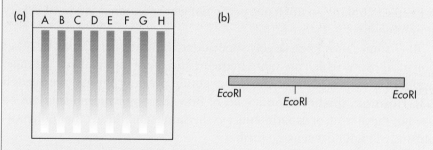

A random fragment (b), is taken from a genomic library of *Eco*RI fragments of DNA, radioactively labeled and used to probe the membrane.

c) In two individuals (C and F), there is clearly an extra *Eco*RI site in this fragment, which has therefore led to two rather than one hybridizing fragment. This is an example of an RFLP. Individual C is homozygous for this polymorphism, and individual F is hetereozygous for it.

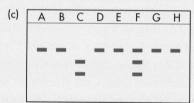

It is important to emphasize that RFLPs generally have no phenotypic effect on the people or organisms that have them, since in the vast majority of cases they are not located in coding regions in the DNA.

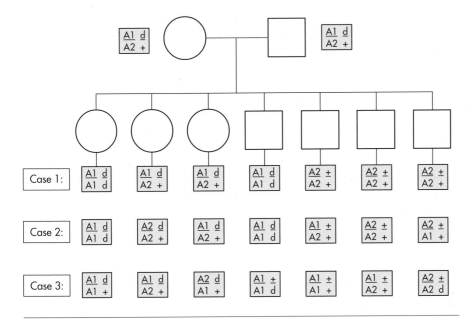

Figure 6.10 Linking an RFLP to a gene of interest. In this hypothetical case, two parents (each heterozygous for a defective gene, labeled d in the defective state) are also both heterozygous for a polymorphism A, which occurs in two forms, A1 and A2. They have seven children, who are analyzed for the presence of the defective gene and for which of the A polymorphisms they carry. Three outcomes are presented: case 1 where A is tightly linked to the defective gene, case 2 where A is weakly linked, or case 3 where A is unlinked. In principle, analyses of this type are done on much larger populations (many families) and using many different polymorphic loci, any of which may have several different forms.

found. The process can then be repeated with this second fragment, and so on, effectively walking along the chromosome. By mapping the position of restriction fragments on these clones, a restriction map of the region of the genome can gradually be built up (Figure 6.11). Fragments isolated in this way may also reveal new polymorphic markers that can be tested for linkage to the target gene. This is important as it can reveal whether the chromosome walk is moving towards, or away from, the gene of interest.

Chromosome walking is laborious and can get stuck in long repetitive regions, which are not uncommon in human and other DNA. Also, some pieces of DNA are intrinsically unclonable, and this can stop a chromosome walk dead in its tracks. To get round these problems, a method called chromosome jumping can be used. Chromosome jumping enables the cloning of stretches of DNA which are genetically linked but not immediately adjacent on the chromosome. The method of chromosome jumping is shown in Figure 6.12. It relies on the existence of restriction enzymes which only cut genomic DNA rarely, usually because they have an eight

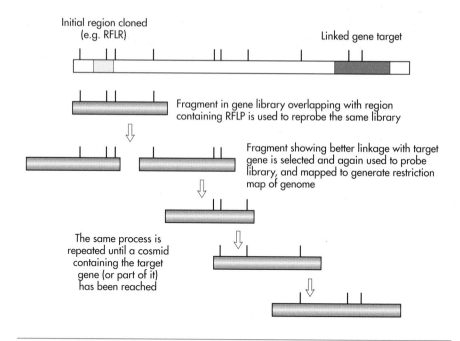

Initial region cloned
(e.g. RFLR)

Linked gene target

Fragment in gene library overlapping with region
containing RFLP is used to reprobe the same library

Fragment showing better linkage with target
gene is selected and again used to probe
library, and mapped to generate restriction
map of genome

The same process is
repeated until a cosmid
containing the target
gene (or part of it)
has been reached

Figure 6.11 Walking along a chromosome. The generation of a restriction map of the genome is also shown, with cut sites for a particular enzyme shown by vertical bars.

base-pair recognition sequence which will be rarer than the more usual six or four base-pair recognition sequences.

Q6.7. On average, how frequently would you expect an enzyme with an eight base-pair recognition sequence to cut DNA?

In a typical experiment, chromosomal DNA is digested with the enzyme *Not*I. This releases large fragments of DNA, which are then ligated with a short piece of DNA that contains a nonsense suppressor gene (Box 6.4) flanked by two *Not*I linkers. Among the products of the ligation will be circles of DNA consisting of individual *Not*I fragments, joined by the linker fragment. The ligation mix is then digested with an enzyme that cuts more frequently than *Not*I, such as *Eco*RI, *Hin*dIII or *Bam*HI. This will cut at many sites within the ligated circle. Some of the fragments will contain the nonsense suppressor gene and this will be flanked by sequences from the two ends of one of the *Not*I fragments. The mix is then ligated into a lambda vector that has a nonsense mutation in one of its essential genes, and hence can only grow and form plaques if it incorporates the fragment containing the nonsense suppressor and the two ends of a *Not*I fragment. This represents regions of the DNA which are linked but not right next to each other on the chromosome. Essentially, we are creating a kind of library where each clone in the library contains the two ends of a large fragment of chromosomal DNA. We now take our clone carrying a marker, which has been shown by linkage analysis to be close to the gene of interest, and use it to probe the

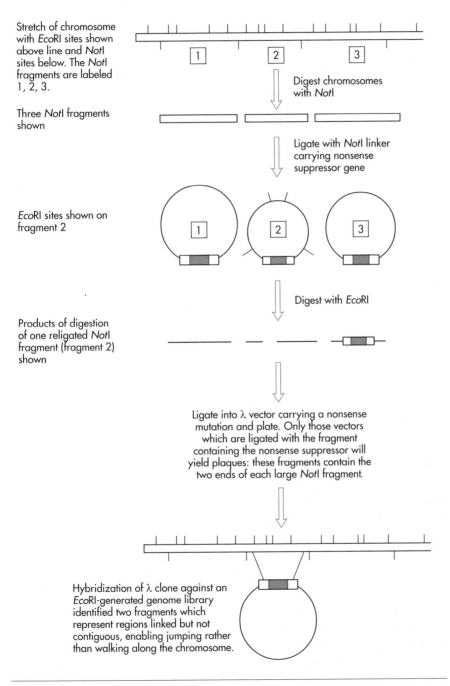

Stretch of chromosome with EcoRI sites shown above line and NotI sites below. The NotI fragments are labeled 1, 2, 3.

Digest chromosomes with NotI

Three NotI fragments shown

Ligate with NotI linker carrying nonsense suppressor gene

EcoRI sites shown on fragment 2

Digest with EcoRI

Products of digestion of one religated NotI fragment (fragment 2) shown

Ligate into λ vector carrying a nonsense mutation and plate. Only those vectors which are ligated with the fragment containing the nonsense suppressor will yield plaques: these fragments contain the two ends of each large NotI fragment.

Hybridization of λ clone against an EcoRI-generated genome library identified two fragments which represent regions linked but not contiguous, enabling jumping rather than walking along the chromosome.

Figure 6.12 The principle of chromosome jumping.

Box 6.4 Nonsense Suppression

If a codon in a gene becomes mutated so that rather than encoding an amino acid it now reads as a stop signal, the protein produced by the gene will terminate prematurely when it is synthesized and is likely to be non-functional. Such mutations are referred to as "nonsense" mutations. Their effect on phenotype can, however, be largely suppressed by the presence of an altered tRNA which instead of recognizing the codon as a stop codon, inserts an amino-acid so that polypeptide chain elongation can continue. This phenomenon can be exploited in devising cloning tools. In the example given here, the presence of the nonsense suppressor gene (that is, the gene for the altered tRNA) on the *Not*I linker means that as the vector itself contains a nonsense mutation in an essential gene and hence cannot normally grow, only those clones which pick up a fragment containing the linker will be able to form plaques on a bacterial lawn. Thus this constitutes a strong selection in favor of the particular clones that we are interested in.

library. It should detect a clone containing not only a piece of DNA with complementary sequence to the probe, but a second piece from the other end of an original large fragment which is linked but not adjacent to the first sequence. This procedure has allowed us to take a jump along the chromosome rather than having to laboriously clone all of the intervening DNA. Of course this jump could be away from the gene of interest rather than towards it, so we will now need to use this new candidate sequence in linkage analysis to determine whether it is nearer to the gene we are trying to find. The method is summarized in Figure 6.12.

Q6.8. How far would this technique enable us to jump?

Using these and other methods, a map can gradually be built up of the region between polymorphic markers which are linked to the gene of interest, and this region can be sequenced. But now we come to the hardest part of all: how do we know we have found the right gene when we get there? Returning to the analogy of the books in the pitch dark room, how do we know when we have found the right book?

6.9 Identifying the Gene of Interest

In practice, several different methods are used to spot when the correct gene has been found, and this is good because several lines of evidence will be needed to be convincing. It must be recalled from Section 2.3 that in eukaryotic organisms, genes are large and can be made up of many introns and exons. Since all the work described above is done using genomic rather than cDNA clones, any candidate sequence believed to be part of the gene of interest may not contain the whole of the coding sequence, or indeed it may be entirely derived from intron sequence. One commonly used method to determine whether the candidate sequence is part of a coding region, rather than being part of an intron, is the zooblot in which the candidate clone is used in a Southern blot (Box 5.4) against genomic DNA from a number of different animals. This works on the principle that evolutionary pressures will ensure that the sequence of coding regions are preserved because a mutation here is likely to result in a protein which does not function as well as the wild-type. On the other hand, the sequence in the noncoding introns is under no such evolutionary pressure and shows considerable variation between closely related species. If the candidate clone detects a homologous sequence in genomic DNA from a number of animals then it is likely that it represents the sequence of an exon. A cDNA clone of the complete gene may need to be isolated, by probing a suitable cDNA library with a probe made from this putative exon, before some of the techniques described below will work.

Once it has been shown that the cloned DNA is likely to be part of a coding region (exon), then the following methods can be used to see whether it is likely to be part of the gene of interest.

1. We can test whether the gene shows patterns of expression which are consistent with its known function, or which are unusual in some way in the disease state.
2. Comparison of the sequence of the complete gene (when it is available) from the defective genome with the same gene from the wild-type should reveal the precise genetic nature of the mutation. This method must be used with some care because sequence variations between individuals may be due to polymorphisms which are not in themselves responsible for the disease. Any putative genetic change must be shown to be 100% linked with the disease state.
3. The sequence of the hypothetical gene can be used to predict the sort of protein which it encodes, and it may be that this will be shown to be consistent with what is known about the nature of the disease.
4. Ultimately, the most convincing experiment is the demonstration that the wild-type gene can correct the genetic defect. This is likely to be done in tissue culture cells, and requires that some defect has already been characterized in cells derived from someone with the defective gene. More recently, methods have been developed for deleting selected genes in mice and other animals, so the effects of doing this to a putative defective gene can be examined. If the mice lacking the gene show symptoms characteristic of the disease, it is highly likely that the correct gene has been identified. Methods for this are described in Section 12.2.

6.10 Cloning of the CF Gene: A Case Study

The steps taken to identify and clone the CF gene, one of the first "disease genes" to be cloned, illustrate the above features very well. It was a laborious and expensive task, and a real landmark in human molecular genetics. For this reason, although many of the methods that were used have been refined or superseded since, it remains an excellent case study to understand the routes used to cloning single genes with known phenotype but no known features.

Traditional genetic methods using known markers and studying the inheritance of CF in human families succeeded in mapping the CF gene to human chromosome 7. As more polymorphic markers became available, they were tested for their degree of linkage to the CF gene. Eventually four markers were identified, called MET, D7S340, D7S122 and D7S8, which were linked to the CF gene. Of these, genetic linkage data showed that D7S122 and D7S340 were likely to be the closest to the gene, so further studies focused on these. Chromosome walking and jumping was used to clone most of the DNA close to these markers. The resulting cosmid clones were checked by various means to ensure that they did indeed represent the correct stretches of DNA. For example, it was shown using Southern blotting that the restriction map derived from the pieces of DNA in the cosmid clones, and the actual restriction map of the genomic DNA, were identical.

To determine where the CF gene was likely to lie within the cloned DNA region, use was made of the fact that coding regions tend to be more highly conserved between species than non-coding regions, since they are subject to stronger selection pressure (i.e. using the technique of the zooblot referred to in Section 6.9). Various cosmid clones were hybridized against genomic DNA from species such as cattle, mice, chickens and hamsters, and certain regions within the area covered by the cosmid clones were identified as being well-conserved. The presence of undermethylated CpG regions in these different areas was then investigated, using the enzyme *Hpa*II which only cuts its recognition site (5'-CCGG-3') when it is unmethylated. Undermethylation is often seen at the 5' end of eukaryotic genes (see Section 2.12) and is another indication that a given region may be part of an actual gene.

After a number of false leads had been pursued, a particular fragment was found which hybridized to a clone from a cDNA library and hence represented part of a transcribed gene. Expression of this gene was elevated in several tissues where the CF gene was thought to be expressed, although there was no difference between levels of expression in individuals homozygous for the defective gene and those with the wild-type gene.

Q6.9. Assuming that the gene was indeed the CF gene, what does this last result imply about the nature of the mutation that causes CF?

The fragment which had been identified was then used to clone a full-length cDNA corresponding to the whole CF gene. This proved tricky as the gene was very unstable when present in *E. coli*, and the sequence had to be determined by sequencing the gene in fragments. Two features in particular made it clear that the gene that had been found was indeed the CF gene. The first was that the predicted structure for the protein encoded by the CF gene was that of a membrane transporter, and it had long been known through studies on tissue culture cells that CF caused an alteration in the ion transport in epithelial cells, although it was not known whether this was itself the cause of the disease. The second piece of evidence, which was conclusive, was that when the same gene was analyzed from the DNA of a large number of individuals who had CF, in every case the gene was found to be mutated, often with the same mutation (deletion of a single amino acid at position 508).

Since the successful cloning of the CF gene, a large number of other genes associated with human genetic diseases have been cloned using similar methods. These methods are also used for cloning genes from many other organisms. Interest is now moving towards a much more complex problem: the identification and cloning of genes where several genes all contribute to a phenotype. This requires a much more sophisticated analysis, which we will not describe here, and to date represents one of the largest challenges in human genetics.

Questions and Answers

Q6.1. Why is it better for transposon tagging to use a transposon which is not normally present in the target genome?

A6.1. *A typical way of using transposon tagging is to generate mutants and then to construct libraries from the mutants which are screened using the transposon as a probe. If the transposon is already present in the genome which is being studied, then some of the clones recovered will be from these endogenous copies of the transposon, not from the one which has been newly introduced. In other words, the presence of copies of the transposon in the target genome can result in a very high background of false positives when the library is screened, which can make identification of the correct clone much harder.*

Q6.2. Why is it important to limit the number of transpositions made by the transposon when attempting to clone genes by transposon tagging?

A6.2. *To answer this question, consider the consequences of not limiting transposition. A given individual may be selected for the presence of the transposon, but if that transposon continues to jump to new locations, there* *are two possible outcomes, depending on the type of transposon used. If the transposon is one which replicates itself and inserts the replicated copy at a new position, then more and more transposition events will accumulate at random in the progeny from that one individual, making it increasingly hard to determine which of the events gave rise to the original phenotype. If the transposon is one which excises from one position and inserts itself randomly elsewhere, the situation is even worse, as individuals will start to arise where the original mutated gene is no longer marked by the transposon and so will not be identifiable.*

Q6.3. Before looking at Figure 6.3, can you see how including a bacterial origin of replication on a transposon which is being used for gene tagging is helpful in recovering the region of DNA containing the transposon after tagging has taken place?

A6.3. *The region of DNA containing the transposon can be identified by simply digesting the DNA using an enzyme that does not cut too frequently, and ligating it under dilute conditions. This will favor self-ligation of individual DNA molecules rather than the formation of concatemers with several DNA molecules ligated together. On transformation of the ligation into E. coli, and selection on the appropriate antibiotic, only those stretches of DNA derived from the transposon and the DNA into which it inserted should be recovered, as only these will contain a functional origin of replication.*

Q6.4. Why is it important to demonstrate that the complementing gene is indeed the one into which Tn5 had inserted in generating the original mutant?

A6.4. *This additional step is important as it is sometimes found that a mutation in one gene can be complemented by expression of a different gene.*

Q6.5. The pool of 96 colonies before infection is referred to as the input pool. The pool of the colonies grown from the animal after infection is referred to as the output pool. If the signature tags from the transposons in the input and the output pools are amplified by PCR and labeled with radioactivity, and are hybridized against the replicate filters, can you predict what you would see with (a) the filter hybridized with amplified signature tags from the input pool and (b) the filter hybridized with amplified signature tags from the output pool?

A6.5. *All the spots of DNA on the filter should be seen to hybridize with signature tags from the input pool, since every colony on the filter is represented in the input pool. However, not all the spots of DNA will hybridize with signature tags from the output pool. Certain spots will represent colonies where the transposon has jumped into a gene required for infection: thus, these bacteria will not be present in the output pool and so will not produce an amplified signature tag to hybridize with the original bacterial DNA.*

Q6.6. Can you think of a reason why some plants were completely normal and others showed patches of cell death in this experiment?

A6.6. *The plants that showed patches of cell death still had the Ac element present in their cells, which allowed the Ds element to re-excise from the cf-9 gene and restore its activity (leading to cell death in those cells descended from cells where the Ds excised). In the stable progeny, where no such variegation was seen, the Ac element was lost during the cross, so the Ds insertion was completely stable.*

Q6.7. On average, how frequently would you expect an enzyme with an 8 base-pair recognition sequence to cut DNA?

A6.7. *As we saw in Section 3.7, an enzyme with a recognition sequence n bases long will cut once every 4^n base pairs. Thus, on average, an enzyme with a 4-base recognition site cuts once every 256 bp, an enzyme with a 6-base recognition site cuts once every 4096 bp, and an enzyme with an 8-base recognition site cuts once every 65,536 bp.*

Q6.8. How far would this technique enable us to jump?

A6.8. *This will depend entirely on the distance apart of the sites which are used to generate the large fragments, whose ends are then cloned. Because restriction enzyme recognition sites occur essentially at random, some jumps will be larger than others.*

Q6.9. Assuming that the gene was indeed the CF gene, what does this last result imply about the nature of the mutation that causes CF?

A6.9. *The fact that the level of expression of the CF gene was the same in affected and unaffected individuals shows that the gene defect does not lead to a large change in gene expression levels. Thus it makes it more likely that the disease is due to some alteration in amino acid sequence of the protein which leads to it becoming unable to carry out its normal function.*

Further Reading

Cloning and characterization of *Pseudomonas aeruginosa fliF*, necessary for flagellar assembly and bacterial adherence to mucin (1996) Arora SK, Ritchings BW, Almira EC, Lory S and Ramphal R, Infection and Immunology, Volume 64 Pages 2130–2136.
This study describes the use of transposon tagging to identify genes in the bacterium Pseudomonas aeruginosa *that are involved in adherence to mucin, the example that was used in Section 6.3. The cloning and further study of these genes is also described*

***In vivo* genetic analysis of bacterial virulence** (1999) Chiang SL, Mekalanos JJ and Holden DW, Annual Review of Microbiology, Volume 53 Pages 129–154.
A detailed review of the topic, including discussion of signature-tagged mutagenesis, and other methods which can be used to identify genes involved in complex biological processes such as bacterial virulence

Isolation of the tomato Cf-9 gene for resistance to *Cladosporium fulvum* by transposon tagging (1994) Jones DA, Thomas CM, Hammond-Kosack KE, Balint-Kurti PJ and Jones JD, Science, Volume 266 Pages 789–793.

Classic paper describing the first cloning of a gene which endows plants (in this case, tomatoes) with resistance to a pest (in this case, a fungus). This was achieved using transposon tagging

Identification of the cystic fibrosis gene: chromosome walking and jumping (1989) Rommens JM, Iannuzzi MC, Kerem B, Drumm ML, Melmer G, Dean M, *et al.*, Science, Volume 245 Pages 1059–1065.

Identification of the cystic fibrosis gene: cloning and characterization of complementary DNA (1989) Riordan JR, Rommens JM, Kerem B, Alon N, Rozmahel R, Grzelczak Z, *et al.*, Science, Volume 245 Pages 1066–1073.

Two landmark papers which described the identification of the gene that, when mutated, leads to cystic fibrosis. These papers were the culmination of much other published work, which is referred to in the papers. The papers are very technical, so do not be too concerned if you do not understand them in their entirety

7 Sequencing DNA

Learning outcomes:

By the end of this chapter you will have an understanding of:

- *the enzymatic sequencing procedure developed by Fredrick Sanger and colleagues*
- *examples of modifications to basic enzymatic sequencing methodology, including PCR-based methods and methods for high-throughput sequencing*
- *how to interpret both gel-based and chromatogram-based DNA sequences*
- *strategies employed for genome sequencing*
- *the methods used to sequence the human genome*

7.1 Introduction

DNA sequencing is one of the lynch pins of modern molecular biology. When DNA sequencing techniques were first developed they allowed scientists to sequence short gene-sized segments of the genome of the organism they were studying. The procedure was intimately linked to library screening since the gene was invariably isolated from either a genomic or cDNA library using procedures that have been described in Chapter 5. Once the sequence of the gene was known a clue to the structure and function of the protein product could be gleaned by comparisons with other known genes and prediction of possible structural motifs. Analysis of DNA and protein sequences will be discussed in Chapter 8. From the DNA, sequence hypotheses could be developed and tested. Often this would involve mutagenesis of the gene and, again, sequencing became a vital tool either to confirm that site-directed mutants had been successfully obtained or to screen random mutants to determine their sequence. More recently, DNA sequencing has been a central part of the "genomic revolution". The complete genome sequence of hundreds of organisms has been determined using the procedures discussed in this chapter, and this has provided a valuable resource that can be exploited to gain an understanding of the biology of these and other organisms.

Two approaches were initially developed for DNA sequencing, which were the chemical method developed by Allan Maxam and Walter Gilbert and the enzymatic method, often referred to as the dideoxy chain termination method, developed by Fredrick Sanger and colleagues. Walter Gilbert and Fredrick Sanger received the Nobel Prize in Chemistry in 1980 "for their contributions concerning the determination of base sequences in nucleic acids". The Sanger strategy is still used today in all sequencing protocols from small-scale sequencing you might do in your laboratory to the high-throughput genome sequencing.

If you are familiar with the protein sequencing strategy, also developed by Fred Sanger, you might think that DNA sequencing will work in a similar manner; i.e. take a DNA fragment, remove one nucleotide from one end and determine which base was present. You would then repeat this process walking your way along the sequence and this would tell you what the sequence was. The problem is that, although this is conceptually very simple, it is not technically possible. DNA sequencing strategies are conceptually more complex, but are technically very straightforward.

7.2 Overview of Sequencing

The sequencing methods of both Sanger and Maxam & Gilbert work by generating four sets of fragments that have a common 5′ end and a base specific 3′ end. It is essential that all four reactions have the same common 5′ end since this will serve as a reference point for reading the sequence. In the case of the Sanger procedure you have four reactions, one for each base. For example, the "G" reaction will contain hundreds of different fragments all with the same 5′ end and all with a "G" at the 3′ end, but the length of the fragment and hence the position of the G in the sequence will vary. So, for example, in Figure 7.1, the G reaction contains five fragments all with the same "C" residue at the 5′ end, but the 3′ end of each fragment contains a different "G" residue. The G reaction therefore generates fragments that are three, four, five, seven and 12 bases in length, and this tells us that the sequence contains G residues at positions 3, 4, 5, 7 and 12 relative to the 5′ end. The A reaction generates fragments that are six, 11, 13 and 14 bases in length and hence indicates the position of A residues within the sequence. The same principle will be true for the C and T reactions. Gel electrophoresis is used to visualize the DNA fragments produced in the sequencing reaction. You could determine the size of each fragment, as discussed in Section 3.11, in the four reactions and map the position of each base within the sequence. However, assuming all four reactions are analyzed in adjacent wells, reading the DNA sequence is simply a matter of "climbing up the gel" jumping from one "rung" to the next across the four ladders to read the DNA sequence 5′ to 3′. The chemical method of sequencing DNA developed by Maxam & Gilbert is no longer used for sequencing.

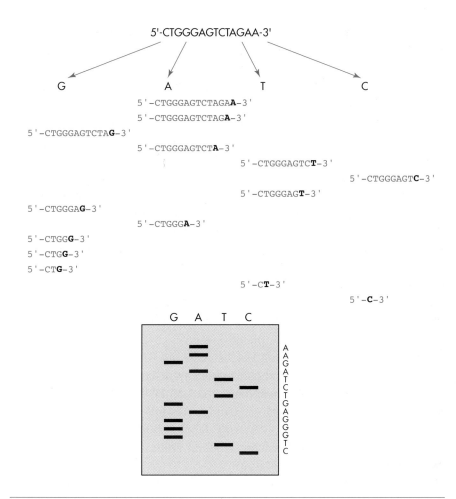

Figure 7.1 A sequencing reaction. Four reactions, one for each base, are used to sequence a DNA molecule. Each reaction produces a set of fragments all with the same 5′ end, but with base specific 3′ ends. The fragments are separated according to size by electrophoresis, the size of fragments in each reaction can be used to map the position of each base in the DNA sequence.

Q7.1. Figure 7.1 shows a schematic representation of a sequencing reaction. The G reaction generates fragments that are three, four, five, seven and 12 bases in length. What size fragments would be generated by the C and T reactions?

7.3 Sanger Sequencing

The Sanger dideoxy method of sequencing DNA works by generating four sets of fragments, all with the same 5′ end, and each with a base-specific 3′ end. The Sanger dideoxy sequencing method uses DNA polymerase to synthesize the DNA to be sequenced. As you know, DNA polymerases cannot synthesize DNA *de novo*, but can only extend an existing nucleic acid chain. DNA polymerases require a primer from which to extend the DNA chain. You will have seen how this has been exploited in the polymerase chain reaction (Section 3.12) to direct the amplification of segments of DNA. In DNA sequencing we again use an appropriately designed primer to direct DNA polymerase to the region of DNA to be sequenced. As with PCR if you anneal a primer to a region of single-stranded DNA in the presence of deoxynucleoside triphosphates (dNTPs) DNA polymerase will synthesize the complementary strand. All strands synthesized will have the same 5′ end as specified by the 5′ end of the primer. This therefore gives us a protocol to synthesize DNA fragments all with a common 5′ end. In order to generate base-specific 3′ ends we take advantage of the nucleotide analogues dideoxynucleoside triphosphates (ddNTPs). They are the same as the normal deoxynucleoside triphosphates (Figure 3.19) but lack the 3′ hydroxyl group on the ribose sugar. When the dideoxy nucleotide is incorporated into a growing DNA chain it results in a chain with no 3′ hydroxyl group and so no more nucleotides can be added and hence chain elongation is terminated, so for example ddGTP gives G-specific ends. DNA fragments are radiolabeled during synthesis, using (in the original protocol developed by Fred Sanger and colleagues) a nucleoside triphosphate that is labeled with ^{32}P at the α phosphate (Box 5.3). DNA fragments are visualized using polyacrylamide gel electrophoresis followed by autoradiography. Later protocols utilized ^{35}S- or ^{33}P-labeled nucleotides, and the very latest protocol utilized fluorescent labeling and capillary electrophoresis and is discussed later in this chapter.

A sequencing reaction specific for G residues is shown in Figure 7.2. The reaction contains a single-stranded DNA template, suitable primer, all four dNTPs, one of which is radiolabeled, ddGTP and DNA polymerase. DNA polymerase will start synthesizing DNA from the primer. Every time the template directs the addition of a "G" to the growing chain it will use either dGTP or ddGTP. If dGTP is incorporated, synthesis of that strand will continue; if, however, a ddGTP is added DNA synthesis will stop. This will result in a series of fragments all starting from the primer and therefore having the same 5′ end, but having a different 3′ end depending on how far DNA polymerase had traveled before incorporating the ddGTP in the growing nucleotide chain. The products of the reaction are then heated so that the double-stranded DNA is denatured to give single-stranded molecules. Because the products are analyzed by denaturing gel electrophoresis followed by autoradiography (Box 7.1) only the newly synthesized single-stranded DNA fragments (which are radiolabeled during synthesis) are visualized. The fragments are, therefore, separated by denaturing polyacrylamide gel electrophoresis followed by autoradiography which should reveal a ladder where each rung represents a G residue in the DNA sequence. The G residues closest to the primer give the shortest fragments,

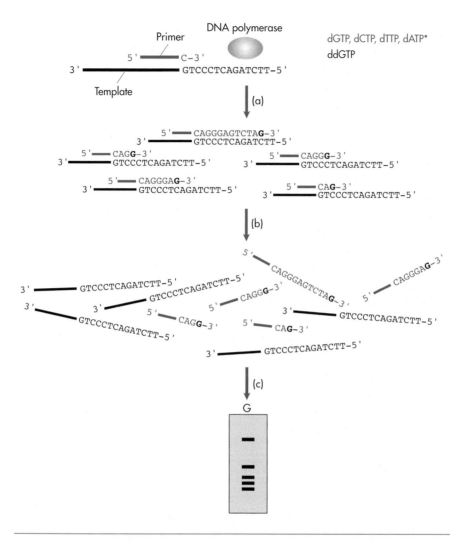

Figure 7.2 The dideoxy chain termination sequencing reaction specific for G residues. DNA synthesis is primed by the short oligonucleotide primer and the reaction contains dGTP, dCTP, dTTP, radiolabeled dATP and ddGTP; the first base is added to the C residue at the 3′ end of the primer. a) The sequencing reaction generates a set of fragments all with the same 5′ end, but the 3′ end of each fragment contains a different G residue. b) The sample is heated to denature the DNA. c) The DNA is separated by denaturing polyacrylamide gel electrophoresis and the newly synthesized radiolabeled DNA visualized by autoradiography.

will run furthest and therefore be at the bottom of the gel. Clearly, you not only want the position of Gs within the sequence, but also the other three bases. A schematic representation of a Sanger sequencing reaction, in which the position of all four bases is determined, is shown in Figure 7.3. The reactions you need to determine the position of the other bases in the

Box 7.1 Denaturing Polyacrylamide Gel Electrophoresis

Sequencing protocols require electrophoresis technology that can separate DNA molecules whose difference in size is only one nucleotide. One common procedure is denaturing polyacrylamide gel electrophoresis, using what is sometime referred to as a "slab" gel. This technique uses large but extremely thin slab gels to deliver the resolving power required. The polyacrylamide gel is poured between two plates, typically 30 cm × 40 cm, separated by 0.4 mm. The presence of 8 M urea within the gel mix provides conditions that denature double-stranded DNA and, in addition, the DNA samples are heated to ensure the DNA is denatured before loading onto the gel. These conditions, therefore, mean that all DNA molecules are analyzed as single-stranded species. The samples are loaded on the gel and a voltage applied. The DNA, being negatively charged, will migrate towards the positive electrode or anode and the polyacrylamide will act as a molecular sieve such that smaller DNA molecules will migrate faster through the gel. On completion of the electrophoresis the two plates are separated, the gel washed in a special "fixing" solution, which retains the DNA in the gel, before being transferred to a piece of filter paper and rapidly dried under vacuum. The dried gel is than placed against a photographic film. The film is developed to produce an autoradiograph. In the case of a sequencing reaction the autoradiograph will contain a set of sequencing ladders that can then be read to reveal the sequence of the DNA being studied.

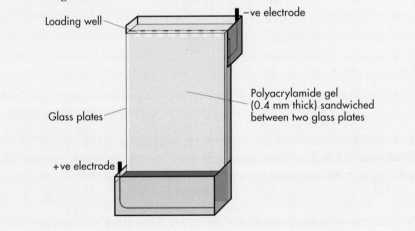

sequence will all contain the DNA template, the same primer, all four dNTPs and DNA polymerase, in addition the A, T and C reactions will contain ddATP, ddTTP and ddCTP respectively. If the sequencing reactions from the G, A, T and C reactions are run side by side it is possible to read the sequence of the autoradiograph by simply working your way up the four ladders.

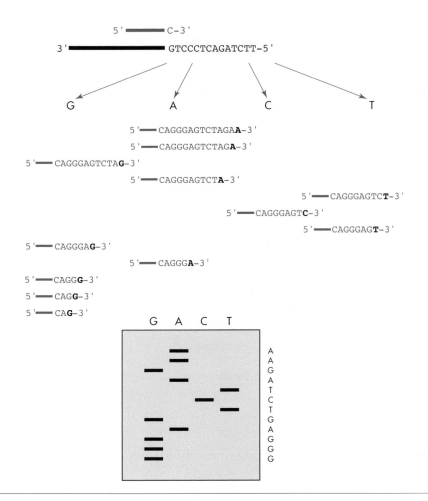

Figure 7.3 The dideoxy chain termination sequencing reaction. Four sequencing reactions, as described in Figure 7.2, one specific for each base are used to sequence a DNA fragment.

Q7.2. Figure 7.4 shows an autoradiograph from a Sanger sequencing procedure. Read the sequence from 5′ to 3′.

7.4 The Sanger Sequencing Protocol Requires a Single-stranded DNA Template

It should be clear from the discussion above that the Sanger chain termination method of sequencing DNA requires a single-stranded DNA template in the reaction. When the method was first developed this meant cloning the DNA to be sequenced into special vectors based on the bacteriophage M13. Bacteriophage M13 is a virus that infects *Escherichia coli*. The virion contains a single-stranded DNA genome encased in a helical capsid. When M13 enters the bacterium, the viral DNA is used as a template to make a

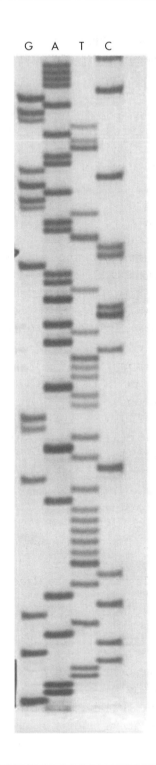

**Figure 7.4 An autoradiograph of a dideoxy chain termination
sequencing reaction.** The sequence is read 5′ to 3′ from the bottom of the
autoradiograph to the top, starting 5′-GAATTC-3′.

double-stranded copy. The double-stranded genome will then direct the expression of viral genes and act as template for the synthesis of single-stranded viral genome that is packaged into the virion prior to release of the virion from the bacteria. Because M13 is a helical virus its genome size is not restricted. So in a culture of *E. coli* infected by bacteriophage M13, the bacterial cells will contain double-stranded viral genome, whereas the culture medium will contain the virion with its single-stranded DNA genome. It is therefore relatively easy to isolate both single and double-stranded forms of the viral genome by separating the bacterial cells from the culture medium by centrifugation. All these properties of M13 make it ideal for producing single-stranded DNA. Plasmid vectors were developed that contain M13 sequences, and these vectors can be purified from host *E. coli* cells in the double-stranded form and are therefore tractable to the standard cloning procedures discussed in Chapter 3. They will, however, also direct the synthesis of a single-stranded copy of the plasmid sequence which is packaged into a viral particle and excreted into the culture medium. Single-stranded DNA can therefore be purified from the growth medium.

Cloning into M13-derived vectors is a very robust procedure for obtaining single-stranded DNA for sequencing. However, assuming that your DNA to be sequenced is already present in a standard plasmid vector, it requires a sub-cloning step to transfer your sequence into the M13-based vector. It was therefore desirable to develop a procedure that would allow the direct sequencing of double-stranded DNA by generating single-stranded template in the reaction by either chemical or heat treatment.

7.5 Modifications of the Original Sanger Protocol

As it is based on *de novo* synthesis of DNA, the Sanger chain termination method of DNA sequencing requires a DNA polymerase; initial protocols used the Klenow fragment (Box 3.5). As sequencing protocols developed the Klenow fragment has been replaced with other DNA polymerases (Box 3.5). T7 DNA polymerase first replaced the Klenow fragment in sequencing protocols because it facilitated the development of more robust protocols in which double-stranded DNA templates could be sequenced. These protocols generate a single-stranded template by alkali denaturation of the template DNA. The primer is added to the denatured template at neutral pH and the sequencing protocol completed as discussed previously and as shown in Figures 7.2 and 7.3.

The development of thermostable DNA polymerase has allowed the use of heat denaturation to generate single-stranded template (as in the polymerase chain reaction). The use of *Taq* DNA polymerase has made possible the use of PCR-based sequencing protocols. Another advantage of the polymerase chain reaction is that it can be exploited to generate sufficient template for sequencing without the need to clone a DNA fragment into a plasmid vector. This does facilitate the direct sequencing of genomic DNA, although other considerations that will be discussed later in this chapter mean that for high-throughput genome sequencing projects the DNA is still fragmented and cloned into vectors.

As you should be aware, the success of the Sanger chain termination method for DNA sequencing relies upon the fact that during synthesis the DNA polymerase will normally add a deoxynucleotide to the growing DNA chain, however, at a low frequency a dideoxynucleotide is added which causes chain termination. The success of the sequencing protocol depends on making sure that the deoxynucleoside triphosphate to dideoxynucleoside triphosphate ratio is such that a readable ladder is produced. If the dideoxynucleoside triphosphate concentration is too high there will be premature termination of all DNA chains and the sequence of only a few bases adjacent to the primer will be determined. If the dideoxynucleoside triphosphate concentration is too low then chain termination will not occur and no ladder will be observed. The dideoxy to deoxy ratio will vary for different DNA polymerases. For example, the Klenow fragment is very discriminatory between deoxy and dideoxy nucleotides, it "knows" that dideoxynucleotides are not the correct building blocks for DNA, and therefore a relatively high concentration of dideoxynucleotides needs to be used to make chain termination possible. In contrast T7 DNA polymerase does not discriminate very well between deoxy and dideoxy nucleotides and therefore a relatively low concentration of dideoxynucleotides is required for chain termination.

Q7.3. (a) What would happen if you added the Klenow fragment to a sequencing reaction in which the dideoxy to deoxy ratio had been optimized for T7 DNA polymerase? (b) What would happen if you added T7 DNA polymerase to a sequencing reaction in which the dideoxy to deoxy ratio had been optimized for the Klenow fragment?

Modern "manual" sequencing exploiting radiolabeled nucleotides, T7 or *Taq* DNA polymerase and denaturing polyacrylamide gel electrophoresis is now a routine, inexpensive procedure. It is relatively easy to obtain 500 bases of sequence information in a single sequencing reaction from a single primer.

7.6 Strategies for Sequencing a DNA Fragment
So how would you sequence your favorite gene? The first task would be to clone the gene, traditionally by generating a cDNA or genomic library and then screening the library for the gene of interest as discussed in Chapters 4 and 5. You now have your favorite gene cloned into a vector. The next step would be to isolate the plasmid as template for the sequencing reaction. For most vectors used in creating DNA libraries, standard sequencing primers are available "off the shelf" which prime DNA synthesis across the

restriction enzyme sites used for cloning. So, if you were sequencing a single gene or cDNA, the DNA would be cloned into a vector and you would use a primer derived from the vector sequence that primed DNA synthesis towards the cloned DNA insert. This would yield approximately 500 bases of sequence information. Most clones would be larger than 500 bp and therefore a strategy is required to sequence further into the DNA fragment. One approach is shown in Figure 7.5. Using this approach, once you had sequenced as far as you could into the gene from the vector primer you would synthesis another primer based on the sequence you had just obtained, then again sequence as far as possible before synthesizing another primer and repeat the process until the whole insert had been sequenced, essentially sequencing by walking along the DNA, a technique called primer walking. Today it is very easy and inexpensive to purchase "tailor-made" oligonucleotide primers and, therefore, determining the sequence of a few kilobases of DNA by primer walking would take only a few weeks.

In the late 1970s when the Sanger protocol was originally developed one of the limitations of the protocol was the difficulty of obtaining a suitable primer and therefore sequencing by walking along the DNA was not as practical as it is today. Sanger therefore developed another strategy that would allow all the sequencing reactions to be primed by the same primer (Figure 7.6). The term "shotgun sequencing" was coined to describe this strategy. Shotgun sequencing involves fragmenting the DNA to be sequenced into random short overlapping fragments and cloning them into a standard vector to generate a library of fragments. Clones are then randomly chosen and sequenced using a primer derived from the vector sequence. The sequencing reads from each fragment are then compared

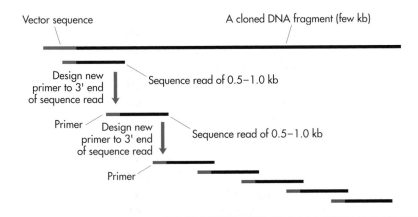

Figure 7.5 Sequencing a cloned DNA fragment by primer walking. The sequence that has been determined in each round of sequencing is indicated by the black line, the primer sequence is indicated in blue.

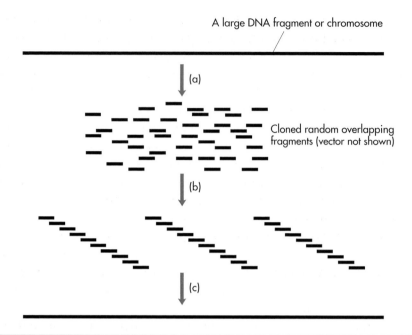

A large DNA fragment or chromosome

(a)

Cloned random overlapping fragments (vector not shown)

(b)

(c)

Figure 7.6 Shotgun sequencing. a) A large DNA fragment, chromosome or complete genome is fragmented into a population of overlapping fragments and cloned into a plasmid vector. b) Clones are chosen at random and the DNA sequenced; sequence overlap between clones is detected. c) The sequence information is assembled into a contiguous sequence (contig).

with each other to find regions of overlap and the sequence assembled into one contiguous sequence. The advantage of this strategy is that the same primer is used for all sequencing reactions. However, more sequence data must be obtained to ensure that overlapping fragments covering the whole region are sequenced. This will also lead to redundancy, i.e. where the same region is sequenced several times. It is also possible that one region is not sequenced, i.e. there are gaps in the sequence, and therefore the sequence has to be "finished" by screening for specific clones and then sequencing to fill the gaps. Redundancy has an advantage in that it should eliminate errors generated by poor sequencing reactions or errors in reading the autoradiograph. Shotgun sequencing, even of relatively short DNA fragments, is heavily reliant on computer analysis to identify the overlapping regions. Although shotgun sequencing is no longer required to sequence a few kilobases, it is now the method of choice for whole genome sequencing.

7.7 High-throughput Sequencing Protocols

Before scientists started sequencing whole genomes, most sequencing projects involved the determination of just a few kilobases of DNA. Scientists would therefore use a "manual" approach to sequencing. They would set up the sequencing reactions on an individual basis, they would manually load their samples on the denaturing polyacrylamide gel, process the gel prior to autoradiography and expose the gel to photographic film overnight, before manually reading the DNA sequence from the autoradiograph. In this procedure a single scientist may be able to sequence a few kilobases of DNA per day, although manually reading the bases of the autoradiograph and then analyzing the sequence was a major undertaking, and the analysis was best completed with the help of a computer. The manual approach is therefore perfectly satisfactory for scientists wanting to sequence their favorite gene, but it is not satisfactory for sequencing the millions or even billions of base pairs that make up whole genome sequences. A modified protocol was therefore required for high-throughput sequencing projects. Such protocols are described next, but it should be emphasized that, currently, genome sequencing is still based upon the technique developed by Fred Sanger and colleagues in Cambridge in the late 1970s. However, we will see in Section 7.10 that new methods not based on the Sanger dideoxy chain termination protocol have been developed.

The limitation in terms of the amount of sequence that can be determined by manual sequencing is not the chain termination reaction itself, but the resolving power of the slab gels used in the denaturing polyacrylamide gel electrophoresis and the bottleneck of manually reading the sequence from the autoradiograph. For a typical sequencing gel you can only resolve a few hundred bases per run, although you can apply a few tricks to extend the read up to 500 bases from each reaction. To sequence the next 500 bases a new primer would need to be designed. In addition the preparation and processing of the slab gels does not lend itself to the scale-up required for high throughput sequencing. A major advance in sequencing has been the development of capillary electrophoresis. In this method, rather than separating the DNA molecules on a traditional slab gel as described in Box 7.2, a gel held in a very fine capillary is utilized (Box 7.2). Another major advance has been the utilization of fluorescent labeling rather than radiolabeling to detect the sequencing products. Modern DNA sequencing protocols utilize dideoxynucleotides that are fluorescently labeled, each of the four dideoxynucleotides with a different fluorescent dye. The products of the sequencing reaction are separated by capillary electrophoresis and as each molecule exits the capillary it is scanned by a laser to excite the fluorescent dyes, the fluorescence then being detected by a CCD camera. The CCD camera is linked to a computer that will generate a file containing the DNA sequence and therefore it is not necessary to manually read the sequence from the electropherogram.

7.8 The Modern Sequencing Protocol

Today, a protocol referred to as "cycle sequencing" is the method of choice for determining DNA sequence. Cycle sequencing exploits the polymerase chain reaction, fluorescent labeling and capillary electrophoresis. The procedure is easily automated and is therefore the protocol used in so called "high-throughput" genome sequencing projects.

A schematic representation of a cycle sequencing reaction is shown in

Box 7.2 Capillary Electrophoresis

The principles of capillary electrophoresis are the same as for polyacrylamide electrophoresis but the slab gel is replaced with a separation matrix held in a capillary tube. A typical capillary is 50 µm in diameter and 50 cm in length. Prior to loading the sample the capillary is filled with a polymer, the precise formula of which will vary depending on the supplier. The polymer will act as the molecular sieve during electrophoresis. The sample is taken up by one end of the capillary and a voltage is applied, the DNA sample will migrate towards the positive electrode or anode. Smaller DNA fragments will migrate through the capillary faster than larger fragments. As with denaturing polyacrylamide electrophoresis, the capillary is capable of resolving DNA fragments that vary in size by just one base pair. At the end of the capillary is a laser beam that will hit each fluorescently labeled DNA fragment as it comes out of the capillary. The laser will cause the fluorescent dyes to fluoresce, this fluorescence is detected by a charged couple device (CCD) camera. The data from the camera is then processed by a computer to generate an electropherogram with the fluorescence plotted against time. Each peak on the electropherogram will represent a single DNA fragment. Capillary electrophoresis can be used to analyze a DNA fragment of over 1000 DNA bases in a single run lasting just 3.5 h.

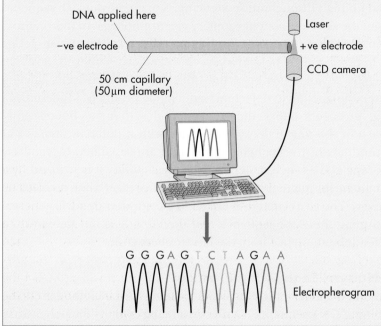

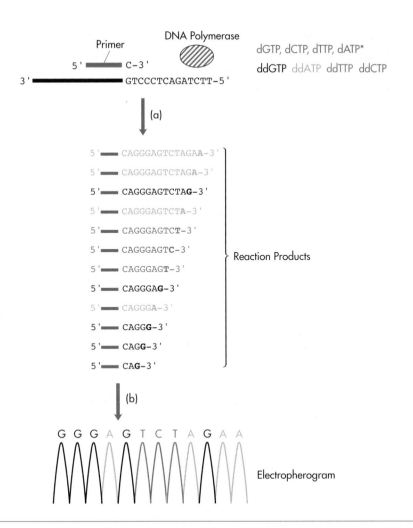

Figure 7.7 The modern sequencing protocol, exploiting differentially fluorescently labeled dideoxynucleoside triphosphates, "PCR"-based reaction and capillary electrophoresis. a) "PCR"-based sequencing reaction in the presence of all four dNTPs and all four ddNTPs; each one of the four ddNTPs has a different fluorescent tag. b) The reaction products are analyzed by capillary electrophoresis.

Figure 7.7. The cycle sequencing reaction is set up in a single tube. The double-stranded template DNA is mixed with primer, deoxynucleoside triphosphates, thermostable DNA polymerase and the four dideoxynucleoside triphosphates. Each dideoxynucleoside triphosphate is labeled with a different fluorescent dye; these modified dideoxynucleotides are often called dye terminators. So when the dye terminator is used by DNA polymerase not only will it lead to chain termination, but it will also label the DNA fragment, the label indicating which base was last incorporated in the growing chain. The reaction is incubated in a thermal cycler and 15 to 30 rounds of amplification are used to give a yield of product that can easily be analyzed (note that the amplification step is not a true polymerase chain reaction because only one primer is used and therefore there will only be linear amplification of one strand of the DNA). The reaction is then analyzed by capillary electrophoresis. Cycle sequencing only requires nanograms of plasmid DNA template, so you would only require one hundredth of the plasmid DNA obtained from a 1 mL overnight culture of *E. coli*.

Q7.4. Why would you not get exponential amplification when using only one primer in a "PCR"?

7.9 Genome Sequencing

A major focus of the international scientific community in the last few years has been whole genome sequencing of bacterial, plant and animal genomes. This has required a vast amount of laboratory work and money and is usually done as collaborative projects. Until the mid 1990s it was not possible to sequence genomes of any organism other than viruses, as the sequencing technology and the computational resources required to piece together and analyze the data were not available. This meant that in the early 1990s sequencing even a bacterial genome was a major undertaking. However, both the sequencing and computational technologies have advanced to such a level that sequencing bacterial genomes has become almost routine, taking only a matter of weeks. Plant and animal genomes can be sequenced in a few years, if not months.

The main sequencing protocol that is used to sequence genomes is still based on the Sanger dideoxy sequencing method but, as discussed earlier, the protocol has been modified to facilitate high throughput. Modifications include labeling the sequencing products with fluorescent tags and analysis by capillary electrophoresis. These modifications make it possible to sequence nearly 1 million bases per day using a single "sequencing machine". The bottleneck for genome sequencing is no longer the sequencing protocol but obtaining the clones to sequence. In Section 7.10 we will discuss a sequencing protocol that does not require cloning and hence has the potential to reduce further the time required to sequence a genome.

So what are the strategies that are used to sequence a whole genome? As discussed above, if you were sequencing a single gene or cDNA, the DNA would be cloned into a vector and you would use primer walking to sequence along the insert (Figure 7.5). Could you apply this approach to sequencing a whole genome? This would not be a sensible approach for two main reasons. First, you cannot clone a whole genome or chromosome as one piece of DNA into a vector that would allow template preparation required for the sequencing reactions. You could not obtain enough template by using the polymerase chain reaction because you do not have the sequence information to design primers for the PCR amplification. Secondly, sequencing by primer walking would be too time consuming, for example given that the *E. coli* genome is 4 million base pairs long, and assuming that you could sequence 1000 base pairs and then synthesize the next primer each day, it would take you at least 10 years to complete the sequencing. To sequence a human chromosome by walking along the sequence would take hundreds of years. Genome sequencing therefore uses the shotgun strategy described earlier (Figure 7.6). However, this is a major undertaking since a library containing tens of thousands of clones is required if sequencing a bacterial genome and millions of clones if sequencing a typical eukaryotic genome. A very large number of these clones then need to be sequenced to give thousands or millions of sequencing reads. Assembly of these sequencing reads into the final genome sequence required an extraordinary amount of computing power. There are two main variations to shotgun sequencing of whole genomes: (1) whole-genome shotgun sequencing and (2) hierarchical shotgun sequencing.

Whole-genome shotgun sequencing

Although there is great debate within the scientific community as to whether whole-genome shotgun (WGS) sequencing is the best method for sequencing large eukaryotic genomes, it is definitely the best method for sequencing bacterial genomes. As you will see from Figure 7.6, the first task is to prepare genomic DNA of the organism to be sequenced, then to fragment this DNA and clone random DNA fragments into a vector. How would you complete this task? In Chapter 4 we discussed how you would construct genomic libraries and this is exactly what we want to do when sequencing genomes. Because the library is going to be used for sequencing rather than screening for a particular gene it will be made in a slightly modified way. The library is normally made from small fragments of around 2 kb. An additional library of larger fragments, 10–20 kb cloned into a bacteriophage λ vector is also constructed. How this library is exploited will be discussed later. A large number of the 2 kb clones are then sequenced using a primer derived from the vector sequence and reading into the inserts. Often primers on both sides of the insert are used so that sequence information is obtained from both ends of the insert. The typical sequence read is 500 bases and therefore only the sequence of the first 500 bases from one or both ends is determined for each clone, leaving 1000 to 1500 bases that have not been sequenced. This is an important point that we will come back to later. The sequence reads are then analyzed by computer to determine where they overlap with each other. The sequence reads that overlap are then assembled into contiguous sequences (referred to as contigs; Figure 7.8). Normally enough fragments are sequenced such that on average each base position in the genome is sequenced six to 10

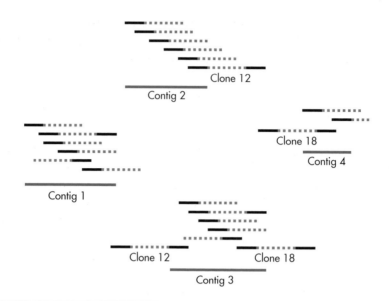

Figure 7.8 Assembly of shotgun sequence reads into contiguous sequences (contigs). The solid black lines indicate the regions of each insert that have been sequenced, the dotted lines represent sequence that has not been determined and is therefore not known. Note some clones have been sequenced from both ends (e.g. 12 and 18); however, there is still some sequence in the middle that has not been determined for these clones.

times to give six- to 10-fold coverage. At this point you will have sequenced tens of thousands of clones, but invariably you will not be able to assemble your sequence reads into one contiguous sequence, i.e. one contig covering the whole genome, as you will have gaps in your sequence information. These gaps will obviously mean that you do not have the complete genome sequence and also you may not know the relative position of each contig or their positions on the genome. This "draft" sequence now has to be "finished". "Finishing" involves determining the sequence across the gaps to generate a complete genome sequence. The sequence gaps will fall into two categories. The first type are called sequencing gaps and these arise because you have often sequenced one clone from both ends and therefore have 500 bases of sequence from both ends but no sequence information from the middle, e.g. clone 12 in Figure 7.8, and it is possible that all or part of this middle sequence has not been determined from any other clones. The two different sequence reads from the same clone will therefore fall into two different contigs because there is no overlapping sequence information to join up the two reads (e.g. the two reads from clone 12 fall into contigs 2 and 3). However, you know that the two sequence reads come from the same clone and so you only need to go back to that clone (clone 12 in this example) and sequence the middle section to fill the gap and join up the two contigs. Sequencing gaps are thus where you know that you have a fragment of DNA that covers the gap, you just have not sequenced it entirely. You can go back to the library and pick out the specific clone you need to sequence to fill the gap. You will see from Figure 7.9 that it is possible to organize adjacent contigs separated by sequencing gaps into a higher order structure called a scaffold.

The second type of gaps are physical gaps where it is not possible to identify a clone in the library that will fill the gap using existing information. Clearly, a physical gap is harder to fill and generally requires screening for suitable clones within the library using traditional library screening techniques such as hybridization (Figure 5.6). The λ library, containing larger (10–20 kb) fragments, is also exploited in gap closure. The ends of the λ clones are sequenced and then the position of the sequence reads determined with respect to the contigs that have been assembled from the short reads (Figure 7.10). This serves three purposes. First, it will help determine the relative position of different contigs or scaffolds. Secondly, if a λ clone spans a physical gap it provides a template for sequencing to fill the gap and therefore to convert a physical gap to a sequencing gap, these sequencing gaps can then be filled by primer walking using the λ clone as template. Finally, the λ clones are used for sequencing repeat structures, for example the rRNA operons which are present in multiple copies and around 5 kb in length. The sequences of the rRNA operons are highly conserved between the different copies, and since the genome is sequenced in 500-base chunks, the rRNA operons are likely to be misassembled. In bacteria that

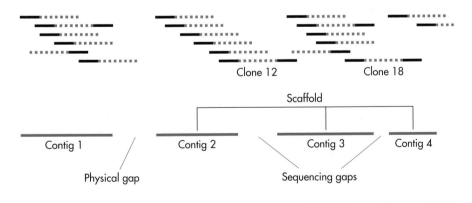

Figure 7.9 Higher order assembly of shotgun sequences. Sequence reads from either end of a clone may be present in two contigs; for example, the sequence read from one end of clone 12 is in contig 2, whereas the sequence read from the other end is in contig 3. This linking information means that the gap between contigs 2 and 3 is called a sequencing gap, sequencing the whole of clone 12 will close the gap. There is no linking information between contig 1 and the other contigs, therefore it is not known where contig 1 lies on the genome: this is a physical gap.

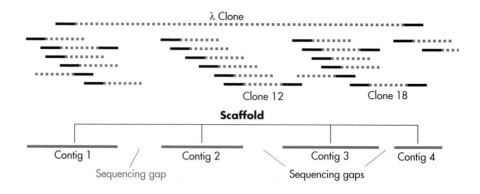

Figure 7.10 Using large clone inserts to help higher order assembly of shotgun sequence and finishing of genome sequences. Sequencing either end of large DNA fragments can give further linking information. In this case it links contig 1 with contigs 2, 3 and 4. The λ clone can be sequenced to close the gap.

have very few repeat sequences, the solution for sequencing the repeats is to screen for λ clones that each contain one complete copy of each of the repeated sequences and the unique flanking sequences, e.g. one complete copy of each rRNA operon flanked by unique sequence, and then to sequence each clone by primer walking from the unique sequences.

Hierarchical shotgun sequencing

This is arguably the best and possibly only method for obtaining the finished genome sequence of a higher eukaryote for which no "related" species has been sequenced. The main reason for the need to use hierarchical shotgun sequencing rather than whole genome shotgun sequencing is the presence of a vast amount of repeat sequences in eukayotes, for many species making up more than 50% of the genome (Sections 2.3 and 2.4), which can lead to misassembly of the finished sequence. Approaches similar to those described above for sequencing through the repeat sequences found in bacterial genomes, i.e. sequencing λ clones by primer walking, would not be suitable for genomes with such a large amount of repeat sequences.

Hierarchical shotgun sequencing involves first fragmenting the genome into relatively large overlapping pieces (approximately 150 kb) (Figure 7.11) which are cloned into a bacterial artificial chromosome (BAC) vector (Section 4.16), to generate a BAC library. A large amount of effort is then required to determine which BAC clones overlap with each other. Having selected a complete set of overlapping BAC clones, which in theory cover the whole genome, these are shotgun sequenced, i.e. clones are broken up into small overlapping fragments that are sequenced and then the sequences from each BAC clone are assembled into a contiguous sequence

(Figure 7.11). Because you know which BACs overlap with each other it is relatively easy to then assemble all the sequence information into one contiguous genome sequence (Figure 7.11).

As an example of hierarchical shotgun sequencing we will discuss how the International Human Genome Sequencing Consortium sequenced the human genome. Several BAC libraries were constructed and used in the sequencing of the human genome. The intention was that approximately 10 libraries, each constructed from DNA obtained from a single anonymous individual, each contributing about 10% of the final sequence would be used. In reality, one BAC library, given the name RPCI-11 and constructed from a male volunteer and therefore containing all 24 chromosomes, was used for the majority of the sequencing effort. The library had an average insert size of 178 kb and contained more than 500,000 clones.

> **Q7.5.** Given the size of the human genome (3×10^9 base pairs), what fold coverage of the genome does the RPCI-11 library have?

A physical map showing the relative positions of a very large number of these BAC clones was constructed by restriction enzyme fingerprinting. As of September 2000, just prior to publication of the draft sequence, 372,264 BAC clones from the RPCI-11 and other libraries had been analyzed by *Hin*dIII digestion. In theory this would give more than 15-fold coverage, i.e. on average each base in the genome would be present in at least 15 BAC clones in the library. The digests were analyzed on large agarose gels containing 242 lanes, 192 lanes being used for BAC restriction digests and 50 lanes for marker DNAs. Clearly, the logistics of setting up thousands of restriction digests, analyzing these digests by agarose gel electrophoresis and then handling the data produced is a major task. In a typical laboratory a single scientist might analyze 10–30 samples by restriction digestion and agarose gel electrophoresis per day, whereas in the laboratories used for the physical mapping of the BAC library 20,000 fingerprints could be completed per week. Overlapping fragments were identified by the fact that they contained common restriction fragments. BACs that were shown to overlap due to similarities in their restriction enzyme digestion patterns were assembled into overlapping contiguous clones called fingerprint clone contigs. Of the 372,264 clones that were fingerprinted, 295,828 could be assembled into 1447 different contigs, leaving 76,436 BAC clones for which no overlap could be detected with another BAC that had been fingerprinted. The number of contigs is continually being reduced as gaps are filled in the map and therefore one contig is joined with another. In a perfect world the map would contain 24 contigs, one for each chromosome, so the physical map was therefore clearly not complete because it contained gaps between the different contigs. There are several reasons why it is

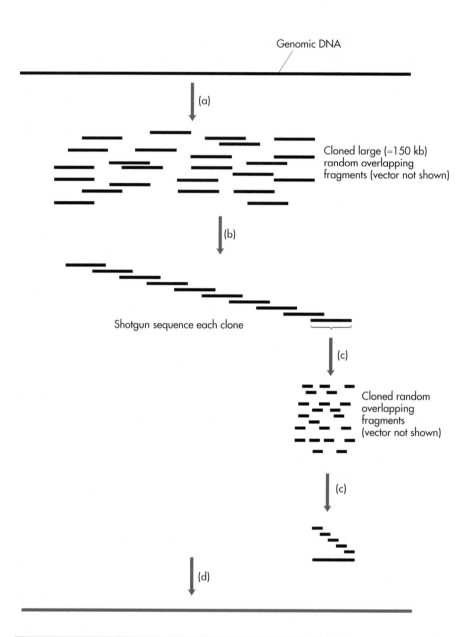

Genomic DNA

(a)

Cloned large (≈150 kb) random overlapping fragments (vector not shown)

(b)

Shotgun sequence each clone

(c)

Cloned random overlapping fragments (vector not shown)

(c)

(d)

Figure 7.11 Hierarchical shotgun sequencing. a) Genomic DNA is fragmented into a random set of overlapping pieces and cloned into a bacterial artificial chromosome (BAC). b) The positions of the clones are mapped with respect to each other to produce a series of large BAC contigs ideally covering the whole genome. The chromosomal position of each BAC contig is mapped. c) The sequence of selected BAC clones, covering the whole genome, is determined by shotgun sequencing (see Figure 7.6). d) Each BAC sequence is assembled into a contiguous sequence covering the whole genome.

almost impossible to obtain an accurate map with just one contig for each chromosome. Firstly, some overlaps between BACs that have been finger-printed are too small to be identified by restriction mapping. In theory these overlaps will be identified once the BACs have been sequenced. Secondly, genome-wide repeats and, in particular, chromosomal duplica-tions will often have similar restriction patterns and will therefore be assembled into one contig rather than two or more separate contigs.

The next task was to determine the chromosomal position of each fin-gerprint clone contig. This again required high-throughput techniques. The chromosomal positions of most BAC clones were determined by screening for the presence of sequence-tagged sites (STSs) within each clone using several different approaches. A sequence-tagged site is any unique sequence in the genome whose location is known. Genes, ESTs and any other unique sequences can be an STS but their chromosomal position must be known. Firstly, the RPCI-11 library had been characterized by many laboratories even before the fingerprinting experiments had been completed and therefore the chromosomal positions of nearly 10,000 of the BAC clones present in fingerprint clone contigs were already known. Secondly, STS sequences were hybridized to filter replicas of the RPCI-11 library, nearly 14,000 STS sequences hybridized to 96,283 different BAC clones within the library. If an STS hybridizes to a BAC clone this will tell you the chromosomal position of the clone and if one STS hybridizes to two different BAC clones this tells you that they overlap. Thirdly, almost 3500 BACs were positioned by fluorescence *in situ* hybridization (FISH), a technique which allows the visualization of the position of a labeled clone on whole chromosomes (Section 13.9). Finally, because the sequencing of the BAC clones was in progress while the BAC mapping was ongoing, some fingerprint clone contigs were sequenced before their chromosomal posi-tion was determined. It was then possible to identify STS sequences from the sequencing data and therefore determine the chromosomal position these sequenced BACs.

Having determined the chromosomal position, where possible, of the fingerprint clone contigs, a minimal overlapping set of clones was selected for sequencing and each BAC clone shotgun sequenced. Because each BAC only contained on average 178 kb of DNA it was relatively easy to assemble the BAC sequences into contigs with relatively few gaps. Because in most cases the chromosomal position of each BAC was known before shotgun sequencing it should have been relatively easy to assemble the genome sequence by slotting the sequenced contigs into the sequence at the mapped position. Because a large number of BACs were sequenced and therefore labeling errors may have occurred, the sequenced BACs were restriction mapped *in silico* and matched to experimentally determined fingerprints. In addition, sequence reads were obtained from the ends of intact BACs, for many of which the chromosomal position was known, and therefore sequence contigs could also be matched to the BAC end reads to determine chromosomal position. The positions of sequenced contigs that could not be determined by either the *in silico* fingerprint or by matching BAC end reads were often resolved because they had overlapping sequence with contigs that had already been sequenced. For some it was necessary to generate an STS from the sequence read and its chromosomal position determined by radiation hybrid mapping, a technique beyond the scope of this book but discussed in many genetics textbooks.

The draft sequence of the human genome was published in February 2001 and covered approximately 87% of the genome, including 90% of the euchromatic (gene-rich) DNA. At the time the genome could be assembled into around 145,000 sequence contigs, i.e. there were around 145,000 gaps in the sequence. Because the assembly contained both a large number of small contigs and also some very large contigs the average contig length was not a good indication of the quality of the assembly, so instead a statistic called the "N50 length" was used. The sequence contigs had an N50 length of 826 kb, which meant that 50% of all bases in the assembly were present in a contig with a length of at least 826 kb. These sequence contigs could be assembled into just over 2000 scaffolds with a N50 length of 2279 kb. In terms of the fingerprint clone contigs, i.e. the physical BAC map, this contained 942 fingerprint clone contigs with a N50 length of 8398 kb. The draft sequence has been an extremely valuable tool for biologists. It has allowed scientists to obtain gene sequences rapidly without the need to screen genomic or cDNA libraries for their gene of interest. The draft sequence has also allowed us to study genome structure and function. However, to exploit the genome information fully the finished sequence is required but sequencing the remaining 10% of euchromatic DNA was a relatively costly and time consuming project. Converting the draft to finished sequence roughly doubled the time and cost of the publicly funded project. Sequencing of the euchromatic sequence of the human genome by the International Human Genome Sequencing Consortium was "finished" in 2004, with 99% of the euchromatic DNA and 93% of the total DNA (euchromatic and heterochromatic) sequenced. The sequence contains only 341 gaps and the N50 length is 38.5 Mb. The sequence has been determined with an accuracy of 99.999%, i.e. less than one error per 100,000 bases. Current sequencing technology cannot sequence the remaining gaps. In most cases the unsequenced parts of the genome either cannot be cloned or, because of the high level of repeat sequences, they could not be assembled following shotgun sequencing and specific primers cannot be made for sequencing via primer walking.

Whole genome shotgun sequencing of the human genome

At the same time that the publicly funded project was sequencing the human genome using the hierarchical shotgun approach, a biotechnology company (Celera) attempted to sequence the genome using a whole-genome shotgun approach. The company invested in a large number of sequencing machines and high-powered computers that would be essential for the assembly. The prospect of sequencing the human genome, at least to draft level, using the whole-genome shotgun approach only became feasible once sequencing technology and computing power had advanced to the stage it was at in the late 1990s. Celera constructed libraries of 2, 10 and 50 kb fragments from five individuals. They sequenced both ends of clones from the three different libraries. Remarkably, Celera completed their part of the sequencing phase of the project in under a year. However, they did not obtain enough data from their own reads to complete the assembly. Instead they exploited the sequence information that had been made available by the publicly funded project. This data they shredded into 550 bp reads and added to their own sequencing reads. In total they obtained 27.27 million reads from their own sequencing reads and 16.06 million 'reads' from the publicly funded project, the combined set giving eight-fold coverage of the genome. Celera then looked for sequence overlap between the 43.33 million reads to assemble the sequence into contigs essentially as described in Figure 7.8. Because they had sequenced from both ends of the clones, especially the 10 and 50 kb clones, it was possible to arrange the sequence contigs into scaffolds (Figure 7.9). The chromosomal positions of the sequence contigs and scaffolds were then determined by exploiting STS marker information and the BAC fingerprint information. Celera published their draft whole-genome shotgun assembly at the same time as the publicly funded project published their draft sequence. The Celera draft assembly contained over 220,000 contigs with over 220,000 gaps. Like the publicly funded draft sequence it included 90% of the euchromatic DNA. The Celera publication did contain a slightly better assembly, which they termed a compartmentalized shotgun assembly, which had exploited positional information from a variety of sources including the publicly funded project. It is clear that the quality of the whole-genome shotgun assembly does not match the quality of the hierarchical shotgun assembly. One of the reasons the publicly funded project had opted for the hierarchical approach was that, although it was agreed that both approaches would yield a large amount of draft sequence information very quickly, the hierarchical approach was far more suitable for determining the sequence of gaps and therefore ultimately finishing the sequence. As discussed above, the hierarchical approach has yielded a finished sequence, whereas Celera have not finished the sequencing, although Celera did subsequently publish a slightly improved assembly with "only" 210,000 gaps.

7.10 High-throughput Pyrosequencing
Recently, a faster and more cost effective high-throughput sequencing protocol has been developed that is not based on the Sanger dideoxy chain ter-

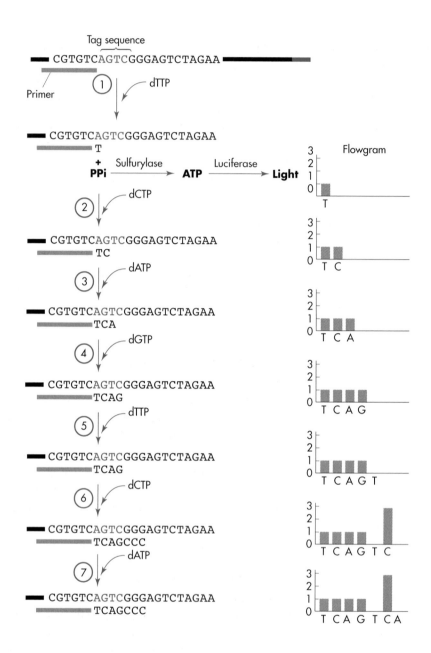

mination protocol. The sequencing reaction was first described several years ago, but only recently has a method been developed to adapt it for high-throughput sequencing. We will discuss the basic sequencing procedure and then look at how it has been developed to allow rapid sequencing of whole genomes.

Pyrosequencing

One of the products of DNA synthesis, in addition to the DNA chain itself, is pyrophosphate which is released every time a nucleotide is added to a

Figure 7.12 Pyrosequencing. The sequencing reaction contains template, primer, DNA polymerase, sulfurylase, luciferase and other co-factors, but no dNTPs. The primer is annealed to the single-stranded template adjacent to a "tag" sequence. Deoxynucleotides are washed over the template sequentially. In step 1, dTTP is washed over the template and "T" is incorporated into the DNA chain, pyrophosphate is released. The pyrophosphate is substrate for a chemical reaction catalyzed by the sulfurylase enzyme which generates ATP, this ATP is used in a second reaction catalyzed by luciferase which results in light emission that is detected by a CCD sensor and indicated in the flow gram. Buffer is passed over the reaction to flush away the dTTP. In steps 2, 3 and 4, dCTP, then dATP and then dGTP are washed over the template, each is incorporated and light is emitted. In step 5, another cycle is started; dTTP is added but is not incorporated so no light is emitted. In step 6, dCTP is added and three "C" residues are added, the light emitted is three times the intensity of that observed for the addition of one nucleotide therefore indicating that a run of three "Cs" has been added to the chain. In step 6, dATP is added but not incorporated. The sequence can be read from the graph as: TCAGCCC.

growing DNA chain. This new DNA sequencing protocol exploits the release of pyrophosphate as a signal for the addition of a nucleotide to a growing DNA chain and hence the technique is called pyrosequencing.

As with the Sanger dideoxy chain termination protocol, pyrosequencing requires a single-stranded DNA template and a primer to initiate the sequencing reaction. Sequencing involves adding a pulse of each deoxynucleotide in turn to the reaction, addition of a nucleotide to the growing chain is signaled by release of pyrophosphate which is detected using coupled enzymatic reactions that emit light. A four-base "tag" is added to all DNA fragments, this is read first and is used to calibrate the sequencing reaction, i.e. to determine the level of light emission due to incorporation of one nucleotide. The tag is added during preparation of the template and this will be discussed in Figure 7.13 when we look at the high-throughput pyrosequencing protocol. The sequencing reaction (Figure 7.12) contains the single-stranded template, primer and DNA polymerase which you will be familiar with from the Sanger dideoxy chain termination method, in addition there are also the enzymes sulfurylase and luciferase. During the sequencing run the four deoxynucleosides are passed over the DNA template sequentially. In Figure 7.12, step 1, dTTP is passed over first and this is incorporated into the growing chain, releasing pyrophosphate which is detected using coupled enzymatic reactions that emit light. In subsequent steps, dCTP, dATP and then dGTP are passed over the template each resulting in chain extension and hence light emission. This tells us the sequence of the first four bases is TCAG and the levels of light emitted are used to calibrate the reaction. In step 5, dTTP is passed over the template. In this example the fragment being sequenced does not have a T at the next

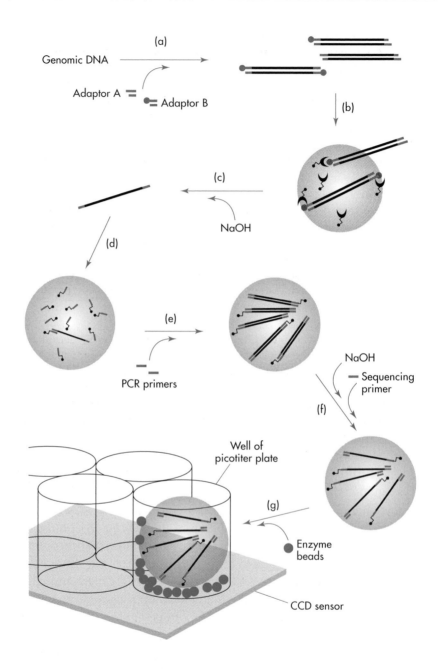

position so no light is emitted. In step 6, when dCTP is added, three nucleotides are added to the growing chain which results in an increase in the intensity of the light emitted. The level of light emission correlates with the number of bases added, in this case the intensity is three times that observed for a single nucleotide addition telling us that the sequence includes a run of three C residues. The cycle of dTTP, dCTP, dATP and dGTP addition is repeated many times, generating a read of 100–200 bases.

Figure 7.13 Template preparation for high-throughput pyrosequencing.
a) Genomic DNA is fragmented into a population of overlapping fragments of
300–800 bp and adaptors are added. b) This will generate two types of fragment;
those with two copies of the same adaptor, either A or B, and those with a
different adaptor at each end. The latter are required by the sequencing
protocol. The B adaptor has a biotin tag at the 5′ end and so will bind to
streptavidin-coated beads. c) Treatment with sodium hydroxide will denature the
DNA, only single-stranded DNA containing an A adaptor at the 5′ end and a B
adaptor at the 3′ end will be released from the streptavidin-coated beads. d) This
single-stranded DNA is collected and bound to beads containing a primer
complementary to the B adaptor sequence; the conditions are controlled such
that each bead binds just one single-stranded DNA fragment. e) PCR is used to
amplify the captured DNA. The PCR is completed in a water–oil emulsion where
each aqueous droplet contains a single bead. In this way the PCR reaction is
"caged" within its own droplet; this ensures that PCR product from each
fragment is compartmentalized. After the PCR reaction there are approximately
10 million copies of a single DNA fragment on each bead, in each case the
strand generated by amplification from the B adaptor is covalently attached to
the bead and remains attached while the other strand is released by denaturing
in alkali conditions. f) This generates a library of around 500,000 beads each
carrying a different single-stranded DNA fragment. The sequencing primer anneals
within the adaptor sequence such that during the sequencing reaction the first
four bases that are read are from within the adaptor sequence and are therefore
the same for all fragments; this acts as the calibration tag. g) The beads are then
deposited into a picotitre plate containing wells that are 44 µm in diameter;
individual wells can only hold one bead. At the same time DNA polymerase and
smaller beads with immobilized ATP sulfurylase and luciferase are also deposited
in each well. The pyrosequencing reaction (Figure 7.12) is completed by passing
the deoxynucleotides over the beads and the light emitted when nucleotide is
incorporated is detected by a CCD sensor at the bottom of the picotiter plate.

High-throughput pyrosequencing (454 sequencing)

High-throughput pyrosequencing is a whole-genome shotgun sequencing
protocol. The method described in Figure 7.13 was developed by 454 Life
Sciences and therefore is often called 454 sequencing and is available commercially. The technique first involves fragmenting the DNA to be
sequenced into short overlapping fragments. These fragments are amplified by PCR and attached to microscopic beads such that each bead contains copies of just one DNA fragment. The DNA is denatured and a
sequencing primer is annealed. The DNA beads are then placed in
"picotiter plates" with each microscopic well containing a single bead and
the enzymes required for the sequencing reaction. The DNA template in
each well is then sequenced using the pyrosequencing protocol.

In a typical pyrosequencing run the cycle of washing dTTP, dCTP, dATP
and dGTP over the plate is repeated 42 times, generating a read of 100–200

bases. The light signal for each of the wells is detected separately giving 300,000 sequence reads. The sequencing procedure takes 5.5 h and generates 20–30 million base reads from a single plate.

The advantage of pyrosequencing is that no cloning step is required and therefore it is very quick, a draft sequence of a bacterial genome can be obtained in a few days. In addition to speed, the method has another advantage over traditional sequencing protocols. Because there is no cloning there is no bias towards sequencing of particular regions of the genome, whereas in traditional library-based shotgun sequencing, certain sequences are hard to clone and therefore sequence. The *Mycoplasma genitalium* genome was resequenced using this approach and a slightly better assembly was obtained than that obtained for the initial clone based shotgun approach (25 and 28 contigs respectively). However, for pyrosequencing, the average read length is only 100–200 bases and sequence scaffolds cannot be built because there is no linking information, so obtaining a finished sequence with no gaps is more difficult. A further drawback with the pyrosequencing reaction is that it is increasingly difficult to accurately determine the length of runs of the same nucleotide when they extend for longer than 6–7 bp.

High-throughput pyrosequencing is ideal for resequencing of bacterial strains, since a single sequence run would generate 8- to 10-fold coverage of a bacterial genome. It has been used successfully to identify point mutations within the genome of *Mycobacterium tuberculosis* responsible for resistance to a novel antibiotic. Pyrosequencing has also been used to sequence samples where the DNA is not of sufficient quantity or quality to clone and generate the genomic library required for traditional sequencing. For example, pyrosequencing was used to sequence samples from an extinct woolly mammoth, 13 million base pairs of sequence information was obtained.

Q7.6. Below is a flowgram from a pyrosequencing procedure. Read the sequence from 5′ to 3′.

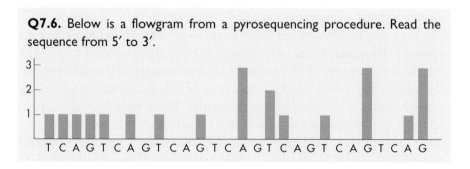

7.11 The Importance of DNA Sequencing

The determination of the nucleotide sequence of genomes from different organisms is an important part of modern biology. Because of the advances discussed in this chapter it is now relatively easy to sequence a bacterial genome, with hundreds being completed to date. These include the genomes of many human pathogens including *Mycobacterium tuberculosis* (which causes TB), *Treponema pallidum* (the cause of syphilis), *Rickettsia prowazekii* (the cause of typhus), *Vibrio cholerae* (the cause of cholera) and *Yersinia pestis* (the cause of plague). The sequencing of genomes from animals and plants is still a major undertaking, but, to date, we have sequenced several such genomes including human, chimpanzee, dog, mouse, rice and the fruit fly, *Drosophila melanogaster*. Genome analysis is a valuable tool to aid understanding of the biology of an organism. It will almost certainly play a vital role in our understanding of pathogenicity. It will help us understand human development and why we are different from other animals such as mice and chimpanzees. Genome sequences will be exploited to develop novel drugs. Sequencing is also a valuable tool in our battle against emerging virus as emphasized in the global response to the coronavirus virus believed to cause severe acute respiratory syndrome (SARS). The genome sequence of this coronavirus was determined within a matter of months of it being identified and has helped greatly our understanding of SARS.

Q7.7. Assuming that your average sequence read is 500 bp, what is the minimum number of reactions needed to sequence the following genomes to give 10-fold coverage: (a) *E. coli* (4×10^6 bp) and (b) human (3×10^9 bp).

Questions and Answers

Q7.1. Figure. 7.1 shows a schematic representation of a sequencing reaction. The G reaction generates fragments that are 3, 4, 5, 7 and 12 bases in length. What size fragments would be generated by the C and T reactions?

A7.1. *The C reaction would generate fragments 1 and 9 bases in length, the T reaction would generate fragments 2, 8 and 10 bases in length.*

Q7.2. Figure 7.4 shows an autoradiograph from a Sanger sequencing procedure. Read the sequence from 5′ to 3′.

A7.2. *The sequence reads:*

5′- GAATTCGCATGCATCTTTTTTATGCTATGGTTATTTCATACCA

TAAGCCTAATGGAGCGAATTATGGAGCAAAA-3′

Q7.3. (a) What would happen if you added the Klenow fragment to a sequencing reaction in which the dideoxy to deoxy ratio had been optimized for T7 DNA polymerase?

(b) What would happen if you added T7 DNA polymerase to a sequencing reaction in which the dideoxy to deoxy ratio had been optimized for the Klenow fragment?

A7.3. (*a*) *The dideoxy concentration would be too low and therefore chain termination would occur at too low a frequency.*

(*b*) *The dideoxy concentration would be too high and therefore virtually all chains will terminate before T7 DNA polymerase has extended past a few bases.*

Q7.4. Why would you not get exponential amplification when using only one primer in a "PCR"?

A7.4. *Since only one strand is copied, the number of templates to which the primer anneals would not increase and so the same number of copies are made in each cycle, e.g. if you have 100 copies of the template, you will only get a maximum of 100 copies of DNA made in each cycle, this would give a maximum of 200, 300, 400, 500 copies after the 1st, 2nd, 3rd, and 4th cycles, respectively, whereas exponential amplification with two primers (one for each strand) would give a maximum of 200, 400, 800, and 1600 copies.*

Q7.5. Given the size of the human genome (3×10^9 base pairs), what fold coverage of the genome does the RPCI-11 library have?

A7.5. *Approximately 30-fold. If there are 500,000 clones each 178 kb in size this is equivalent to 8.9×10^{10} bp, which is $(8.9 \times 10^{10}) / (3 \times 10^9) = 30$.*

Q7.6. Below is a flowgram from a pyrosequencing procedure. Read the sequence from 5′ to 3′.

A7.6. *The sequence reads:*

  5′- TCAGTATGAAATTCTGGGAGGG-3′

Remember that runs of two or more of the same base are represented by an increase in the intensity of the light emitted, this is shown on the flowgram by bars of greater height.

Q7.7. Assuming that your average sequence read is 500 bp, what is the minimum number of reactions needed to sequence the following genomes to give 10-fold coverage:

(a) *E. coli* (4×10^6 bp), and

(b) Human (3×10^9 bp).

A7.7. (*a*) *Ten-fold coverage means that on average every base in the genome has been sequenced 10 times, so you would need to read* $(4 \times 10^6) \times 10$ *bases, i.e.* 4×10^7. *If the reads are 500 bp you require* $(4 \times 10^7) \div 500$ *reads, so the answer is 80,000.*

(*b*) *For the human genome you would need* $(3 \times 10^{10}) \div 500$ *reads, so the answer is 60 million.*

Further Reading

Determination of nucleotide sequences in DNA (1980) Sanger F, From *Nobel Lectures, Chemistry 1971–1980*, Editor-in-Charge: Tore Frängsmyr; Editor: Sture Forsén, World Scientific Publishing Co., Singapore, 1993
This article describes the development of the Sanger dideoxy method of sequencing DNA. This article can be accessed using the following url:
http://nobelprize.org/chemistry/laureates/1980/sanger-lecture.html

Whole-genome random sequencing and assembly of *Haemophilus influenzae* Rd (1995) Fleischmann RD *et al.*, Science, Volume 269 Pages 496–512.
This is a primary research paper describing the sequencing of the first bacterial genome using a whole-genome shotgun approach

Initial sequencing and analysis of the human genome (2001) International Human Genome Sequencing Consortium, Nature, Volume 409 Pages 860–921.
The primary research paper reporting the "draft" human genome sequence

Clone by clone by clone (2001) Olson MV, Nature, Volume 409 Pages 816–818.
A commentary on the hierarchical shotgun sequencing approach used to sequence the human genome

Finishing the euchromatic sequence of the human genome (2004) International Human Genome Sequencing Consortium, Nature, Volume 431 Pages 931–945.
The primary research paper reporting the "finished" human genome sequence

End of the beginning (2004) Stein LD, Nature, Volume 431 Pages 915–916.
A commentary on the publication of the "finished" human genome sequence

The Sequence of the human genome (2001) Venter JC, *et al.*, Science, Volume 291 Pages 1304–1351.
The primary research paper reporting the "draft" human genome sequence as determined by the whole-genome shotgun approach

Revisiting the independence of the publicly and privately funded drafts of the human genome (2003) Cozzarelli NR, Proceedings of the National Academy of Sciences of the United States of America, Volume 100 Page 3021.
This is an editorial with links to papers that discuss the advantages and disadvantages of whole-genome shotgun approach to sequencing the human genome

Genome sequencing in microfabricated high-density picolitre reactors (2005) Margulies M. *et al.*, Nature, Volume 437 Pages 376–380.
The primary research paper discussing 454 pyrosequencing

8 Bioinformatics

Learning outcomes:

By the end of this chapter you will have an understanding of:

- *how a DNA sequence can be searched for regions which could code for genes (open reading frames) and how these can be evaluated in both prokaryotic and eukaryotic organisms*
- *how to interpret a pair-wise alignment between related sequences, showing an understanding of similarity and identity as applied to amino acids*
- *the types of information stored in primary sequence databases, secondary pattern databases and structure databases*
- *the two main programs used for similarity searching and how to interpret the output from them*
- *how to interpret a multiple sequence alignment including identifying conserved regions*
- *how to interpret regular expressions describing consensus patterns*
- *what the three-dimensional structure of proteins can add to the understanding of the protein*
- *how phylogenetic trees depict the relationships between organisms or species*

8.1 Introduction

Determining DNA sequences has become a routine procedure; specialist laboratories can sequence vast stretches of DNA each day and new whole-genome sequences are published regularly. Why are biologists so keen to know the sequence of the genes they study? Determining DNA sequence is not an end in itself but a milestone along the path to understanding a particular biological phenomenon. Unless you can interpret the DNA sequence, work out which parts of it are genes and what sorts of proteins these encode, then the information is a meaningless series of Gs, As, Ts and Cs. There are basically two approaches that you can take to interpreting a DNA sequence. Firstly, you can look at the piece of DNA sequence that you have determined and use what is known about the characteristics of a typical gene to work out from first principles which parts of the sequence are likely to code for proteins. The second and very powerful way of looking at DNA sequence uses homology searching, comparing the sequence with other DNA sequences, which are stored in vast public databases. Using a combination of these approaches can unlock an immense amount of information from a DNA sequence. Both of these approaches require the use of computers to help in the analysis of immense amounts of data, and together form the basis of the discipline which is called bioinformatics. Bioinformatics is a theoretical discipline in that it allows you to make predictions about DNA and protein sequence, but these need to be tested in the laboratory.

Remember the flow of genetic information in a cell is from DNA to RNA and finally to proteins. The DNA is transcribed into discrete, essentially gene-sized, pieces of RNA which are then translated into a sequence of amino acids. These will fold into a complex three-dimensional protein structure and may be further modified before they are fully matured into functioning proteins. However, despite the fact that proteins are not translated directly from DNA, most sequence analysis is performed on the DNA sequence and so we will adopt the conventions of DNA rather than RNA in the discussions in this chapter, referring to thymine rather than uracil.

Q8.1. What is the DNA triplet codon for methionine? What would the codon be in an RNA molecule? Refer to Table 8.1, which shows the genetic code.

Q8.2. What do the codons UAA, UAG and UGA encode?

8.2 What Does a Gene Look Like?

The DNA sequence of a gene is the sequences which encode proteins together with the sequences which regulate gene expression (Figure 8.1). In bioinformatics terms any region of DNA sequence without a stop codon could potentially encode a protein; these regions are called open reading frames (ORFs). If you examine the sequence in Figure 8.2, you should be able to see that even in this very short sequence stop codons occur in frames 1 and 3. The sequence in frame 3, however, could encode part of a gene. Most computer algorithms designed to look for ORFs use the simple principle of looking for regions without stop codons.

Q8.3. Why are three alternative translations given below the sequence in Figure 8.2?

Q8.4. Are the three alternative translations given in Figure 8.2 the only possible translations of this sequence?

Table 8.1 The genetic code

UUU	Phe	UCU	Ser	UAU	Tyr	UGU	Cys
UUC		UCC		UAC		UGC	
UUA	Leu	UCA		UAA	STOP	UGA	STOP
UUG		UCG		UAG	STOP	UGG	Trp
CUU	Leu	CCU	Pro	CAU	His	CGU	Arg
CUC		CCC		CAC		CGC	
CUA		CCA		CAA	Gln	CGA	
CUG		CCG		CAG		CGG	
AUU	Ile	ACU	Thr	AAU	Asn	AGU	Ser
AUC		ACC		AAC		AGC	
AUA		ACA		AAA	Lys	AGA	Arg
AUG	Met	ACG		AAG		AGG	
GUU	Val	GCU	Ala	GAU	Asp	GGU	Gly
GUC		GCC		GAC		GGC	
GUA		GCA		GAA	Glu	GGA	
GUG		GCG		GAG		GGG	

The table shows the RNA triplet codon for each of the 20 amino acids. As it is RNA which is translated into protein in cells the genetic code is written as RNA. However most sequence analysis is performed on DNA rather than RNA sequence. When translating between protein and DNA sequence you need to read T for U.

Looking for the sequence typical of the beginning of a gene can further refine the ORF. The synthesis of all proteins begins with a methionine residue, consequently the triplet codon ATG is found at the beginning of most protein coding regions and is referred to as the start codon. There are some exceptions to this rule as in rare circumstances the initial methionine residue can be encoded by GTG or TTG. Also, as the ATG start codon encodes methionine, which does not only occur at the beginning of proteins, it is useful to have some additional evidence to support the identification of the beginning of the gene. Looking for conserved sequences which occur upstream of the start of a gene, such as promoter sequences and ribosome binding sites can firm up the identification of the start of the gene.

Not every stretch of DNA sequence between a start and stop codon actually encodes a protein. Very small ORFs can be excluded on the basis of size alone. Most computer algorithms allow you to choose the threshold above which you consider it likely that an ORF will encode a gene, a threshold of 100 nucleotides is a common starting point. Figure 8.3 gives a representation of some sequence from *Pseudomonas aeruginosa*, showing potential ORFs, the largest of these in frame 2 is too big to have arisen by chance and is likely to encode a gene.

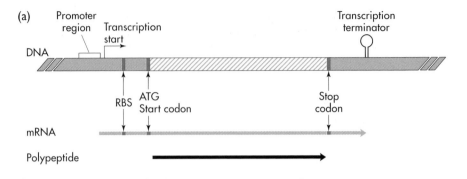

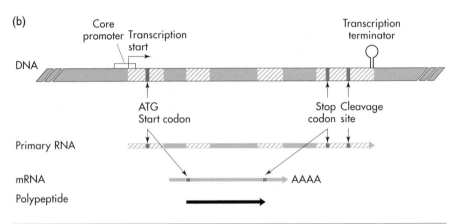

Figure 8.1 Structure of a gene. A gene is the fundamental unit of heredity. In terms of DNA sequence this consists of the sequences which are eventually translated into protein by the ribosomes, and those involved in regulation of expression of the gene. a) In the prokaryotic gene the promoter and terminator are sequences on the DNA molecule which are recognized by RNA polymerase. RNA is synthesized using the DNA between the transcription start and terminator as template. The ribosome binding site (RBS) (or Shine–Dalgarno sequence), start and stop codons (gray boxes) are part of the DNA sequence of the gene but do not become operational until they have been transcribed into RNA. They then become the signals which indicate where the ribosome should bind and which part of the sequence the ribosome should translate into protein sequence. b) In the eukaryotic gene the promoter region overlaps the transcription start, additional sequences, discussed in Chapter 11, are involved in regulation of gene expression. The ATG start codon is normally in the first exon and the final exon contains the stop codon and the cleavage site (shown as gray boxes). Transcription initially produces a primary RNA molecule which includes exons (hatched) and introns (solid), this is spliced to remove introns, cut at the cleavage site and the poly A tail added to give a mature mRNA molecule. The sequence between the ATG start codon and the stop codon is translated into protein.

Q8.5. How many amino acids would you expect a protein to consist of if it was encoded by a 100 nucleotide stretch of DNA?

```
5'accgcgcatggtgctaggcaaaccgcaaacagacccgactctcgaatggttcttgtctga3'
+1 ThrAlaHisGlyAlaTrpGlnThrAlaAsnArgProAspSerArgMetValLeuValEnd
+2  ProArgMetValLeuGlyLysProGlnThrAspProThrLeuGluTrpPheLeuSerGlu
+3   ArgAlaTrpCysEndAlaAsnArgLysGlnThrArgLeuSerAsnGlySerCysLeuIle
```

Figure 8.2 Nucleotide sequence translated in the three forward reading frames. The translation is shown in the three letter code beneath the triplet codon. End indicates a stop codon.

There are other pieces of information that you can use to analyze ORFs that you have identified to help you to decide whether they are likely to encode genes. One of the most useful of these is to look at codon usage (Box 8.1). The lower graph in Figure 8.3 (middle panel) shows how codon usage varies in the three forward reading frames that we have been looking at. You can see that in the ORF in frame 2, which we have already identified, the line in the graph rises well above the other lines indicating the average value. This indicates that most of the codons used in this region are

---

Box 8.1 The Degeneracy of the Genetic Code, and Codon Usage

The genetic code (Table 8.1) is a triplet code, each codon made up of three bases encodes a specific amino acid. There are 64 possible triplet codons, three of which are stop codons indicating the ends of genes, the remaining 61 encode amino acids. As there are only 20 amino acids there is redundancy in the genetic code with some amino acids being encoded by up to six codons. The degeneracy of the genetic code gives rise to considerable variation in the overall GC content of the genomes of different species (Section 2.9), this variation stems from a difference in the frequency with which different species use particular codons.

For example, threonine can be encoded by any of the possible triplets starting AC, if you look at the codon usage table for *Pseudomonas aeruginosa* (Table 8.2) (gray shading) you can see that in this organism ACC is used in preference to any of the other possibilities. This preference is not always as marked, in the case of valine, codons starting GT (blue shading), the third position is likely to be a G or a C but there is no clear preference between them. In the *P. aeruginosa* genome as a whole there is a strong preference for GC base pairs, this is reflected by an overall 67% GC bias. This bias becomes even more striking if you look at the third position of the triplet where *P. aeruginosa* uses a G or a C in 88% of cases. The biological basis of this preference for particular triplet codons is the availability of tRNAs for these particular codons. Codon preference is seen throughout the animal and plant kingdoms as well as in bacteria

Table 8.2 Codon usage table for *Pseudomonas aeruginosa*

Amino Acid	Triplet Codon	Frequency per 1000	Amino Acid	Triplet Codon	Frequency per 1000	Amino Acid	Triplet Codon	Frequency per 1000	Amino Acid	Triplet Codon	Frequency per 1000
Phe	TTT	1.7	Ser	TCT	0.8	Tyr	TAT	5.3	Cys	TGT	1.0
	TTC	33.7		TCC	12.1		TAC	20.0		TGC	9.0
Leu	TTA	0.3		TCA	0.6	Stop	TAA	0.3	Stop	TGA	2.4
	TTG	8.7		TCG	13.0	Stop	TAG	0.3	Trp	TGG	14.8
Leu	CTT	3.1	Pro	CCT	2.1	His	CAT	6.3	Arg	CGT	7.9
	CTC	27.8		CCC	13.0		CAC	15.4		CGC	49.2
	CTA	1.4		CCA	2.2	Gln	CAA	6.2		CGA	2.4
	CTG	82.9		CCG	33.3		CAG	36.2		CGG	14.1
Ile	ATT	2.9	Thr	ACT	1.7	Asn	AAT	3.7	Ser	AGT	2.6
	ATC	37.8		ACC	32.8		AAC	22.6		AGC	25.9
	ATA	0.9		ACA	0.8	Lys	AAA	3.6	Arg	AGA	0.5
	ATG	20.2		ACG	6.3		AAG	25.0		AGG	2.0
Val	GTT	2.7	Ala	GCT	4.8	Asp	GAT	10.5	Gly	GGT	8.3
	GTC	28.8		GCC	67.7		GAC	42.6		GGC	61.9
	GTA	4.0		GCA	4.9	Glu	GAA	23.4		GGA	4.2
	GTG	33.4		GCG	38.8		GAG	37.3		GGG	9.9

The table shows the frequency with which each of the 64 possible triplet codons is used in *Pseudomonas aeruginosa*. The data is calculated by analyzing DNA sequences in the GenBank database.

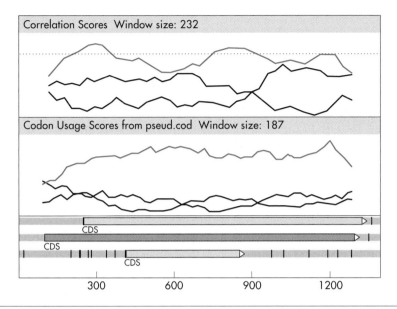

Figure 8.3 Potential open reading frames in a 1400bp sequence from *Pseudomonas aeruginosa*. The top panel is a plot showing how well the sequence in each reading frame correlates with the typical pattern of amino acids in a globular protein. The middle panel is a Gribscov plot showing how well the sequence in each reading frame conforms to the typical codon usage for *P. aeruginosa*. In the bottom panel potential open reading frames are shown as bars. In each case the blue line indicates the plot for the second reading frame. This diagram is redrawn from the output from the Artemis program. http://www.sanger.ac.uk/Software/Artemis/

those favored by *Pseudomonas aeruginosa* the organism from which this gene was cloned, this suggests that this ORF is likely to be translated into a protein.

Another useful technique is to look at how well the amino acid composition of the translated sequence correlates with the composition of typical proteins. The top graph in Figure 8.3 (top panel) shows the degree of correlation between the translated sequence of each of the three forward reading frames and that of globular proteins. Again you can see that the sequence of the ORF that we have identified correlates well with that of typical globular proteins.

Q8.6. Can you think of any circumstances in which a functioning gene would not show a pattern of codon usage typical of that of the organism from which the DNA was cloned?

Q8.7. Are there any circumstances where the translated sequence of a protein might deviate from that of a typical globular protein?

The fact that the answer to both of the above questions is that there are circumstances where a protein can be encoded by an ORF, which would not score well in terms of codon usage or correlation score, raises an important point. These methods of analysis are not infallible; they simply allow you to accumulate evidence in support of the proposition that an ORF is in fact a gene. If these were applied rigidly to exclude ORFs which did not match up to the criteria, many important genes would be missed.

In this example we have identified a region of our DNA sequence which is an open reading frame between a start codon and a stop codon. We have supporting evidence that this ORF is indeed translated into a protein because the codon usage conforms with the preferred pattern of codon usage in *P. aeruginosa*, and the amino acid sequence correlates well with that of known globular proteins. This type of analysis from first principles works quite well with bacterial DNA sequences. However, the situation is not so straightforward with eukaryotes.

8.3 Identifying Eukaryotic Genes

Whereas in prokaryotic organisms, genes exist as a continuous sequence, this is not the case in eukaryotic organisms where genes are often interrupted by introns (Section 2.3). If the DNA sequence that you wish to analyze is a cDNA clone, then it is derived from mRNA and the introns will have been spliced out from the primary transcript and the exons joined together in one continuous sequence during processing. In this case the process of identifying the protein coding region and translating the DNA sequence into an amino acid sequence is very similar to that for prokaryotic genes. In the case of cDNA sequence from a eukaryotic organism you have the added benefit of knowing that the sequence represents the product of a single gene and as long as the cDNA was complete, the gene will be too.

If on the other hand you want to analyze genomic DNA from a eukaryotic organism, for instance as part of one of the eukaryotic genome sequencing projects, the presence of introns may make the analysis much more difficult. Introns disrupt the protein coding region of the gene and consequently make them more difficult to identify. Simply looking for open reading frames of a particular size will not work if the exons, which constitute the gene, are smaller than this size. Considering that the average size of an exon in the human genome is only 145 bp the threshold would have to be set quite low to be sure of picking up most exons. Even if you know where the beginning and end of the gene are, in order to derive the sequence of the protein encoded it is necessary to be able to detect the intron–exon boundaries. The sequence of nucleotides at the intron–exon boundary is used by the cell's splicing mechanism to recognize the part of the sequence to be removed. In humans there is a conserved motif GT–AG which is surrounded by a more variable, but nonetheless conserved sequence (Figure 8.4).

```
A G G T R A G T                              C A G G
.7 .8  1  1 .6 .7 .8 .5                      .8  1  1 .6
```

Figure 8.4 The consensus sequence found at the intron exon boundary in a typical human gene. The invariant GT–AG motif is shown in bold. Numbers underneath the sequence indicate the proportion of sequences in which the base is found. R indicates either of the purine residues, adenine or guanine.

Because dinucleotides like GT and AG occur very frequently in genome sequences and the surrounding motifs are not highly conserved, and vary between species, predicting these boundaries is not an easy task.

There are a number of software packages available which attempt to detect intron–exon boundaries by scanning the DNA sequence for patterns, some also make use of information from related cDNA, EST (see Section 4.20) and protein sequences to refine their analysis. Some of these packages are "trained" using well-characterized regions of the genomes of particular species to enhance their ability to detect the unique features of exons in these species.

It is also difficult when analyzing eukaryotic DNA sequence to distinguish the first exon from internal exons and hence it is difficult to define the 5′ ends of eukaryotic genes. In order to detect the initial exon, which will begin with the start codon, you need to look for sequences characteristic of the 5′ ends of eukaryotic genes, these include promoter sequences (Section 11.1), and in the case of mammalian sequences, CpG islands.

Q8.8. Why does the dinucleotide CpG occur less frequently in a genome than would be expected from the GC content of the organism?

Software packages which look for CpG islands usually allow the user to define the minimum criteria for a CpG island, but regions of at least 500 bp with a GC content of at least 55% and at least 65% as many CpG dinucleotides as you would expect from the GC content of the organism, are associated with the 5′ ends of many mammalian genes.

In the example shown in Figure 8.5 a portion of human genomic DNA has been scanned for conserved sequences found at intron–exon boundaries in vertebrates, and eight potential exons have been detected. The presence of putative promoter sites in addition to an ATG start codon has been used to locate the initial exon. Further analysis, using protein homology information (see Section 8.9), results in the elimination of one small exon and refinement of the intron–exon boundaries. In this case the analysis has been very effective and the results match the known gene structure.

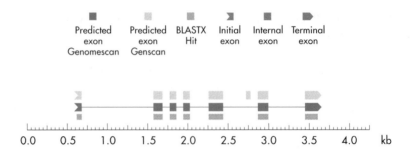

Figure 8.5 The arrangement of introns in the human chymotrypsin gene. The top panel (light blue) indicates the introns detected by scanning the sequence for known intron–exon boundary consensus sequences using Genscan. The next panel (dark blue) shows the refinement possible when incorporating protein homology information using Genomescan. The third panel (medium blue) shows that the regions of the sequence are similar to known protein sequences as detected by BLAST.

This type of analysis works well with genes belonging to well-characterized groups, and where the results are assessed by a knowledgeable experimenter. However, with the large amounts of sequence data generated by genome projects it is essential to be able to perform this process automatically, at least in the first instance. In February 2001 when the analysis of the draft human genome sequence was published the International Human Genome Sequencing Consortium estimated that there were between 30,000 and 40,000 genes in the human genome; in 2004 they revised this estimate to between 20,000 and 25,000. As gene-finding algorithms are refined it is possible to draw the definitions more tightly, reducing the number of false positives without running the risk of missing true positives. It will be many years before we know exactly how many genes there are in the human genome, but this example demonstrates the problem with automated gene finding. None of the algorithms currently available for the location of eukaryotic genes is yet good enough to be relied on entirely, although by using a combination of different approaches it is possible to detect a significant proportion of genuine genes.

A further complicating factor in identifying genes in higher eukaryotes is that many genes can be spliced in a number of different ways to give different gene products. This alternative splicing (Section 2.3) is a major factor in producing the diversity of proteins expressed by the human genome, with some genes being able to produce up to a thousand different forms of the protein. The possibility of alternative splicing inevitably clouds any analysis based on looking for splice sites and makes gene discovery more difficult.

It is important when using these tools for gene identification to understand that this is all that they are, and that they need to be used in conjunction with everything else that you know about the genes you are looking for. In the final analysis this type of bioinformatics analysis can only help you to formulate a theory which can then be tested in the laboratory. This type of sequence analysis can only take us so far in interpreting the sequence data. The next step in the analysis is to compare our sequence with that of other known sequences. This can give us information about the evolutionary relationship between sequences, as well as about their structure and function.

8.4 Sequence Comparisons

Sequence comparison is a very powerful tool; its use is based on the assumption that proteins, which have evolved from a common ancestor, will have similar sequences. This is a phenomenon called homology. If you can show that your sequence belongs to a group of similar sequences, then you can deduce that it has evolved from the same ancestral sequence; it is also likely that it has a similar function to the other members of the group.

8.5 Pair-wise Comparisons

A pair-wise comparison basically involves lining up two sequences with each other and observing how similar they are. Have a look at the two amino acid sequences in Figure 8.6a; you can probably see some similarities between the two sequences. You should be able to spot three runs of at least three amino acids that are identical in the two sequences. The sequences are not exactly the same length and the spacing between the regions that match up vary so it is necessary to introduce two gaps to make an optimal alignment of the sequences Figure 8.6b. With short sequences like these it is reasonably easy to align them by eye, with longer sequences you need to use an objective and automatable method to do this. The first step in this process is to prepare a grid like that in Figure 8.7. Each filled-in square in the grid represents a position where the two sequences are identical. The regions where the two sequences are identical show up as diagonal lines.

Because there are only 20 amino acids you would expect to see some positions where any two randomly chosen sequences were identical. These matches are not relevant in an evolutionary sense and introduce noise into

(a) SVKLQDRKLQDRMKSGKVSVKLQD   (b)   SVKLQD---RKLQDRMKSGKVSVKLQD
                                                |||||| ||| | ||||||||
     VSVKLQDQLSQRLQDNSKMVSVKLQD   VSVKLQDQLSQRLQDNSK--MVSVKLQD

Figure 8.6 Pair-wise alignment of amino acid sequences. a) Shows two arbitrary amino acid sequences. b) The best alignment of the two sequences, vertical bars represent identical amino acids.

(a)

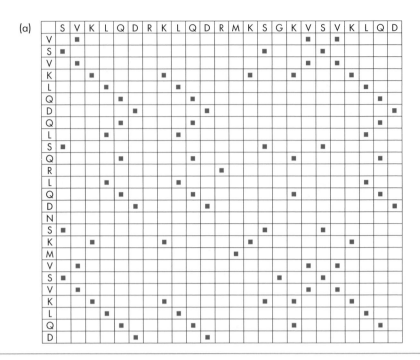

Figure 8.7 Grids for pair-wise sequence comparison. a) The first sequence is written along the top of the grid and the second sequence down the left hand side. Wherever the two sequences are identical the corresponding square on the grid is filled in.

(b)

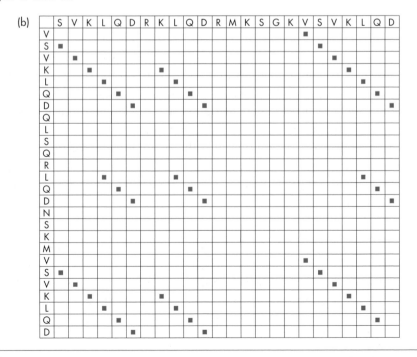

Figure 8.7. b) The grid shows positions where there are at least three consecutive amino acids which are identical in both sequences.

(c)

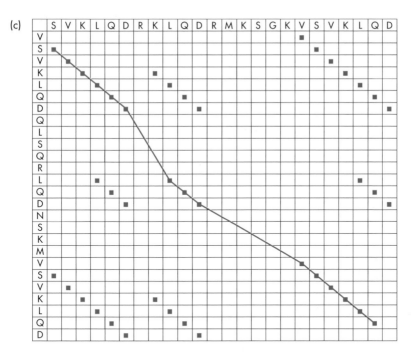

Figure 8.7. c) The line shows the best path through the grid, maximizing the number of matched amino acids and minimizing the number of gaps which have to be inserted to enable the sequences to be aligned.

the system. One way of filtering out this noise is to look for regions of three or more consecutive identical amino acids. This is referred to as using a word size of 3 (Figure 8.7b). This process makes the relationship between the two sequences clearer; a single pathway through the grid can now be determined which maximizes the number of matches and introduces the smallest number of gaps (Figure 8.7c). This alignment can be represented by the sequence alignment shown in Figure 8.6b. When a grid of this type is created for two longer sequences it is referred to as a dot-plot, an example of which is shown in Figure 8.8.

Q8.9. Although sequence comparisons can be made either between two DNA sequences or between two protein sequences, it is usual, wherever possible, to use amino acid sequences in comparisons. Can you explain why this is? *Hint.* Remember that because of the degeneracy of the genetic code, many amino acids can be encoded by more than one DNA sequence.

If a pair-wise comparison of two amino acid sequences reveals a significant number of identical amino acids in the same positions in the two

Figure 8.8 Pair-wise comparison of the amino acid sequences of the EIA protein from two different strains of adenovirus. The comparison was generated using a programme called dotlet http://www.isrec.isb-sib.ch/java/dotlet/Dotlet.html.

sequences this suggests that the two proteins have evolved from a common ancestor, in other words that they are homologous. It also suggests that it is likely that they have the same, or a related function, depending on the degree to which the two sequences match. The same is true for a comparison between two DNA sequences, but as the amino acid sequence is more information rich most sequence comparisons are performed using amino acid sequences.

8.6 Identity and Similarity

Where two related sequences differ at a particular position this is referred to as a substitution. Not all substitutions in amino acid sequences are likely to be as significant in their effect on the protein product. Some amino acids have very similar biochemical and physical properties and can be interchanged with relatively little effect on the protein. These are referred to as conservative substitutions, as they are less likely to be selected against by evolutionary pressures they are seen more frequently. For example the three amino acids valine, leucine and isoleucine, which are termed aliphatic amino acids, all have very similar side chains. Substitutions of one of these amino acids for another are not likely to have a significant effect on the structure or the protein. However, a substitution of any of these three amino acids

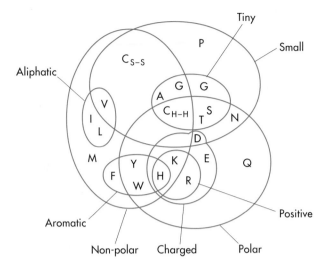

Figure 8.9 Venn diagram showing a classification of amino acids. The amino acids are classified according to chemical and physical properties thought to be important in determining protein structure. The diagram was taken from W.R. Taylor, The Classification of Amino Acid Conservation. *J Theoret Biol* 1986; 119: 205–218.

with histidine, which has an aromatic, positively charged side chain, is likely to have quite a drastic effect.

There are several classification systems that attempt to group amino acids together according to their biochemical and physical properties, one of which is shown in the Venn diagram in Figure 8.9.

Sequence comparison programs can take into account whether or not an amino acid substitution is likely to affect the structure of the protein. Instead of using a particular classification of amino acids to decide which substitutions are significant, most programs use what is called a substitution matrix (Figure 8.10). This is a representation of how often particular substitutions actually occur in related proteins. Substitution matrices were developed as an objective way of deciding which substitutions are conservative and which are not.

Q8.10. Have a look at the blosum 62 matrix in Figure 8.10. Which amino acid is most often substituted for tyrosine (Y)? Which amino acids are least often substituted for tyrosine?

Q8.11. Which group of amino acids are most interchangeable and what do they have in common?

	C	S	T	P	A	G	N	D	E	Q	H	R	K	M	I	L	V	F	Y	W	
C	9																				C
S	-1	4																			S
T	-1	1	5																		T
P	-3	-1	-1	7																	P
A	0	1	0	-1	4																A
G	-3	0	-2	-2	0	6															G
N	-3	1	0	-2	-2	0	6														N
D	-3	0	-1	-1	-2	-1	1	6													D
E	-4	0	-1	-1	-1	-2	0	2	5												E
Q	-3	0	-1	-1	-1	-2	0	0	2	5											Q
H	-3	-1	-2	-2	-2	-2	1	-1	0	0	8										H
R	-3	-1	-1	-2	-1	-2	0	-2	0	1	0	5									R
K	-3	0	-1	-1	-1	-2	0	-1	1	1	-1	2	5								K
M	-1	-1	-1	-2	-1	-3	-2	-3	-2	0	-2	-1	-1	5							M
I	-1	-2	-1	-3	-1	-4	-3	-3	-3	-3	-3	-3	-3	1	4						I
L	-1	-2	-1	-3	-1	-4	-3	-4	-3	-2	-3	-2	-2	2	2	4					L
V	-1	-2	0	-2	0	-3	-3	-3	-2	-2	-3	-3	-2	1	3	1	4				V
F	-2	-2	-2	-4	-2	-3	-3	-3	-3	-3	-1	-3	-3	0	0	0	-1	6			F
Y	-2	-2	-2	-3	-2	-3	-2	-3	-2	-1	2	-2	-2	-1	-1	-1	-1	3	7		Y
W	-2	-3	-2	-4	-3	-2	-4	-4	-3	-2	-2	-3	-3	-1	-3	-2	-3	1	2	11	W
	C	S	T	P	A	G	N	D	E	Q	H	R	K	M	I	L	V	F	Y	W	

Figure 8.10 Blosum 62 substitution matrix. The acronym blosum stands for **blo**cks **su**bstitution **m**atrix. The table shows how likely a substitution is between pairs of amino acids. A positive number indicates that the substitution occurs relatively frequently, whereas a negative number indicates that it is a rare substitution. The commonly used blosum 62 matrix is generated using sequences which have no more than 62% identity between them and is useful for making comparisons between distantly related sequences. There are other blosum matrices compiled from different data sets which can be used in other types of comparison.

Programs that produce sequence alignments often report the percentage of identical and of similar amino acids in the alignment. Have a look at the alignment in Figure 8.11, 39.4% of the amino acids in the two sequences are identical but nearly 50% are described as being similar.

Q8.12. What criteria are used to decide which amino acids are substituted with similar ones? You might find it useful to look up some of the pairs of similar amino acids in the blosum 62 substitution matrix.

8.7 Is the Alignment Significant?

Most alignment algorithms are designed to find the best alignment between two sequences, and they will always report an alignment. This does not mean that the alignment is of any significance. One of the simplest ways of checking to see whether an alignment is significant is to jumble up one of the sequences and run the alignment again. This new alignment is not meaningful because one of the sequences is essentially nonsense. If the similarity score is similar to that of your initial alignment then

```
Matrix: EBLOSUM62
GAP_penalty: 10.0
Extend_penalty: 0.5

Length: 325
Identity:      128/325 (39.4%)
Similarity:    162/325 (49.8%)
Gaps:           95/325 (29.2%)
Score: 513.0


E1A05    1 MRHIICHGGVITEEMAASLL-----DQLIEEVLADN----LPPPSHFEPP     41
             :..||...:|       |.::|. |.||    :|.......|
E1A12    1           MRTEMTPLVLSYQEADDILEH-LVDNFFNEVPSDDDLYVP     39


E1A05   42 TLHELYDLDV-TAPEDPNEEAVSQIFPDSVMLAVQEGIDLLTFPPAPGSP     90
             :|:||||||| :|.||.||:||::.||:|::||..||:.|        |
E1A12   40 SLYELYDLDVESAGEDNNEQAVNEFFPESLILAASEGLFL---------P     80


E1A05   91 EPPHLSRQPEQPEQRALGPVSMPNLVPEVIDLTCHEAGFPPSDDEDE---    137
             |||.||      |....:|...||.|.||.:||.|:|.|||.||.|||
E1A12   81 EPPVLS-----PVCEPIGGECMPQLHPEDMDLLCYEMGFPCSDSEDEQDE    125


E1A05  138 ------------------EGEEFVLDYVEHPGHGCRSCHYHRRNTGDPDIM    170
             |.|||.||:.|.|||.|:||.:||.:||.:||:.|:|
E1A12  126 NGMAHVSASAAAAAADREREEFQLDHPELPGHNCKSCEHHRNSTGNTDLM    175


E1A05  171 CSLCYMRTCGMFVYSPVSEPEPEPEPEPEPARPTRRPKMAPAILRRPTSP    220
             |||||:|...||:||||||:.||||
E1A12  176 CSLCYLRAYNMFIYSPVSDNEPEP--------------------------    199


E1A05  221 VSRECNSSTDSCDSGPSNTPPEIHPVVPLCPIKPVAVRVGGRRQ-AVECI    269
             ||:.|. |..||   ||:::...||...||||..||.|||: |||.|
E1A12  200 -----NSTLDG-DERPS--PPKLGSAVPEGVIKPVPQRVTGRRRCAVESI    241


E1A05  270 EDLLNEPGQ----PLDLSCKRPRP         289
             .||:.|..:    |:|||.||||.
E1A12  242 LDLIQEEEREQTVPVDLSVKRPRCN        266
```

Figure 8.11 Pair-wise alignment of the E1A protein from two different strains of adenovirus produced using the EMBOSS program needle. Exact matches between the two sequences are indicated by |, conservative changes (with a blosum value of 1 or greater) by :, non-conservative changes by • and gaps introduced to improve the alignment are indicated with a –.

it is likely that your alignment was also meaningless. It is usual to compare one sequence with 10 random sequences created in this way; this will give you an average and a standard deviation for the identity and similarity scores. If your original alignment has a score which does not fall within the range indicated by the standard deviations, then it is at least statistically significant. Random alignments generated by jumbling up the second of the two sequences in Figure 8.11 gave an average identity score of 14.9 ± 2.6 and an average similarity score of 22.7 ± 3.5. The alignment shown in Figure 8.11 is therefore statistically significant. It is important to remember that, when you are evaluating an alignment, you need to use your knowledge of the biological and evolutionary implications of the alignment as well as the statistical test. The example used in the alignment in Figure 8.11 is explained in the next section and you will be able to see that the alignment fits very well with what is known about the two proteins.

8.8 What Can Alignments Tell Us About the Biology of the Sequences Being Compared?

Pair-wise comparison of two related sequences can be a very powerful tool. The following example illustrates the sort of information that can be deduced from this type of comparison. It involves the comparison of a protein from different strains of a human adenovirus. Some types of this virus are oncogenic, capable of causing tumors to develop, while others are not. The protein E1A is the product of the virus "early region" and has been shown to be involved in the induction of tumors. The dot-plot in Figure 8.8 and the alignment in Figure 8.11 show a comparison of the amino acid sequences of the E1A protein from an oncogenic strain (12) with that of a non-oncogenic strain (05). The strong diagonal line on the dot-plot (Figure 8.8) indicates that these sequences are very similar along large parts of their length, which is to be expected since they are variations of the same protein; you can see these regions of similarity on the alignment (Figure 8.11). There are two large gaps which have to be introduced into the alignment, marked with arrows in Figure 8.8. It is the extra sequence, starting at position 123 in the A1E12, which is not present in the non-oncogenic strain which is responsible for this virus causing tumors. This observation from sequence alignments has been tested in the laboratory by cloning this extra sequence into the non-oncogenic strain and observing that it is then capable of causing tumors in mice.

8.9 Similarity Searches

Imagine that you could do a pair-wise alignment with your query sequence and every known sequence and then look at, say, the best 50 alignments. This would tell you which of the proteins, for which sequence data was available, were most similar to your query. If the structures and functions of these proteins are known then they will give clues as to the structure and function of your query sequence. DNA and protein sequence data are stored in large publicly available databases (Box 8.2); in addition to the sequence data these databases contain varying amounts of annotation. For example, a typical entry in the protein database, SwissProt, gives a brief description of the protein and its origin, references to papers describing the derivation and analysis of the sequence, a comments section which will identify any key features of the protein such as its activity and whether it belongs to a characterized family of proteins, hyperlinks to relevant entries in other databases and an outline of the key features of the sequence. Hence searching this type of database provides access to an immense amount of information.

Box 8.2 DNA and Protein Sequence Databases

Over the past few years, as DNA sequencing has become a routine and automated procedure, vast amounts of sequence data have become available. Most of the data is stored in publicly accessible databases, although some, particularly the genome sequences of commercially useful species, is held by commercial companies. The primary repository for DNA sequence data is one of the three members of the International Nucleotide Sequence Database Collaboration (INSDC). INSDC consists of EMBL (Europe), GenBank (USA) and DDBJ (Japan) and they exchange data on a daily basis. Data submitted to these databases is annotated by the scientists who have sequenced it; they supply information such as the name of the organism from which it came, the positions of any genes that they have located and any publications where they have referred to it.

The central database of protein sequence data is the UniProt knowledge base, so-called because in addition to sequence data it contains information about the functions of the proteins as well as links to other databases. UniProt consists of two parts, the most comprehensive of these is the TrEMBL database which contains translations of the nucleic acid sequences automatically extracted from the EMBL database. The main drawback with TrEMBL is that entries are generated by an automated procedure and any errors in the original EMBL submission are perpetuated in TrEMBL. SwissProt is a sister database of TrEMBL; it is what is known as a curated database. This means that the annotations accompanying the entries in SwissProt are more comprehensive and have been checked by experts before being entered into the database. SwissProt is considerably smaller than TrEMBL but it gives access to a wide range of information about the sequences that it contains.

Secondary sequence (or pattern) databases contain information derived from the primary databases described above. The Prosite database of protein families and domains is an example of a secondary database. Prosite consists of entries for recognized families of proteins and gives sequence patterns and profiles, which can be used to identify members of these families. As with primary databases, secondary protein sequence databases can be generated automatically as in the case of Blocks and ProDom, or can be curated: Prosite, Pfam and Prints. These databases contain extensive links between themselves and to the primary databases, the curated databases also provide useful, authoritative summaries of the properties of the protein families.

When a DNA or protein sequence is submitted to these databases it is assigned a unique accession number. The major biological databases all use these numbers to refer to sequences. A given DNA sequence will be found in all of the INSDC databases but each entry will refer to the same accession number; accession numbers are also used for cross referencing for instance between databases.

The problem with trying to perform a pair-wise alignment, between your query sequence and all of the sequences in one of the databases, is that the databases are too big for it to be practical to perform such an exhaustive search. However, a number of computational methods have been developed which will perform a very good approximation to this type of search. The two most commonly used programs are Basic Local Alignment Search Tool (BLAST) and Fasta (it stands for FAST-All). Each of these programs uses a short cut to allow it to produce a list of similar sequences without performing an exhaustive alignment between the query sequence and every entry in the database. There are numerous web-based resources which allow you to run Fasta and BLAST searches. The addresses for some of these can be found in the Further Reading section at the end of this chapter.

A classical example, which demonstrates the power of this technique, is the case of the gene responsible for cystic fibrosis. In this inherited disease, thick mucus accumulates in the lungs and intestines, which results in frequent lung infections and malnutrition. Patients also have high levels of salt in their sweat and are often sterile. Intensive efforts to clone and sequence the gene responsible culminated in 1989 in the publication, by a large team led by Lap-Chee Tsui and John Riordan, of the sequence of the *CFTR* gene; mutations in this gene give rise to cystic fibrosis. Performing a similarity search with the *CFTR* gene today will yield a huge number of similar sequences, as it belongs to a much studied family of membrane transporter proteins. However, in the late 1980s the situation was very different because there were far fewer known sequences. Homology searches at the time showed that, somewhat surprisingly, the gene responsible for this genetic disease in humans was similar to a family of proteins responsible for resistance of bacteria to a wide range of antibacterial agents. These ABC multidrug transporter proteins function by pumping the drugs out of the bacteria. This was an important clue as to the function of the CFTR protein, which is now known to be involved in pumping sodium and chloride ions across membranes in mammalian cells.

8.10 Fasta

Typically, the output from a Fasta search produces a summary table such as that shown in Figure 8.12a. The table lists the sequences from the database which most closely match that of the query sequence. The first column shows the name of the entry in the database, or database identifier, followed by an extract from the description of the protein. Many web interfaces for the Fasta program provide hyperlinks to the database entries so that you have immediate access to a wealth of information about the proteins. The other columns give statistical information, which enables you to assess the match between your query and the database entry. In the example shown you can see that most of the sequences are of a similar length to

(a)

DB:ID	Source	Length	Identity%	Ungapped%	Overlap	E
SW:CTR2_CANFA	Chymotrypsinogen 2 precursor (EC	263	86.692	86.692	263	1.8e-103
SW:CTRB1_HUMAN	Chymotrypsinogen B precursor (EC	263	85.932	85.932	263	1.33-101
SW:CTRB1_RAT	Chymotrypsinogen B precursor (EC 3	263	84.411	84.411	263	2.8e-101
SW:CTRB_BOVIN	Chymotrypsinogen B (EC 3.4.21.1).	245	81.224	81.224	245	2.4e-91
SW:CTRA_BOVIN	Chymotrypsinogen A (EC 3.4.21.1).	245	76.735	76.735	245	1.3e-84
SW:CTRA_GADMO	Chymotrypsin A precursor (EC 3.4.	263	67.755	67.755	245	2.0e-76
SW:CTRB_GADMO	Chymotrypsin B (EC 3.4.21.1).	245	67.073	67.623	246	1.6e-74
SW:CTRL_HUMAN	Chymotrypsin-like protease CTRL-1	264	55.682	55.894	264	9.0e-68
SW:ELA2_PIG	Elastase 2 precursor (EC 3.4.21.71)	269	44.444	47.619	270	1.0e-42
SW:ELA2_RAT	Elastase 2 precursor (EC 3.4.21.71)	271	43.333	46.063	270	8.2e-42
SW:CLCR_HUMAN	Caldecrin precursor (EC 3.4.21.2)	268	42.471	44.898	259	1.1e-41
SW:ELA2A_HUMAN	Elastase 2A precursor (EC 3.4.21	269	43.911	47.410	271	1.9e-40
SW:ELA2_MOUSE	Elastase 2 precursor (EC 3.4.21.7	271	41.481	44.094	270	5.9e-40
SW:CLCR_RAT	Caldecrin precursor (EC 3.4.21.2)	268	41.699	44.082	259	1.1e-39
SW:ELA2_BOVIN	Elastase 2 precursor (EC 3.4.21.7	269	43.333	46.429	270	3.9e-39
SW:TMPS9_MOUSE	Transmembrane protease, serine 9	1065	40.784	42.623	255	8.5e-38
SW:ELA2B_HUMAN	Elastase 2B precursor (EC 3.4.21	269	41.544	45.200	272	1.7e-37
SW:TMPS9_RAT	Transmembrane protease, serine 9	1061	40.000	41.803	255	5.6e-37
SW:TMPS3_MOUSE	Transmembrane protease, serine 3	453	40.711	44.017	253	7.7e-37
SW:PLMN_CANFA	Plasminogen (EC 3.4.21.7)(Fragme	333	38.314	41.667	261	1.2e-36

(b)

```
>>SW:CTR2_CANFA_PO4813 Chymotrypsinogen 2 precursor (EC   (263 aa)
 initn: 1607 initl: 1607 opt: 1607  Z-score: 1984.4  bits: 374.8 E(): 1.8e-103
Smith-Waterman score: 1607;  86.692% identity (86.692% ungapped) in 263 aa overlap (1-263:1-263)

              10        20        30        40        50        60
Query  MAFLWLLSCFALVGAASSCGVPAIEPVLNGLSRIVNGEDAVPGSWPWQVSLQDSTGFHFC
       :::::::::::::.:.: .:::::.:::::.:::::::::::::::::::::::::::::::
SW:CTR MAFLWLLSCFALLGTAFGCGVPAIQPVLSGLSRIVNGEDAVPGSWPWQVSLQDSTGFHFC
              10        20        30        40        50        60

              70        80        90       100       110       120
Query  GGSLISEDWVVTAAHCGVRTSHLVVAGEFDQSSSEENIQVLKIAEVFKNPKFNMFTVRND
       :::::::::::::::::::::.: :::::::::.: :.:::::::::.:::::::::::. ::
SW:CTR GGSLISEDWVVTAAHCGVRTTHQVVAGEFDQGSDAESIQVLKIAKVFKNPKFNMFTINND
              70        80        90       100       110       120

             130       140       150       160       170       180
Query  ITLLKLATPARFSETVSPVCLPQATDEFPPGLMCVTTGWGRTKYNANKTPDKLQQAALPL
       :::::::::::::.::: :::::::::.:: : .::::::: ::..  .::::::::::::
SW:CTR ITLLKLATPARFSKTVSAVCLPQATDDFPAGTLCVTTGWGLTKHTNANTPDKLQQAALPL
             130       140       150       160       170       180

             190       200       210       220       230       240
Query  LSNAECKKFWGSKITDVMICAGASGVSSCMGDSGGPLVCQKDGAWTLVGIVSWGSGTCST
       ::::::::::::::::.:::::::::::::::::::::::::::::::::::::::::::::
SW:CTR LSNAECKKFWGSKITDLMVCAGASGVSSCMGDSGGPLVCQKDGAWTLVGIVSWGSGTCST
             190       200       210       220       230       240

             250       260
Query  SVPAVYSRVTELIPWVQEILAAN
       :.:.::.:::.::::::.:: ::
SW:CTR STPGVYARVTKLIPWVQQILQAN
             250       260
```

Figure 8.12 Excerpts from a Fasta output showing a search with a canine chymotrypsinogen amino acid sequence against the SwissProt database. a) The summary table lists, in order, the sequences from the SwissProt database which most closely match the query sequence. This excerpt shows the top 20 matches although the full output would normally list 50 or 100. b) The first pair-wise alignment showing the similarity between the query sequence and the most similar sequence in the database. Exact matches between the two sequences are indicated by :, conservative changes, as defined by a blosum 50 matrix by • .

the query sequence, and that the region of similarity between the two sequences is over most of the sequence. Between 87% and 38% of the amino acids are identical in each pair of sequences.

While the percentage of identical amino acids gives you some idea of how significant the matches are, the E value in the final column is the most important piece of statistical data. The "expect" or E value for a match between two sequences is a measure of how likely it is that the match has occurred by chance when searching a database of a particular size. In this example there is a 1.8×10^{-103} chance that the first match could have occurred by chance, in other words the alignment between these two sequences is statistically highly significant.

There are some useful rules of thumb which can help in evaluating E values although because these are statistical measures they are only a tool and as always need to be used in conjunction with knowledge of the biology of the proteins involved. In general, an E value of 0.02 is considered to be significant although, as in the example in Figure 8.12, E values are often very much smaller. An E value of 1 indicates that, when searching a database of this size, you would expect one match, this good, just by chance; it is therefore very unlikely to have any biological or evolutionary significance. E values of between 0.02 and 1 fall in what is known as the twilight zone, within which you will need your knowledge of the biology of the genes and proteins involved in order to assess the significance of the match. It is in the twilight zone that exciting discoveries showing relationships between proteins previously thought to be unrelated are being made.

In addition to the summary table, Fasta also displays pair-wise alignments between the query sequence and each of the sequences that match it (Figure 8.12b). Examination of this comparison shows that the query sequence and the match with the smallest E value are very similar along their entire lengths with no gaps in the alignment and very few substitutions, most of these being with amino acids with similar properties.

8.11 BLAST

This program performs much the same task as Fasta although the computational approach is different. The BLAST output also gives a summary of the sequences which most closely match the query sequence (Figure 8.13a). Again, E values indicate how significant the matches are and hyperlinks point to other databases. Some implementations of BLAST, such as at NCBI (the US National Center for Biotechnology Information) also provide a graphical overview (Figure 8.13c) which can be particularly useful when, as in the example shown, some matches are only to part of the sequence. Another useful feature, which is often combined with BLAST, is the ability to look for the occurrence of conserved domains in the sequence (Figure 8.13d). Conserved domains will be discussed in more detail in Section 8.15.

(a)

```
Sequences producing significant alignments:                    (bits) Value
gi|130316|sp|P00747|PLMN_HUMAN  Plasminogen precursor [Conta...  1652  0.0
gi|114062|sp|P08519|APOA_HUMAN  Apolipoprotein(a) precursor ...  1035  0.0
gi|123114|sp|P26927|HGFL_HUMAN  Hepatocyte growth factor-lik...   472  e-133
gi|123116|sp|P14210|HGF_HUMAN   Hepatocyte growth factor prec...  459  e-129
gi|137119|sp|P00750|TPA_HUMAN   Tissue-type plasminogen activ...  236  2e-62
gi|123057|sp|P05981|HEPS_HUMAN  Serine protease hepsin (Tran...   197  2e-50
gi|13124582|sp|P57727|TMPS3_HUMAN  Transmembrane protease, s...   192  3e-49
gi|18203319|sp|Q9NRR2|TRYG1_HUMAN  Tryptase gamma precursor ...   191  7e-49
gi|18202734|sp|Q9BQR3|PRS27_HUMAN  Pancreasin precursor (Mar...   191  9e-49
gi|547643|sp|Q04756|HGFA_HUMAN  Hepatocyte growth factor act...   191  9e-49

-------------
gi|9296990|sp|Q9UKR3|KLK13 HUMAN  Kallikrein 13 precursor (K...   143  3e-34
gi|19860140|sp|P48740|MASP1_HUMAN  Complement-activating com...   137  2e-32
gi|3915626|sp|P00746|CFAD_HUMAN  Complement factor D precurs...   133  2e-31
gi|7387859|sp|O00187|MASP2_HUMAN  Mannan-binding lectin seri...   129  4e-30
gi|3914477|sp|O43240|KLK10_HUMAN  Kallikrein 10 precursor (P...   106  2e-23
gi|20978519|sp|Q96MU8|KREM1_HUMAN  Kremen protein 1 precurso...    69  5e-12
gi|6226950|sp|Q99767|APBA2_HUMAN  Amyloid beta A4 precursor ...    38  0.012
gi|38258267|sp|Q9UK17|KCD3_HUMAN  Potassium voltage-gated ch...    33  0.40
gi|45477126|sp|Q86X10|K1219_HUMAN  Protein KIAA1219            30  3.4
gi|2497674|sp|Q16527|CSR2_HUMAN  Cysteine and glycine-rich p...    29  5.8
```

(b)

```
>gi|130316|sp|P00747|PLMN_HUMAN  Plasminogen precursor [Contains: Angiostatin]
           Length = 810

 Score = 1652 bits (4278), Expect = 0.0
 Identities = 760/810 (93%), Positives = 760/810 (93%)

Query: 1    MEHKEVVXXXXXXXXKSGQGEPLDDYVNTQGASLFSVTKKQLGAGSIXXXXXXXXXXXXFT 60
            MEHKEVV       KSGQGEPLDDYVNTQGASLFSVTKKQLGAGSI            FT
Sbjct: 1    MEHKEVVLLLLLFLKSGQGEPLDDYVNTQGASLFSVTKKQLGAGSIEECAAKCEEDEEFT 60

Query: 61   CRAFQYHSKEQQCVIMAENRKSSIIIRMRDVVLFEKKVYLSECKTGNGKNYRGTMSKTKN 120
            CRAFQYHSKEQQCVIMAENRKSSIIIRMRDVVLFEKKVYLSECKTGNGKNYRGTMSKTKN
Sbjct: 61   CRAFQYHSKEQQCVIMAENRKSSIIIRMRDVVLFEKKVYLSECKTGNGKNYRGTMSKTKN 120

Query: 121  GITCQKWSSTSPHRPRFSPATHPSEGLEENYCRNPDNDPQGPWCYTTDPEKRYDYCDILE 180
            GITCQKWSSTSPHRPRFSPATHPSEGLEENYCRNPDNDPQGPWCYTTDPEKRYDYCDILE
Sbjct: 121  GITCQKWSSTSPHRPRFSPATHPSEGLEENYCRNPDNDPQGPWCYTTDPEKRYDYCDILE 180
```

(c)

Distribution of 92 Blast Hits on the Query Sequence

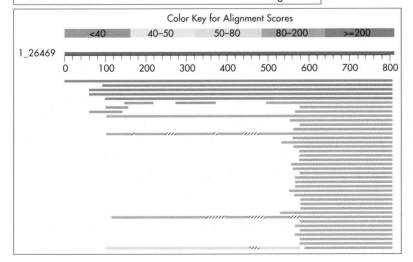

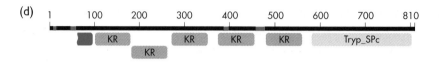

Figure 8.13 Excerpts from a BLAST output showing a search with the blood clotting protein precursor plasminogen against the human sequences in the SwissProt database. a) The summary table lists the best matches between the query sequence and the human sequences in SwissProt. BLAST would normally list up to 100 matches, the first and matches 41–50 are shown here. b) Part of the first pair-wise alignment showing the similarity between the query sequence and the most similar sequence in the database. Sequences of low complexity are excluded from the analysis and are marked by X in the query sequence. c) Graphical overview showing the alignments between the database sequences and the query. d) Graphical representation showing putative conserved domains detected in the query sequence.

Most scientists have a preference for using either Fasta or BLAST; some may have good reasons for this preference and for some it may just be habit. In practice, for the student, both programs will do a good job and there is probably little to choose between them. However, it is good practice to use both programs, particularly if you are looking at matches in the twilight zone. Because both programs use methods designed to speed up what would otherwise be a very computationally intensive process, they may miss matches in this region. Using both programs is one way to minimize the risk of missing important matches.

8.12 What Can Similarity Searches Tell Us About the Biology of the Sequences Being Compared?

As with pair-wise alignments, when we find a match between our query sequence and other sequences we can infer that the proteins are likely to have a similar structure and consequently a similar function. We also infer an evolutionary relationship: proteins with similar sequences are likely to be homologues, and to have evolved from a common ancestor. Their evolutionary relationship is evident from the similarity observed in their sequences and reinforced if it can be demonstrated that they have similar structures and functions. In the case of a sequence about which nothing is known, for instance a sequence generated during one of the genome sequencing projects, a similarity search, in conjunction with techniques which look for the characteristics of genes, will help to identify the protein.

The query sequence used in the comparison shown in Figure 8.12 is that of canine chymotrypsinogen, the precursor of an important digestive enzyme chymotrypsin. In this case the sequence is taken from a canine genome sequencing project. You can see that the sequence at the top of the list is also of chymotrypsinogen, and if you look at the database identification you can also probably guess that it is also of canine origin. Had the query sequence been an unknown it would be safe to conclude from this search that it was a chymotrypsinogen.

In fact the first seven matches in Figure 8.12 are to chymotrypsinogen or chymotrypsin precursor in a range of animals from humans to *Gadus morhua* the atlantic cod. By following the links to the SwissProt entries for these sequences we find that chymotrypsin is an extracellular enzyme, that is produced as an inactive precursor (chymotrypsinogen), it belongs to a family of enzymes with similar properties (the serine proteases), and that it preferentially cleaves peptide bonds after a tyrosine, tryptophan, phenylalanine or leucine residue. These chymotrypsin precursors are all homologues and because they all perform the same function and are evolutionarily related, but because their sequences have diverged from each other as a result of mutations accumulated after speciation, they are referred to as being orthologous.

> **Q8.13.** Examine Figure 8.12a. What other proteins are similar to chymotrypsinogen?

As well as telling us that our query sequence is chymotrypsinogen, we now know that it is similar to members of a large family of related proteins called trypsin-like serine proteases, which cleave peptide bonds in polypeptides. The mechanism of action of these enzymes involves a highly reactive serine and two other amino acids, an aspartate and a histidine; together these amino acids make up what is known as the catalytic triad. Our bioinformatics analysis has thus given us access to information about both the structure and the function of the protein.

All of the proteins that belong to the trypsin-like family of serine proteases are thought to have evolved from a common ancestor and consequently are homologous. However, as they have diverged from each other to perform different functions in the same organism, they are known as paralogues. The mechanism which is thought to make this type of evolutionary divergence possible is gene duplication. Once a chance event results in duplication of a gene, one copy can continue to fulfil its original function while the other is able to accumulate mutations and adapt to another related function.

One way of studying paralogues is to limit your search to one particular organism, in this way the multiple examples of proteins with the same function in different organisms are eliminated from the comparison. The BLAST search in Figure 8.13 uses only the human sequences in the SwissProt database. The query sequence is another serine protease, the human precursor of the protein plasmin which is involved in blood clotting.

Q8.14. Examine Figure 8.13a. Which of the matches listed are likely to be of biological and evolutionary significance? Which are in the twilight zone and which are not significant? Are these matches with paralogues or orthologues?

The BLAST output also gives pair-wise alignments of the high scoring comparisons (Figure 8.13b), a graphical overview of the matches (Figure 8.13c) which can help to identify the regions of the proteins that match the query sequence and a graphic showing any conserved domains that have been detected in your query protein (Figure 8.13d).

Q8.15. Examine Figure 8.13c. Most of the proteins detected by this search are serine proteases. Which part of the protein is likely to be responsible for this protease activity?

Q8.16. Examine Figure 8.12d. Is the location of the trypsin-like serine protease conserved domain consistent with the location of this activity in the C-terminal 200 amino acids.

In summary, similarity searches give access to a large amount of information about the identity, structure, function and evolution of the query sequence. Because the databases that can be searched contain such vast numbers of sequences, including those from whole genomes, it is likely that there will be significant matches to any protein coding sequence. Depending on how close the match is between query and database sequences it may result in identification of the protein. With matches of lower significance, or matches along part of the sequence, the analysis may not identify the protein but may suggest possible structures that the amino acid sequence could adopt and also possible functions for the whole or parts of the protein.

Similarity searches also give us information about the evolutionary relationships both between organisms and between proteins and by implication the genes which encode them. Genes that encode proteins which have similar sequences are assumed to have evolved, by accumulating mutations, from a common ancestral gene. In the case of the serine proteases used in the example here, the proteins have evolved into a very diverse family of proteins which have important roles in processes as distinct as digestion and blood clotting. However, all of these proteins have a common enzymic activity, which is that they cleave polypeptide chains in specific ways. As a result they all have a common conserved domain, the trypsin family serine protease domain. Grouping proteins into families in this way allows us to understand the structures that they adopt and how this relates to their function. By comparing members of one of these families we can begin to understand which parts of the sequence are important and whether their roles are structural or more directly related to function. These relationships can be further explored by performing a multiple sequence alignment of a group of related proteins (see Section 8.13). Comparisons of the amino acid sequence of orthologous proteins can help to elucidate the evolutionary relationships between organisms. The globin molecules of hemoglobin have been used for this purpose in mammals and the nucleic acid sequence for the 16S ribosomal RNA for bacteria (see Section 8.17).

As we mentioned in Section 8.5, sequence comparisons are almost always carried out using amino acid sequence data; this is also true for similarity searches. This is because DNA sequences may vary but still encode the same amino acid sequence; consequently there may be differences between two DNA sequences which have no effect on the amino acid sequence. Exceptions to this rule would be when comparing sequences which do not encode proteins, such as rRNA genes or when studying non-coding regions such as regulatory sequences.

Both BLAST and Fasta offer the possibility of translating query sequences in all six reading frames so that searches can be performed even if you have not managed to conclusively identify open reading frames. In fact, finding a similar sequence may give important clues to deciding whether a DNA sequence encodes a protein, and where the open reading frame is.

8.13 Multiple Sequence Alignments
In order to analyze the groups of sequences which make up gene families it is necessary to be able to do more than look at pair-wise comparisons. Often, by looking at a group of related sequences it is possible to pick out conserved sequences and to deduce which are the important structural and functional components of the proteins. From these observations it may be possible to formulate hypotheses which can be tested in the laboratory.

In a multiple alignment, sequences are aligned in such a way that similarities and differences can be seen (Figure 8.14). ClustalW is a good general-purpose multiple alignment program, which performs alignments between each pair of sequences; it then creates a guide tree (not a phylogenetic tree) from which it creates the multiple alignment. Various methods are used to indicate the conserved regions in a multiple alignment, including shading the alignments and calculating a consensus sequence.

A consensus sequence is a sequence derived from a multiple alignment of a group of related sequences and represents the positions where there is agreement between the sequences. This helps to identify regions where

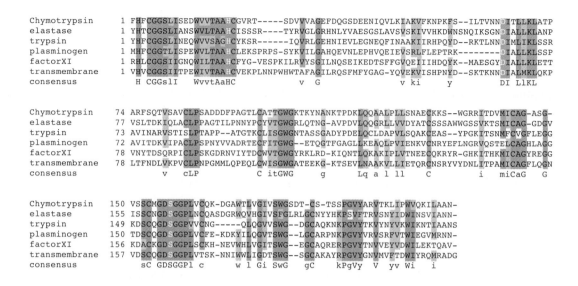

```
Chymotrypsin    1 FHFCGGSLISEDWVVTAAHCGVRT-----SDVVVAGEFDQGSDEENIQVLKIAKVFKNPKFS--ILTVNNDITLLKLATP
elastase        1 YHTCGGSLIANSWVLTAAHCISSSR----TYRVGLGRHNLYVAESGSLAVSVSKIVVHKDWNSNQIKSGNDIALLKLANP
trypsin         1 YHFCGGSLINEQWVVSAGHCYKSR------IQVRLGEHNIEVLEGNEQFINAAKIIRHPQYD--RKTLNNDIMLIKLSSP
plasminogen     1 MHFCGGTLISPEWVLTAAHCLEKSPRPS-SYKVILGAHQEVNLEPHVQEIEVSRLFLEPTRK--------DIALLKLSSP
factorXI        1 RHLCGGSIIGNQWILTAAHCFYG-VESPKILRVYSGILNQSEIKEDTSFFGVQEIIIHDQYK--MAESGYDIALLKLETT
transmembrane   1 VHVCGGSIITPEWIVTAAHCVEKPLNNPWHWTAFAGILRQSFMFYGAG-YQVEKVISHPNYD--SKTKNNDIALMKLQKP
consensus         H CGGslI  WvvtAaHC            v G              v ki    y        DI LlKL
```

```
Chymotrypsin   74 ARFSQTVSAVCLPSADDDFPAGTLCATTGWGKTKYNANKTPDKLQQAALPLLSNAECKKS--WGRRITDVMICAG-ASG-
elastase       77 VSLTDKIQLACLPPAGTILPNNYPCYVTGWGRLQTNG-AVPDVLQQGRLLVVDYATCSSSAWWGSSVKTSMICAG-GDGV
trypsin        73 AVINARVSTISLPTAPP--ATGTKCLISGWGNTASSGADYPDELQCLDAPVLSQAKCEAS--YPGKITSNMFCVGFLEGG
plasminogen    72 AVITDKVIPACLPSPNYVVADRTECFITGWG--ETQGTFGAGLLKEAQLPVIENKVCNRYEFLNGRVQSTELCAGHLAGG
factorXI       78 VNYTDSQRPICLPSKGDRNVIYTDCWVTGWGYRKLRD-KIQNTLQKAKIPLVTNEECQKRYR-GHKITHKMICAGYREGG
transmembrane  78 LTFNDLVKPVCLPNPGMMLQPEQLCWISGWGATEEKG-KTSEVLNAAKVLLIETQRCNSRYVYDNLITPAMICAGFLQGN
consensus           v   cLP         C itGWG    g      Lq a l ll     C         i  miCaG  G
```

```
Chymotrypsin  150 VSSCMGDSGGPLVCQK-DGAWTLVGIVSWGSDT-CS-TSSPGVYARVTKLIPWVQKILAAN-
elastase      155 ISSCNGDSGGPLNCQASDGRWQVHGIVSFGLRLGCNYYHKPSVFTRVSNYIDWINSVIANN-
trypsin       149 KDSCQGDSGGPVVCNG-----QLQGVVSWG--DGCAQKNKPGVYTKVYNVVKWIKNTIAANS
plasminogen   150 TDSCQGDSGGPLVCFE-KDKYILQGVTSWG--LGCARPNKPGVYVRVSRFVTWIEGVMRNN-
factorXI      156 KDACKGDSGGPLSCKH-NEVWHLVGITSWG--EGCAQRERPGVYTNVVEYVDWILEKTQAV-
transmembrane 157 VDSCQGDSGGPLVTSK-NNIWWLIGDTSWG--SGCAKAYRPGVYGNVMVFTDWIYRQMRADG
consensus         sC GDSGGPl c       w l Gi SwG    gC    kPgVy  V  yv Wi    i
```

Figure 8.14 Multiple sequence alignment. The alignment was performed using ClustalW and the output was shaded and the consensus sequence calculated using Boxshade. Dark blue boxes indicate identical amino acids; light blue indicates conservative substitutions. In the consensus line upper case letters indicate identical amino acids and lower case indicates the most common amino acid where conservative substitutions have occurred.

there is most agreement between sequences, and which are hence likely to be important in the structure and function of the proteins.

Q8.17. Examine the multiple alignment in Figure 8.14 and identify regions where the sequences are conserved.

8.14 What Can Multiple Sequence Alignments Tell Us About the Structure and Function of Proteins?

It is assumed that regions of proteins which are important for either the structure or the function of a protein are likely to accumulate fewer mutations than regions of less importance, as mutations in these areas are likely to have a detrimental effect on the protein. In this case it should be possible to pick out these important regions because they should be conserved in otherwise divergent sequences. In this way multiple alignments can be used to locate important features of sequences.

The multiple alignment shown in Figure 8.14 is constructed from proteins which belong to the serine protease family; three are digestive enzymes, chymotrypsin, trypsin and elastase, two are involved in blood clotting, plasmin and factor XI and one is a protease located in cell membranes. Although the biological roles of these proteins are diverse they all act by cleavage of polypeptide chains at specific points and all have a common functional domain. Plasmin, factor XI and the membrane protease are all larger proteins than the digestive enzymes, but the serine protease activity resides in a region towards the C termini, and it is these regions which have been used in the alignment. The three key amino acids (Figure 8.14 shown in blue) that make up the catalytic triad can be seen to be present in all of the sequences. However, it is also clear that the sequence of amino acids around these key residues is also conserved. The high degree of conservation of sequence in the region around the active site serine suggests that this is an important structural component of the protein and plays a key role in the proteolytic function of the protein. There are also seven conserved cysteine residues, and an eighth which is found in five out of six proteins. These cysteines are involved in forming disulfide bridges and are important for the tertiary structure of the proteins.

As with similarity searching, multiple alignments are usually performed on amino acid sequences. The real strength of this analysis is in identifying conserved regions within sequences which allows predictions to be made about the structures they are likely to adopt and about possible functions. As with any bioinformatics analysis these are merely predictions, which are very useful in designing experiments to test hypotheses about the function of proteins or to modify the function of proteins. Multiple sequence alignments are also useful in helping to identify new members of protein families.

8.15 Consensus Patterns and Sequence Motifs

Where a conserved region of consecutive nucleotides or amino acids is recognized and is thought to have some biological significance it is often referred to as a sequence motif. These motifs are described by "regular expressions" or "consensus patterns". It is possible to derive a consensus pattern from a multiple alignment, which represents the characteristic signature of a family of proteins. The consensus pattern can then be used to assess whether other proteins are likely to belong to the family.

A consensus pattern describing the region around the active site serine in the proteins in our alignment could be written as dsCxGDSGGPlv. In this representation upper case letters indicate invariant amino acids, lower case letters indicate positions where the majority of sequences agree and x represents degenerate positions. A more informative way of expressing this information is in the form of a sequence logo such as that shown in Figure 8.15. In the sequence logo it is possible to indicate which amino acids are found in the partially conserved positions. In either case it is clear that there are key amino acids which do not vary amongst these sequences. Six sequences is, in fact, rather a small set from which to derive a reliable consensus pattern. These are usually derived from much larger data sets, and the resulting patterns are stored in what are known as secondary databases (Box 8.2).

Figure 8.15 Sequence logo. This shows the region around the conserved active site serine of the alignment from Figure 8.14. A sequence logo is a graphical representation of a multiple alignment, the height of the letter indicates both the frequency of an amino acid at a particular position and the degree of conservation at that position. This means that the height of the column where a single residue is always present is greater than that in less well-conserved positions.

One such database, Prosite, gives a consensus pattern describing the conserved region around the active site serine of the trypsin-like serine proteases in the form of a regular expression as:

```
       1              2        3,4  5   6  7 8  9        10
[DNSTAGC]-[GSTAPIMVQH]-x(2)-G-[DE]-S-G-[GS]-[SAPHV]-
       11              12
[LIVMFYWH]-[LIVMFYSTANQH]
```

In this representation a single letter indicates positions where only one amino acid is ever found in multiple alignments, such as positions 5, 7 and 8. In this example the serine at position 7, shown here in blue, is the active site serine. Positions 6 and 9 allow one of two possible residues, namely aspartate or glutamate at position 6 and glycine or serine at position 9. Positions 1, 2, 10, 11 and 12 allow any one of the amino acids from the group in square brackets. Positions 3 and 4 can be occupied by any residues, as indicated by x(2). If you compare this regular expression to the consensus pattern derived from our multiple alignment you should see that it describes the conserved region from our multiple alignment but also allows for greater variation. This is because it was generated from a much larger data set that included more distantly related members of the trypsin-like serine protease family.

Q8.18. Can you locate the sequence described by the following regular expression on the multiple alignment in Figure 8.14?
[LIVM]-[ST]-A-[STAG]-H-C

These two regular expressions, describing the motifs for both the serine and histidine active sites are diagnostic of trypsin-like serine proteases. If sequences matching both are present then the protein can be unambiguously assigned to the family. In situations like this where several motifs are required to define a protein family these can be grouped together into what are known as fingerprints. The fingerprint provides a signature for the protein family. The PRINTS protein fingerprint database is a collection of these fingerprints which can be searched in a variety of different ways. The prints entry for the trypsin-like serine proteases consists of three elements including both of the Prosite patterns that we have already discussed and a third motif representing the active site aspartate.

Both of the above methods of representing the characteristic patterns of amino acids typical of specific protein families are a compromise between a very restricted expression, which is likely to miss some of the more divergent examples, and a very fuzzy expression which may result in false identifications of family members. These expressions are refined by an iterative process of scanning the primary protein databases and evaluating the families identified. An alternative solution is to preserve the full alignment used to identify the family in the form of a matrix in which the frequency of each amino acid at each position is recorded. These are known as profiles and are found in addition to the patterns in the Prosite database.

The real power of these secondary or pattern databases is that they offer a fast track to identify important structural and functional regions of proteins and in doing so provide a link to a wealth of information about the biological function of proteins. These patterns which are derived from extensive alignments of many sequences may in some cases be able to identify more distant relationships, and more divergent members of families, than similarity searching. As the primary databases expand at ever increasing rates, secondary databases can also help to overcome some of the problems of "noise" created by the occurrence of many very similar sequences. We have already come across an example of this when looking at BLAST searches (Section 8.10). A search of the conserved domain database (CDD) (Figure 8.13d) is run by default at the same time as a protein BLAST at NCBI. This makes it possible to study the domain architecture of the query protein even before the BLAST search is complete.

8.16 Investigating the Three-dimensional Structures of Biological Molecules

The logical extension of studying the sequence of DNA and protein molecules is to understand the three-dimensional structures that they adopt. These three-dimensional structures must be determined experimentally by techniques such as X-ray crystallography and nuclear magnetic resonance (NMR) spectroscopy. This is a much more involved process than determining sequence and is beyond the scope of this book. However, once the structures have been resolved they are deposited as a series of coordinates in one of the structural databases and these can be displayed in three-dimensions on a desk-top computer using free software such as Chime or Rasmol. Where three-dimensional structures are available you will see links from SwissProt entries and BLAST search outputs. Even where there is no structure available for a particular protein it is possible to model the structure if a homologous sequence can be found for which a structure is available, although this is really a job for the experts.

The power of comparing three-dimensional structures rather than sequences is demonstrated by returning to the example of the serine proteases. The trypsin-like serine proteases are composed largely of β sheets (Figure 8.16a). There is another class of serine proteases, which are found mainly in bacteria, called the subtilases. These enzymes are very different in overall structure from the trypsin family enzymes, being mostly composed of α helices (Figure 8.16b). However, like the trypsin family enzymes, their catalytic activity is provided by the catalytic triad of serine, histidine and aspartate. The sequence around the amino acids of the catalytic triad is different in the two proteins and you would not detect the subtilases using a profile or pattern derived from the trypsin family. However, the two different protein structures bring the amino acids of the catalytic triad into almost exactly the same positions in three-dimensional space, creating conditions for the same catalytic mechanism to operate. Comparing the

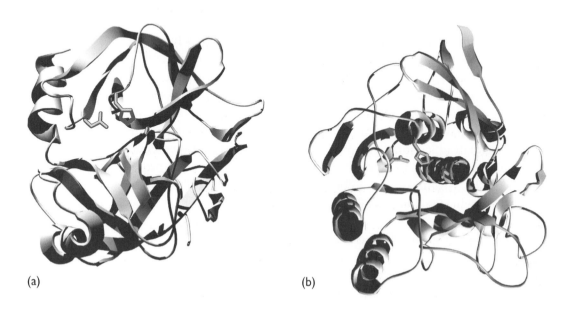

(a) (b)

Figure 8.16 Three dimensional structure of serine proteases. a) Chymotrypsin, this member of the trypsin-like family of serine protease is largely composed of β sheets. b) Subtilisin from *Bacillus subtilis*; this bacterial serine protease is largely composed of α helices.

three-dimensional structures of the two proteins (Figure 8.16a and b) is the only way of detecting this similarity. This is thought to be an example of convergent evolution, where two different evolutionary pathways lead to the same solution to a problem.

8.17 Using Sequence Alignments to Create a Phylogenetic Tree

In Section 8.12 we discussed how similarity searches and multiple alignments can provide information about evolutionary relationships as well as about the structure and function of proteins. This type of information is often represented in the form of a branched diagram or phylogenetic tree (Box 8.3). Traditionally these are used to depict evolutionary relationships but they can also be useful as a method of organizing what you know about a group of proteins and their relationships with each other. There are two fundamentally different evolutionary questions which can be addressed by looking at sequence data. The first is to ask questions about how a group of paralogous proteins, for instance the serine proteases, are related to each other. These proteins are thought to have evolved by a process of gene duplication and subsequent accumulation of mutations resulting in the evolution of proteins with new functions. In theory it should be possible to draw a phylogenetic tree depicting the evolutionary relationships in this group of proteins. In practice this is a very complex example and outside the scope of this book. The second is to use the accumulation of mutations in a key molecule as a molecular clock. In this way the evolutionary relationship between species can be examined.

A phylogenetic tree can be calculated from a multiple alignment by a number of different methods. The simplest of these are the distance methods such as neighbor joining in which a table is constructed representing the degree of difference between pairs of sequences. The clustering of similar sequences and the lengths of the branches of the tree are computed from this information. A neighbor-joining tree is shown in Figure 8.17. Other methods involve tree searching in which many possible trees are evaluated. Different criteria can be used to assess which is the most likely tree depending on which evolutionary model you choose to use, or which tree is the most parsimonious (requires the fewest steps).

Bacteria provide a very good example of the way sequence data can be used to investigate evolutionary relationships (Box 8.4). In this case the sequences of the RNA of the small subunit of ribosomes (16S rRNA) are compared to investigate evolutionary relationships. The tree shown in Figure 8.17 was constructed, using the neighbor-joining method, from a multiple alignment of the 16S rRNA sequences from a set of seven Gram negative bacteria. It shows that the two most closely related organisms in the set are *E. coli* and *Shigella dysenterie*, and that *Salmonella typhimurium* is also closely related to these two organisms. This fits with what is known

Box 8.3 Phylogeny, Parsimony and Dendrograms

Mutations, which arise by chance, give rise to slightly different versions of genes and proteins. While some of these mutations may affect the fitness of the organism most will be neutral. Mutations which confer a selective advantage and mutations whose effect is neutral will be inherited by subsequent generations. One of the ways in which we can study evolutionary relationships is to compare the sequences of proteins or nucleic acid molecules in living organisms, using the accumulation of mutations as a "molecular clock" reflecting the evolutionary distance between the molecules, and hence the organisms which carry them.

Imagine a short amino acid sequence such as (a). It would require three mutations, of one amino acid at a time, to change it to sequence (c).

(a)	M	S	A	T	H	C		(a)	M	S	A	T	H	C
	I	S	A	T	H	C			I	S	A	T	H	C
(b)	I	T	A	T	H	C		(b)	I	T	A	T	H	C
(c)	I	T	A	G	H	C			L	T	A	A	H	C
								(d)	L	T	A	A	H	C

This process is a bit like one of those word games where you have to get from the word "fish" to the word "bike" in the fewest possible steps. (I can do it in five.) This analogy works well because in the word game each intermediate has to be a real word. In the world of amino acid sequences each intermediate needs to be a functioning protein. In this example it is also possible to arrive at sequence (d) in four mutational steps, and in both pathways there is a common sequence (b). There are of course many other pathways between sequences (a) and (c), many just as plausible as the one shown. However, some involve non-conservative substitutions and are likely to result in proteins which do not function, these will not be part of the evolutionary pathway. Others involve more steps and so are less parsimonious; evolutionary theory assumes that the shortest or most parsimonious route is the most likely unless there is evidence to the contrary.

Branching diagrams or dendrograms are used to represent the evolutionary relationship between sequences. Both of the dendrograms shown below are representations of the relationship between the sequences used in this example. Sequences are linked by a line with those to which they are most similar and the length of the line indicates the number of intervening sequences. Diagram a) makes no assumptions about the direction of evolution and is referred to as an unrooted tree. Diagram b) assumes that the sequence ISATHC is the ancestral sequence and that the others have evolved from it; this is called a rooted tree.

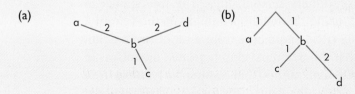

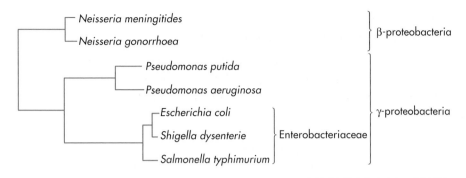

Figure 8.17 Neighbor-joining tree showing the relationship between the 16S rRNA sequences of a group of Gram negative bacteria. The most closely related of the bacteria shown are *E. coli* and *S. dysenterie* these are separated by the shortest distances on the horizontal axis of the diagram. *S. typhimurium*, the other member of the Enterobacteriaceae is next most closely related to this group, these three organisms form a subgroup on the tree. The two species of *Neisseria* form a second subgroup and are also closely related to each other but not as closely as *E. coli* and *S. dysenterie*. The two *Pseudomonas* species form a third subgroup being more closely related to each other than to any of the other species but not as closely related to each other as the members of the other subgroups are to each other. The Enterobacteriaceae are further grouped with the *Pseudomonas* species and this correlates with the taxonomic group the γ-proteobacteria.

about the relationship between these bacterial species, which belong to a group of closely related organisms, the Enterobacteriacae. The tree also suggests two other groupings, with the two *Pseudomonas* species in one group and the two *Neisseria* species in another. The lengths of the branches are assumed to be an indication of evolutionary distance. In this example the evolutionary distance between the two *Pseudomonas* species is greater than between organisms in either of the other two groups. Finally, although all of these bacteria belong to a taxonomic group called the proteobacteria, the Pseudomonads and the Enterobacteriacae belong to a subgroup designated γ while the *Neisseria* species are β-proteobacteria; this is also reflected in the tree. This example shows how sequence comparisons can be used to study evolutionary relationships, in the case of the sequence of an unclassified bacterial species it can be placed in the overall phylogeny based on its 16S sequence.

Q8.19. If amino acid sequences were used to analyze phylogeny, what class of mutations would be omitted from the analysis?

The best way to understand the various bioinformatics tools described here, and to make sure that you understand how to interpret the analyses

> ## Box 8.4 Importance of Comparison of Ribosomal RNA Sequences
>
> A major breakthrough in our understanding of the relationship between the main groups of organisms, the bacteria and the eukaryotes, and the identification of a new major taxonomic grouping the Archea came about in the early 1980s as a result of sequence comparisons and molecular phylogeny. Carl Wose and his colleagues made a study of evolutionary relationships using a molecule that was common to all living things; they used the RNA molecule from the small subunit of ribosomes (16S rRNA in prokaryotes and 18S rRNA in eukaryoutes). These analyses led to the now widely accepted proposal that the Archea, a group of small single-celled organisms, are in fact more closely related to eukaryotes than to the Eubacteria. Further developments of this technique have led to the analysis of 16S rRNA from the ribosomes of bacteria being adopted not only as the definitive method for understanding bacterial taxonomy but also as an approach to routine identification of bacteria.
>
> One major advantage of 16S rRNA sequencing in the identification of bacteria is that there is no requirement for the bacteria to be grown in laboratory culture; this can be a significant problem especially for bacteria isolated from unusual environments. The 16S rRNA gene can be amplified by PCR from environmental samples and sequenced directly. This can then be used to identify the bacterium and determine its place in the taxonomy of bacteria.

is to try it out for yourself. All of the analyses that have been described in this chapter can be run using freely available resources which are either web-based or free to download. The DNA and protein sequences, together with their accession numbers, and the URLs for the web resources are listed below. As this chapter is intended as an introduction to the important concepts in sequence analysis, rather than a step-by-step guide you will need to read the help which accompanies the resources in order to run them successfully. You might also like to use the tutorials provided as part of the 2Can Bioinformatics Educational Resource at the European Bioinformatics Institute. Don't be afraid to have a go: the best way to learn is to try things out for yourself.

Questions and Answers

Q8.1. What is the DNA triplet codon for methionine? What would the codon be in an RNA molecule? Refer to Table 8.1, which shows the genetic code.

A8.1. *The triplet codon ATG represents methionine in DNA, however in RNA thymine is replaced by uracil so that the RNA codon that encodes methionine is AUG.*

Q8.2. What do the codons UAA, UAG and UGA encode?

A8.2. *These codons do not code for an amino acid, they signal the end of the part of a gene that is translated into protein; they are referred to as stop codons.*

Q8.3. Why are three alternative translations given below the sequence in Figure 8.2?

A.8.3. *Because the genetic code uses three nucleotides to encode each amino acid, there are three possible places where you can begin to translate a sequence; these are called the three forward reading frames.*

Q8.4. Are the three alternative translations given in Figure 8.2 the only possible translations of this sequence?

A8.4. *No, there are also three reverse reading frames. Remember that DNA consists of two anti-parallel complementary strands, the translations shown in Figure 8.2 are from one of these strands, and it is just as likely that genes will be present on the other strand. In order to look at the reverse frames you need to read the reverse complement of the DNA sequence and then translate it.:*

```
   5'tcagacaagaaccattcgagagtcgggtctgtttgcggtttgcctagcaccatgcgcggt3'
-1   SerAspLysAsnHisSerArgValGlySerValCysGlyLeuProSerThrMetArgGly
-2   GlnThrArgThrIleArgGluSerGlyLeuPheAlaValCysLeuAlaProCysAla
-3    ArgGlnGluProPheGluSerArgValCysLeuArgPheAlaEndHisHisAlaArg
```

Q8.5. How many amino acids would you expect a protein to consist of if it was encoded by a 100 nucleotide stretch of DNA?

A8.5. *Because the genetic code uses triplet codons, 100 nucleotides could potentially encode 33 amino acids.*

Q8.6. Can you think of any circumstances in which a functioning gene would not show a pattern of codon usage typical of that of the organism from which the DNA was cloned?

A8.6. *There are a number of such circumstances. Genes encoded on pathogenicity islands (Section 2.9) and other regions acquired by horizontal transfer of genetic material show patterns of codon usage typical of the organism from which they originate rather than the current host organism. Also, while there is strong evolutionary pressure on genes which are expressed at high levels to use the codons for which there are plenty of tRNAs, the selective pressure is not so strong on the sequence of genes which are expressed at low levels; these are often seen to diverge from the norm in their codon usage.*

Q8.7. Are there any circumstances where the translated sequence of a protein might deviate from that of a typical globular protein?

A8.7. *Many proteins do not conform to this pattern, as they may have specialized functions which require unusual patterns of amino acids. Proteins which span membranes would be an example.*

Q8.8. Why does the dinucleotide CpG occur less frequently in a genome than would be expected from the GC content of the organism?

A8.8. *In Section 2.12 you learned that the cytosine in most CpG dinucleotides is methylated, and that spontaneous deamination of this base generates thymine which is not easily detected by the cell's repair mechanisms as an error. As a result CpG nucleotides are lost from the genome except in areas where the cytosine residues are not methylated such as the 5' ends of genes in mammalian genomes, these areas are referred to as CpG islands.*

Q8.9. Although sequence comparisons can be made either between two DNA sequences or between two protein sequences, it is usual, wherever possible, to use amino acid sequences in comparisons. Can you explain why this is?
Hint. Remember that because of the degeneracy of the genetic code, many amino acids can be encoded by more than one DNA sequence.

A8.9. *Because of the degeneracy of the genetic code it is possible for two DNA sequences which differ by as much as one third of the bases in the sequence, to encode the same amino acid sequence. It is the similarity in the amino acid sequence, and hence in the protein products, which is important in evolution; consequently you will produce a more meaningful comparison with two amino acid sequences. Another way of looking at this is to say that the amino acid sequence is more information rich. A difference between two amino acid sequences represents a real difference in the two proteins, which may consequently function differently. Differences between DNA sequences may simply be noise not reflected in the protein product. Often differences between nucleic acid sequences will be due to differences in codon usage between different organisms.*

Q8.10. Have a look at the blosum 62 matrix in Figure 8.10. Which amino acid is most often substituted for tyrosine (Y)? Which amino acids are least often substituted for tyrosine?

A8.10. *The largest positive value in the substitution matrix for tyrosine is 3 for the substitution with phenylalanine, this substitution occurs relatively frequently in related proteins. This also makes sense in terms of what we know about the properties of the two amino acids as they both have aromatic ring structures in their side chains. Proline, aspartate and glycine are not commonly substituted for tyrosine; they each have a value of –3 in the substitution matrix.*

Q8.11. Which group of amino acids are most interchangeable and what do they have in common?

A8.11. *The aliphatic amino acids, isoleucine, leucine, and valine are commonly substituted with each other in related proteins; all of the values in the substitution matrix are positive values. These amino acids are hydrophobic and have non-reactive side chains. The other group of amino acids which is easily interchangeable are the aromatic amino acids tryptophan, tyrosine and phenylalanine.*

Q8.12. What criteria are used to decide which amino acids are substituted with similar ones? You might find it useful to look up some of the pairs of similar amino acids in the blosum 62 substitution matrix.

A8.12. *Amino acids with a value in the substitution matrix that is greater than 0 are considered to be similar.*

Q8.13. Examine Figure 8.12a. What other proteins are similar to chymotrypsinogen?

A8.13. *All of the proteins in the list in Figure 8.12 have similar amino acid sequences to chymotrypsin, including elastase, caldecrin, transmembrane protease and plasminogen; these proteins are all serine proteases.*

Q8.14. Examine Figure 8.13a. Which of the matches listed are likely to be of biological and evolutionary significance? Which are in the twilight zone and which are not significant? Are these matches with paralogues or orthologues?

A8.14. *The first two sequences, plasminogen and apolipoprotein have E values of 0, these are highly statistically significant matches. All of the other proteins in the first part of the list are also significant. The second part of the excerpt shows matches 41–50. The first six (41–46) matches with low E values are clearly still significant matches. The amyloid beta A4 protein has an E value of 0.012, only just below the cut-off value of 0.02. The potassium voltage gated channel with an E value of 0.4 falls in the twilight zone and the remaining proteins have high E values and the matches are unlikely to be of significance.*
All of the significant matches in this list are paralogues as they are all found in humans and have evolved from an ancestral protein to perform a wide variety of differing roles.

Q8.15. Examine Figure 8.13c. Most of the proteins detected by this search are serine proteases. Which part of the protein is likely to be responsible for this protease activity?

A8.15. *You can see from Figure 8.13c that many of the similar proteins are much shorter than plasminogen and match at the C terminus of the protein;*

as these proteins are all serine proteases the C-terminal 200 amino acids are likely to be responsible for the protease activity.

Q8.16. Examine Figure 8.12d. Is the location of the trypsin-like serine protease conserved domain consistent with the location of this activity in the C-terminal 200 amino acids.

A8.16. *Yes the diagram in Figure 8.12d shows a conserved serine protease domain at the C terminus of the protein.*

Q8.17. Examine the multiple alignment in Figure 8.14 and identify regions where the sequences are conserved.

A8.17. *There are two main regions where the sequences are conserved. The first is at the beginning of the alignment, at amino acids 2–20, and the second at the beginning of the third line, 152–164 (numbered according to chymotrypsin). There are other smaller regions such as amino acids 65–71 and 98–104.*

Q8.18. Can you locate the sequence described by the following regular expression on the multiple alignment in Figure 8.14?
[LIVM]-[ST]-A-[STAG]-H-C.

A8.18. *This regular expression occurs in the region of the active site histidine at amino acids 15–20 in the multiple alignment.*

Q8.19. If amino acid sequences were used to analyze phylogeny, what class of mutations would be omitted from the analysis?

A8.19. *Changes in the DNA sequence that did not result in a change in the amino acid sequence, because of the redundancy of the genetic code, would be lost from the analysis. These so-called silent mutations are nonetheless important indicators of the evolutionary distance between sequences.*

Further Reading

Instead of recommending journal articles that you can read if you want to deepen your knowledge in this area we recommend that you try out some of the analyses for yourself. To help you to do this the DNA and protein sequences used are listed below together with their accession number. To locate a DNA or protein sequence use the "Quick text search" facility of the Sequence retrieval system at EBI and search for the accession number.

The 2Can bioinformatics educational resource is an excellent tutorial for the resources provided by the European Bioinformatics Institute.

The journal Nucleic Acids Research *publishes a special database issue at the beginning of each year with articles about all of the major DNA and protein databases. The following two papers, both of which have many authors, are examples.*

EMBL Nucleotide Sequence Database: developments in 2005 (2006) Cochrane G *et al.*, Nucleic Acids Research, Volume 34 Pages D10–D15.

The Universal Protein Resource (UniProt): an expanding universe of protein information (2006) Cathy H *et al.*, Nucleic Acids Research, Volume 34 Pages D187–D191.

General bioinformatics web sites

European Bioinformatics Institute	http://www.ebi.ac.uk/
National Center for Biotechnology Information (USA)	http://www.ncbi.nlm.nih.gov/
2can Bioinformatics Educational Resource	http://www.ebi.ac.uk/2can/

Sequences used in the analyses presented in this chapter

Human adenovirus 5 Early E1A protein	P03255
Human adenovirus 12 Early E1A protein	P03259
Human plasminogen precursor	P00747
Human chymotrypsin	P17538
Human trypsin	P07477
Human elastase	P08217
Human factorXI	P03951
Human transmembrane protease	O15393
Canine chymotrypsin	P04813

Web resources

Sequence retrieval system at EBI	http://srs.ebi.ac.uk/
Artemis DNA sequence viewer and annotation tool	http://www.sanger.ac.uk/Software/Artemis/
Genscan	http://genes.mit.edu/GENSCAN.html
Genomescan	http://genes.mit.edu/genomescan/
Dotlet	http://www.isrec.isb-sib.ch/java/dotlet/Dotlet.html
EMBOSS Pairwise Alignment Algorithm	http://www.ebi.ac.uk/emboss/align/
Fasta at EBI	http://www.ebi.ac.uk/fasta33/
BLAST at NCBI	http://www.ncbi.nlm.nih.gov/
ClustalW at EBI	http://www.ebi.ac.uk/clustalw/
WebLogo for generating sequence logos	http://weblogo.berkeley.edu/logo.cgi
Prosite database of protein families and domains	http://www.expasy.org/prosite
Boxshade	http://www.ch.embnet.org/software/box_form.html

9 Production of Proteins from Cloned Genes

> **Learning outcomes:**
>
> **By the end of this chapter you will have an understanding of:**
> - *the reasons for producing proteins from cloned genes*
> - *some of the more common methods that are used to express proteins at high levels*
> - *some of the problems that may be encountered in protein production, and how they can be overcome*

9.1 Why Express Proteins?

There are many reasons why you might want to clone a gene. Similarly, there are many reasons why you might want to be able to express, and ultimately purify, the protein that it encodes. The aim of molecular biology is to understand life at the molecular level, and the aim of the range of industries encompassed by the term "biotechnology" is the exploitation of this understanding. Proteins have many essential roles in living organisms and so the study of proteins is central to molecular biology, while their production is important in many industries.

Generally, proteins are best studied in pure form. The great enzymologist Efraim Racker once said "Don't waste clean thinking on dirty enzymes". It is a central concept of much molecular biology that living processes can be broken down into individual steps that are much easier to study than entire processes. This means that studying the properties of individual proteins can help us to understand the processes in which they participate. Examples of the kinds of studies that can be done with purified or partially purified proteins are many (some are listed below) and, while some of these studies were possible long before molecular biology came along, cloning techniques have greatly enhanced our ability to study and understand proteins. This is mostly because we can now express and purify them in high quantities in organisms which are easily grown in the laboratory.

What sorts of studies can we do with pure proteins? Obviously, one thing to do with a pure protein is to determine its three-dimensional structure, by using structural methods such as X-ray crystallography or nuclear magnetic resonance (NMR) spectroscopy. The structure of a protein gives very important information about how it functions in the cell, whatever its role. From structures, hypotheses about the role of particular amino acids within the protein can be formulated, and these can be tested by specifically mutating the gene which codes for the protein, and then expressing, purifying and analyzing the mutant protein. If we have some idea of the activity of the protein, we can try to reproduce that activity in a test tube. Experiments on purified cellular components such as proteins or DNA are referred to as *in vitro* experiments (as opposed to experiments on living cells or organisms, which are called *in vivo* experiments). Our ability to study proteins *in vitro* depends on two things: the availability of a suitable assay, and the availability of the pure protein. The activity may be a relatively straightforward one such as an enzyme activity, or a more complex one such as binding to DNA and regulating gene expression, or membrane transport or signaling. Protein purification can vary from easy to difficult depending on the particular protein, but our ability to express many proteins at high levels by cloning the genes that encode them is very helpful in devising successful ways of purifying them.

Q9.1. Why do you think it is generally best to study proteins in pure form?

Proteins have many practical uses in the laboratory. With molecular biology itself, of course, most of the key reagents are proteins: restriction enzymes, DNA ligase, various DNA polymerases such as the thermostable polymerases used in the polymerase chain reaction (PCR), and so on. These enzymes were originally purified from a diverse range of organisms, but now nearly all of them have been cloned and expressed in *Escherichia coli*, and are purified from there.

Proteins also have numerous applications, many of which produce goods with which we are all familiar. Enzymes, in particular, are widely used: in brewing and wine making, in paper manufacture, in the production of detergents, in making cheese, in making soft-center mints, and in many other examples. Proteins also have very important roles as pharmaceuticals. Insulin for diabetics, factor VIII for hemophiliacs, and the active component of the vaccine against the hepatitis B virus for children, are some of the many examples of proteins which are drugs. Many more drugs are active against particular proteins. The search for compounds which are active against particular proteins is helped enormously by having the pure proteins available for study in the laboratory. (To give an example, the anti-cancer drug Glivec® is a highly specific inhibitor of a protein called bcr-abl

Table 9.1 Examples of proteins approved for clinical use which are produced using recombinant DNA methods

Name of Protein	Organism or Cell Type in Which it is Produced	Used in treatment of:
Insulin	*E. coli*	Diabetes mellitus
Human growth hormone (hGH)	*E. coli*	hGH deficiency in children
Various interferons	*E. coli*	Types of leukemia, chronic hepatitis C, genital warts, multiple sclerosis
Hirudin	*S. cerevisiae*	Thrombosis
Glucagon	*S. cerevisiae*	Hypoglycemia
Hepatitis B surface antigen (HBsAg)	*S. cerevisiae*	Component of vaccine against disease caused by infection with Hepatitis B virus
Factor VIII	Chinese hamster ovary cells (CHO cells)	Hemophilia
Erythropoietin	CHO cells	Anemia
Tissue plasminogen activator (tPA)	CHO cells	Heart attacks and strokes

tyrosine kinase which itself is implicated in the development of chronic myeloid leukemia. The availability of this protein, expressed from the cloned gene, was essential in the development of the drug.) Again, in the vast majority of cases, the proteins in these examples are produced by cloning the gene for the protein from its original organism and expressing it in a different ("heterologous") organism, often (though not exclusively, as we shall see) *E. coli*. A few of the many examples of pharmaceutical proteins which are already approved for clinical use, their roles, and the organism or cell type from which they are produced, are listed in Table 9.1.

In summary, there are numerous examples stretching from pure research, through the production of mundane and everyday items, to the development of advanced pharmaceuticals, which illustrate the use of different proteins. In an increasing number of cases, the genes for these proteins have been cloned through the use of the sorts of techniques discussed in the earlier chapters of this book. In the sections that follow, we are going to look at how proteins can be produced in both bacterial and eukaryotic cells, focusing on cases where the proteins are needed in reasonably large amounts and in an active form. The issue of how these proteins are then extracted and purified from the organisms in which they are expressed will not be covered here in detail, but we will return to it briefly at the end of this chapter.

9.2 Requirements for Protein Production From Cloned Genes

The basic concept of protein production in a heterologous organism (one which is not where the gene for the protein first originated) is simple. The aim is usually to get the organism to produce as much of that protein as possible. To do this, the processes which lead to the production of the protein (transcription and translation) must be made to occur at as high a level as possible. The gene for the protein has to be placed downstream from a strong promoter, to maximize the amount of mRNA which is produced for translation into protein. Ideally, there will be many copies of this gene present in each cell, which will increase the amounts of mRNA and hence of protein which are produced. This is done by placing the gene and its promoter into a suitable vector, typically a plasmid. The protein needs to be produced in such a way that it does not harm the host, at least until high levels of the protein have accumulated. Finally, the protein ideally needs to be in a soluble, active form which can be easily purified away from the other components of the organism, such as other proteins, lipids, metabolites and nucleic acids, all of which might interfere with the uses to which the protein is going to be put. The ways in which these outcomes are achieved will be considered in the sections that follow.

9.3 The Use of *E. coli* as a Host Organism for Protein Production

The commonest organism used for the expression of proteins is *E. coli*. Many of the issues that arise with getting high protein expression from *E. coli* can also occur with other expression systems, and you should bear this in mind while reading the description of protein production in *E. coli* that follows. *E. coli* is an easy organism to grow: it grows rapidly and in very large numbers in fairly cheap media. The strains used for protein production are not harmful to humans or to the environment – indeed, most of them have been grown in the laboratory for so long that they would struggle to survive outside the laboratory or production plant. Most importantly, a large number of biological "tools" have been developed for maximizing the efficiency of protein production in *E. coli*, and it is these that we will discuss now.

Vectors for gene expression

For regulated protein expression to be possible, plasmid vectors are needed that contain the key features of cloning plasmids, plus a suitably strong promoter for expressing the gene of interest. There must also be a place where the gene of interest can be ligated, downstream from the promoter. Often, this will be a multiple cloning site (also called a polylinker), which consists of a DNA sequence where the restriction sites for several enzymes which cut only at this site in the plasmid are present very close to each other (see Section 3.7). Promoters are the most important feature of these vectors, and

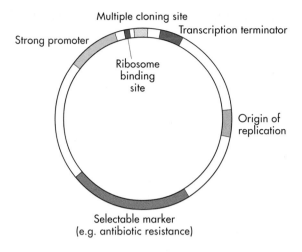

Figure 9.1 The key features of a typical vector for protein production.

are discussed in detail in the next section. Several other features important in these plasmids, which are often referred to as expression vectors, are summarized in Figure 9.1 and discussed in more detail below.

Promoters for expression
Protein production requires the use of a promoter which can be used to drive expression of the gene to high levels. This promoter will be upstream of the gene of interest in a suitable plasmid vector, such that it directs transcription of the gene which encodes the protein of interest. A number of different promoter systems are widely used in *E. coli*, with the choice for any particular experiment depending on a variety of factors. The two most important features of a promoter are its strength and the degree to which expression from it can be regulated. The first is important in achieving high levels of expression: in general, the stronger the promoter, the more mRNA will be synthesized, and the more protein will result from the translation of this mRNA. The ability to control expression is also important, however, particularly as it is often the case that high levels of expression of a heterologous protein can lead to a decrease in growth of *E. coli*, or even death. Thus the ideal promoter for most expression experiments is one which is strong, but easy to turn down or off.

The simplest examples of promoters for protein expression are those which are derived from operons of *E. coli*, with the *lac* promoter of the *lac* operon, and the *ara* promoter from the arabinose operon (also sometimes referred to as the pBAD promoter), being widely used (Figure 9.2). You may already know that expression from the *lac* promoter is regulated by the *lac* repressor protein, LacI (see Chapter 11 for more information on regulation of gene expression). In the absence of lactose LacI is bound to the DNA,

(a) The *lac* promoter

(i) Glucose absent, IPTG present: promoter ON

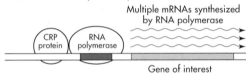

(ii) Glucose present, IPTG absent: promoter WEAKLY ON

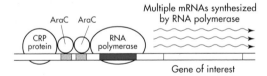

(b) The *ara* (or pBAD) promoter

(i) Glucose absent, arabinose present: promoter ON

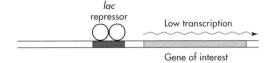

(ii) Glucose present, arabinose absent: promoter OFF

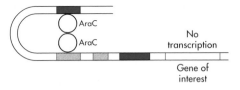

Figure 9.2 The *lac* and *ara* promoters. The diagram illustrates their modes of action when on and off. CRP protein: a positive activator that binds in the absence of glucose.

and hence prevents expression of the genes of the *lac* operon in the absence of lactose, which encode proteins for lactose uptake and cleavage. The *lac* promoter DNA which contains the binding sites for RNA polymerase can be cloned upstream of any gene, thus placing the expression of those genes under the control of the signals which normally direct *lac* gene expression. Expression from the *lac* promoter is usually induced experimentally by the gratuitous inducer IPTG (Box 3.4) which prevents repression of the *lac* promoter but is not itself metabolized by the products of the *lac* operon, and so does not have to be continuously replenished in the medium as the cells grow. Maximum activity of the *lac* promoter also requires the binding of another protein, called the CRP or CAP protein, which activates expression from the *lac* promoter but only binds in the absence of glucose. As we discussed in Section 3.9, many plasmid vectors exist which have the *lac* promoter in them upstream from a stretch of DNA with a large number of restriction sites in it, into which DNA encoding any gene can be inserted. If these vectors are placed in a host strain of *E. coli* which has a functional *lac* repressor, expression of these genes will be fairly low until IPTG is added, whereupon expression will reach maximal levels.

Although it is widely used, the *lac* promoter is not always the best choice for expression of a heterologous protein in *E. coli*. First, it is weaker than some other promoters, and so maximal levels of protein expression will not be achievable if it is used. Second, it is rather "leaky", which means simply that even under conditions where no IPTG or lactose is present in the medium, a fair amount of transcription can still occur from this promoter. This is not necessarily a problem. However, certain proteins are toxic to *E. coli*, which cannot tolerate their presence for long. In these cases, it is usually best to keep expression of the gene encoding them switched off until the cells have grown to a fairly high density, and then turn expression of the gene on, to cause protein production, for an hour or two. The leakiness of the *lac* promoter can be reduced to a certain degree by over-expressing the *lac* repressor, but even this will not always reduce the level of expression from the promoter to an acceptably low level. An alternative approach is to use the *ara* promoter for gene expression. This is the promoter for the *E. coli ara* operon which encodes the proteins for arabinose utilization. This promoter is regulated by the AraC protein, which has the interesting property that it can act both as a repressor and an activator of gene expression. When cells are grown in the presence of glucose, it represses expression from the *ara* promoter to a very low level. However, if arabinose is introduced into the growth medium, AraC switches to being an activator, and high levels of expression occur from the *ara* promoter. As is the case with the *lac* promoter, high levels of expression also require the binding of CRP protein, and hence only occur in the absence of glucose.

Many other promoter systems have been developed over the years for use in *E. coli*. For example, the *tac* promoter is an artificial promoter consisting of DNA sequences from both the *trp* and the *lac* promoters (the *trp* promoter is upstream from the *E. coli trp* operon, which produces the enzymes needed for the biosynthesis of tryptophan). To understand why this promoter was constructed, it is important to understand something about the sequences which determine the activity of a promoter; this is

```
lac:
CCAGGCTTTACACTTTATGCTTCCGGCTCGTATAATGTGTGGA

trp:
GAGCTGTTGACAATTAATCATCGAACTAGTTAACTAGTACGCA

tac:
GAGCTGTTGACAATTAATCATCGGCTCGTATAATGTGTGGAA
```

Figure 9.3 Sequences of *lac*, *trp*, and *tac* promoters. The underlined regions are the −35 and −10 sequences which are important for promoter strength.

covered in Section 11.1 and Figure 11.1. Essentially the *tac* promoter combines the "best" features of the *lac* and *trp* promoters, namely, an ideal −10 and −35 sequence (see Figure 9.3). The *tac* promoter is still repressed by the *lac* repressor (since it still contains the *lac* repressor binding site) but is stronger than the *lac* promoter and hence gives higher level expression of the genes that are downstream from it when derepressed by the addition of IPTG.

Not all the promoters used in *E. coli* are derived from *E. coli* operons. Some of the most useful are those derived from promoters of genes expressed by bacteriophage which infect *E. coli*, since these promoters have evolved to direct very high levels of gene expression. Two examples, shown in Figure 9.4, will illustrate different useful features of these promoters.

The P_L promoter from bacteriophage λ is a promoter which is highly active very early in the life cycle of bacteriophage λ, and which is normally repressed by a protein called the cI repressor protein (which is also encoded by bacteriophage λ). A useful genetic variant of the cI protein is called cI857, a temperature sensitive mutant which is active as a repressor at 37°C but not at 43°C. Thus, genes for expression can be placed under the control of a P_L promoter in a suitable vector which, when present in cells that contain the cI857 repressor protein, will give rise to very little protein at 37°C. When the cells are shifted in their growth temperature to 43°C, the cI857 repressor protein is inactivated and high level expression from the P_L promoter ensues (see Figure 9.4a).

Q9.2. What do you think would happen to the expression of a gene under the control of the λ P_L promoter if the plasmid it was on was transformed into a cell that did not contain a gene for the λ repressor protein cI?

The P_L promoter from bacteriophage λ uses the RNA polymerase encoded by *E. coli* to initiate mRNA synthesis. However, some bacterio-

(a) The λP$_L$ system

37°C: promoter OFF

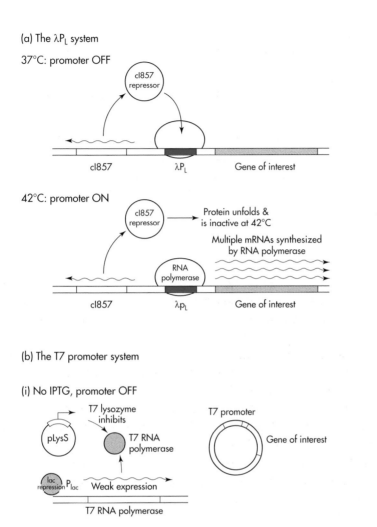

42°C: promoter ON

(b) The T7 promoter system

(i) No IPTG, promoter OFF

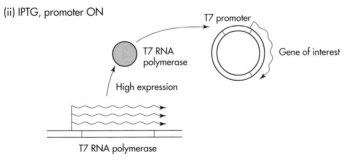

(ii) IPTG, promoter ON

Figure 9.4 The λP$_L$ (a) and T7 (b) promoter systems for expression.

phage (such as T7) produce their own RNA polymerases, which recognize their promoters but not those of *E. coli* genes. Because these are very strong promoters, which are only active when the bacteriophage RNA polymerase is present, this gives us another useful way of tightly regulating promoter expression. In such a system, the gene of interest is cloned downstream of the bacteriophage promoter in a suitable vector. In the absence of the RNA polymerase encoded by the same bacteriophage, this promoter will be inactive. The vector is then introduced into a strain of *E. coli* which has a copy of the gene for the bacteriophage RNA polymerase integrated into its chromosome, itself under the control of a *lac* promoter. As long as the strain is grown under conditions where the bacteriophage RNA polymerase is not present, or present at low amounts, the cloned gene will not be transcribed. If IPTG is added to the culture, the *lac* promoter is derepressed, T7 RNA polymerase is synthesized, and the gene is highly expressed (Figure 9.4b). This gives us the ability to regulate expression from the T7 RNA polymerase promoter by adding different levels of IPTG to cells. However, in the case of very toxic proteins, even the low level of expression of the T7 RNA polymerase that can occur from the *lac* promoter in the absence of IPTG due to its inherent leakiness can be too high. To cope with such proteins, the expression system can be improved by including another plasmid called pLysS which encodes T7 lysozyme, which has the property of inhibiting the T7 RNA polymerase. Thus, until the T7 RNA polymerase expression is increased by adding IPTG, expression from the T7 promoter is extremely low.

To conclude this section, it is worth mentioning another couple of features that are often important when considering the design of expression vectors. First, although strong expression of the cloned gene is needed for high level protein production, the use of strong promoters can cause problems for plasmids. This is because transcription beyond the cloned gene can run into the region controlling plasmid stability and replication, which often leads to loss of the plasmid. For this reason, expression vectors usually have a region of DNA, referred to as a transcriptional terminator, downstream from the multiple cloning site where the cloned gene will be inserted. RNA polymerase reaching this site ceases transcription, so only the cloned gene itself will be transcribed from the strong promoter. Second, in order to get high levels of protein, it is important not only to produce large amounts of mRNA (from the strong promoter) but also that the mRNA is efficiently translated into protein. Good translation of mRNA in *E. coli* and other organisms requires the presence of a sequence which can be recognized by the ribosomes, shortly upstream of the ATG which will encode the first methionine residue of the translated protein (Figure 8.1). This site, referred to as a ribosome binding site, or (after its discoverers) as a Shine–Dalgarno sequence, is present in many expression vectors shortly upstream of the multiple cloning site that will be used to insert the cloned gene. A "perfect" ribosome binding site is completely complementary to the sequence at the 3′ end of the 16S rRNA of the ribosome, AUUCCUCCA. The distance between the final complementary T residue in the DNA and the A of the ATG initiator codon is quite critical; a spacing of five bases is optimal. Good expression can be obtained from promoters with ribosome binding sites which are not a perfect match to the 16S rRNA, and the spacing can vary somewhat, but if expression of a cloned gene is low, checking the sequence and distance from the ATG of the ribosome binding site is always good practice.

Q9.3. List the key features of a plasmid used for expression.

Monitoring protein expression
A method is needed, when attempting to over-express a protein using methods such as those described here, to monitor whether or not expression is occurring. The simplest way to do this, as long as the protein is expressed at a sufficiently high level, is to prepare crude extracts of total protein from the expressing cells at different times after induction of expression, and analyze them by SDS PAGE electrophoresis (Box 5.1). A typical gel of such an induction profile, with clear evidence of a new protein being synthesized after induction, is shown in Figure 9.5. Note that only the band corresponding to the induced protein increases in intensity after induction, showing that the amounts of the other proteins detected on this gel are not altered during the experiment.

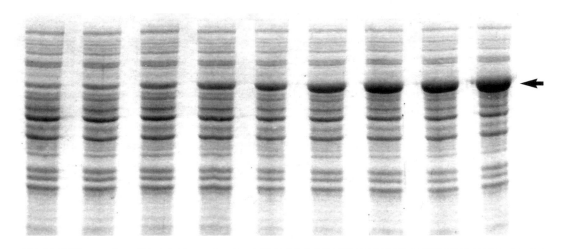

Figure 9.5 Protein profile of an induced protein on an SDS PAGE gel after induction. Protein samples were taken at equal time intervals from a culture expressing a protein under the control of the *lac* promoter, after the addition of IPTG. The induced protein is shown by the arrow.

9.4 Some Problems in Obtaining High Level Production of Proteins in *E. coli*

Although the basic idea of expressing genes at a high level in *E. coli* is straightforward, there are many pitfalls that can occur between obtaining the cloned gene of interest and producing large amounts of protein from it. Understanding how to deal with these often requires an understanding of the underlying biology that causes the problem in the first place. Problems can arise in essentially three areas: the protein may not be produced at high levels, or it may be produced at high levels but in an inactive form, or it may be produced but is toxic to the cells. Toxicity, as discussed in Section 9.3, can be dealt with by keeping the promoter for the gene which encodes the toxic protein fully repressed until cells have grown to a high concentration and then inducing expression for a short period of time to build up a high level of protein before the cells die or become damaged. The other two problems have several possible origins, and it is these that we will discuss now.

Low levels of protein

When the gene for a particular protein is highly expressed in *E. coli*, levels of the protein can reach astonishingly high levels, to the extent of even becoming the most abundant protein in the host. However, it is sometimes the case that even with a protein encoded by a gene which is expressed from a strong promoter with a good ribosome binding site on a high copy number plasmid, final levels of the protein are found to be disappointing. A number of things can cause this, but two in particular are worthy of note and are easily remedied.

First, the gene may have been cloned from an organism with a different codon usage to that of *E. coli*. (For a discussion of codon usage, see Box 8.1). The levels of tRNAs which recognize rare codons are low within the cell. If a gene from an organism which frequently uses codons which are rare in *E. coli* is introduced into *E. coli*, it may be poorly expressed simply because the tRNAs required for its efficient translation are present in low amounts. There are two potential solutions to this problem.

One is to alter the rare codons in the cloned gene to more commonly used ones, using site-directed mutagenesis (Section 10.5). This can be very effective, but the drawback with this method is that if a particular gene has a large number of rare codons within it, it quickly becomes very laborious. The other solution is to use a strain of *E. coli* which has been specifically engineered to express the rare tRNAs at higher levels. Such strains are commercially available, and can be remarkably successful at increasing the yields of protein from genes with rare codons.

> **Q9.4.** Would either of these methods alter the amino acid sequence of the final protein produced?

Second, the problem may not be one of insufficient translation, but one of too efficient turnover. The final amount of any protein in the cell depends upon the rate at which the protein is synthesized, and the rate at which it is degraded. Increasing the latter has the same effect as decreasing the former. As an analogy, imagine a bath filling up with water. The level of the water depends on the rate at which water enters the bath through the taps and the rate at which it drains out through the plug-hole, so if we want to increase the amount of water in the bath we can either turn the taps up more, or put the plug in. Increasing promoter strength is like turning the taps on more. *E. coli* has within it a number of proteases which recognize and degrade proteins, and the activity of these may in some cases lead to low overall levels of a particular protein in the cell. So to decrease the rate of protein degradation (equivalent to putting the plug in the bath) we can use a strain of *E. coli* where the genes for the proteases have been deleted. All the major protease genes in *E. coli* have been identified, and strains where the genes for them are deleted retain viability, so this is a very feasible option.

Insolubility of the expressed protein

Another problem which can arise when trying to obtain maximum amounts of protein is that although a clear band is seen on a protein gel when the protein is over-expressed, it turns out subsequently that the protein when purified is inactive. This is usually because it has not reached its correct folded form. Often it is the case that an over-expressed protein forms a dense aggregate of inactive protein which is poorly soluble; such aggregates are referred to as inclusion bodies (see Figure 9.6).

Why does over-expression often lead to protein aggregation, and what can be done about it? The reasons why some proteins and not others form inclusion bodies when over-expressed are not fully understood, but it is known that the same forces that drive protein folding also lead to protein aggregation. These forces are principally hydrophobic ones, and arise because of the strong tendency for the hydrophobic side chains of certain amino acids to interact with each other rather than with water. Under normal circumstances this leads to these amino acids orientating their side chains towards the interior of the protein, and is a key step in the folding of proteins, but when protein concentrations are artificially high (as in an over-expression experiment) the side chains on different protein molecules may interact, leading to aggregation and insolubility.

In some cases, the fact that the protein is produced as an insoluble lump is actually an advantage. This is because it can considerably simplify the purification of the protein, which will be dense and easily collected by centrifugation after growing up cells and breaking them open. The problem then becomes one of getting the protein into its active form, which has to be done by denaturing the insoluble protein and then letting it refold

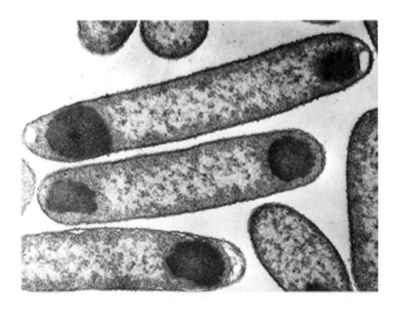

Figure 9.6 Inclusion bodies in *E. coli*, viewed by transmission electron microscopy (TEM). Image courtesy of Professor Jonathan King, MIT.

under carefully controlled conditions that allow it to reach its native and active state. Many commercially produced proteins are in fact made in this way. However, the refolding process can be difficult and expensive, and sometimes it is preferable to find a way of getting the protein soluble and correctly folded while still inside the cell. Sometimes the solution to this is straightforward: the problem arises from very high protein expression, so the answer is to lower protein expression levels. This may be done by using a weaker promoter, a lower copy number plasmid vector, or simply by growing the strains at a lower temperature or in a less rich medium. While all these have the undesirable effect of also decreasing yield, it is better to have a small amount of active protein than a large amount of inactive protein. However, other approaches can also be used.

One approach to the inclusion body problem is to manipulate the folding capacity of the cell. *E. coli* (and all other organisms) contains a number of proteins called molecular chaperones whose role it is to assist protein folding in various ways. By co-expressing these molecular chaperones at higher levels when the heterologous proteins are being produced, in some cases the formation of inclusion bodies is lowered or even stops altogether, without any decrease in the level of expression of the protein concerned.

Another approach which is sometimes successful is to change the gene for the protein by adding to it some codons which code, in the same reading frame as the protein of interest, for a part of another protein which is very soluble. This is a specific example of a more general method, which has many applications in biological research, of fusing the genes for two proteins or parts of proteins together so that they are translated as a single polypeptide. These are called translational fusions, and are described in Box 9.1. When transcribed and translated this will produce a protein called a hybrid or fusion protein which contains both the protein of interest plus the sequence for the other protein. Remarkably, if the added protein is very soluble, it often makes the fusion protein soluble as well, thus avoiding the problem of inclusion body formation.

A potential problem with fusion proteins of the type described here and in Box 9.1 is that once the protein is expressed and purified, it now has some extra amino acids attached to it, which may alter the properties of the protein in undesirable ways. One method which is often used to deal with this problem is to incorporate codons for a sequence of amino acids between the protein of interest and the protein or peptide that it is fused to, the sequence being recognized by a protease which can cleave the peptide bond. The most widely used protease is factor Xa, a protease whose normal role is in the blood coagulation cascade. Factor Xa recognizes the amino acid sequence Ile-Glu-Gly-Arg and cleaves the peptide bond which is on the C-terminal side of the arginine residue. Thus if codons for these four amino acids are included between the protein of interest and the fusion partner, the two can be separated by factor Xa cleavage after purification.

> **Q9.5.** If you wanted to obtain a protein identical in sequence to its naturally occurring form, and had to do this by expressing it initially as a fusion protein with a factor Xa cleavage site separating it from its cleavage partner, would the fusion partner need to be at the N-terminal or the C-terminal end of the protein?

Another problem that can arise when expressing heterologous proteins, and can lead either to inclusion body formation or to inactive proteins, is the protein folding in the incorrect redox state. Proteins which are normally expressed in an environment which is oxidizing will tend to form disulfide bonds between cysteine residues, and these bonds are of great importance in determining the ways in which the protein folds and, hence, its activity and stability. Examples of proteins which are highly likely to form cysteine bonds are proteins in eukaryotes which are synthesized on ribosomes bound to the rough endoplasmic reticulum, and which will eventually be secreted outside the cell or inserted into the plasma membrane. The bacterial cytoplasm, however, is a reducing environment where disulfide bonds cannot generally form. Thus those eukaryotic proteins which depend on disulfide bond formation for their activity are rarely active if expressed in the cytoplasm of *E. coli*.

Again, two approaches have been taken to deal with this problem. The

Box 9.1 Translational Fusions and Protein Tags

One of the great powers of recombinant DNA methods is that they allow you to construct novel genes that encode proteins which do not occur in nature. An example of this is the construction of so-called translational fusions. In a translational fusion the promoter, ribosome binding site, start codon and part of the coding sequence from one gene are fused in the same reading frame to all or part of another protein. It is important that the proteins are fused so that the coding sequences are in the same reading frame otherwise the second protein will not be translated. When this new gene is expressed, it will be translated into a polypeptide chain containing the encoded amino acids from the two original proteins, and remarkably this will often fold up into a protein that now has properties derived from both of the starting proteins. Translational fusions have a range of applications including fusion to reporter proteins for analyzing gene expression (Section 10.2), adding short stretches of amino acids to proteins to make them easier to purify, and fusing them with soluble partners to increase solubility.

Adding "tags" (short stretches of amino acids from one protein) onto the N-terminal or C-terminal ends of another protein can be useful in a number of situations. First, it is often the case that you wish to identify a protein in a complex mixture, but you have no way of picking out that particular protein. However, if the protein is expressed as a translational fusion with part of another protein for which you have an antibody available, the new protein will now be recognized by the antibody. This process is called "epitope tagging", an epitope being the name given to the part of the protein which is recognized by the antibody. A commonly used "tag" is a domain from a human gene called "myc", for which there are highly specific antibodies available, which importantly will not cross-react with other proteins in an *E. coli* or yeast cell extract. Second, many proteins are insoluble when expressed at high levels in a heterologous host such as *E. coli*. In many cases, this problem can be solved by tagging the protein with a domain from a highly soluble protein such as maltose binding protein (MBP). Finally, tags can be used to help purify the protein. An example of this is the use of "his" tags. Short stretches of DNA that code for several histidine residues can be fused to the N or C terminus of the protein of interest, and this will often enable the protein to be purified in one or a few steps on a column with a high affinity for histidine residues, as described in the final section in this chapter.

Often the fusion is constructed in such a way that the fused amino acids can be cleaved off from the other protein after purification has taken place, for example with a specific protease. This step, however, can be problematical as it may be difficult to remove all of the tag amino acids.

first is to direct the protein to a compartment in *E. coli* where disulfide bond formation does occur, namely the bacterial periplasm (i.e. the area between the inner and outer membranes of the bacterial cell). Unlike the cytoplasm, the periplasm is more or less in equilibrium with the medium in which the organism is being grown, and this will usually be oxidizing, thus allowing disulfide bonds to form. Moreover, the bacterial periplasm (like the eukaryotic endoplasmic reticulum) contains enzymes which help the formation of disulfide bonds. As discussed in Box 9.2, selected proteins made in *E. coli* are transported across the inner membrane to the periplasm. These proteins are identified as being destined for the periplasm by the presence of an N terminal "signal sequence". Fusing the codons for the signal sequence from a typical bacterial periplasmic protein to the gene for a heterologous protein will thus cause that protein to be directed to the periplasm, where disulfide bond formation can occur. Using precisely this approach, several eukaryotic proteins which contain large numbers of disulfide bonds have been successfully synthesized in their active form in *E. coli*.

The second approach requires an ingenious manipulation of *E. coli* to bring about a change in the redox state of its cytoplasm. The details of this are beyond the scope of this book, but essentially the *E. coli* cytoplasm is normally reducing (meaning that the side chains on the cysteines in proteins will be in the reduced thiol form, -SH) due to the effect of two different pathways which can be removed by mutating particular genes. When this is done, the cytoplasm becomes more oxidizing, and so cysteines can then form disulfide bonds even when the protein is in the cytoplasm.

The points above summarize just some of the issues that can arise when attempting to maximize protein expression in *E. coli*, and how to deal with them. These are also shown in flow diagram format in Figure 9.7.

9.5 Beyond *E. coli*: Protein Expression in Eukaryotic Systems

Versatile and powerful though *E. coli* is for expressing proteins, it does have limitations. Its major drawback arises from the fact that the properties of proteins often depend not only on the primary amino acid sequence of the protein but also on modifications to the protein which are made to it after it has been translated. This is particularly true of eukaryotic proteins, where the problem is that the modifications made are often ones which cannot be made in *E. coli*, simply because it lacks the enzymes to do them.

Probably the most significant post-translational modification is N-linked glycosylation, which involves the addition of chains of specific sugar molecules to specific asparagine residues in certain proteins. It takes place in the endoplasmic reticulum in eukaryotic cells. The chains are further modified, often very significantly, by the action of different enzymes in the endoplasmic reticulum and in the Golgi body (Figure 9.8). The molecular mass of these chains (referred to as glycans) can be quite large, accounting

Box 9.2 Signal Sequences and Protein Secretion

In prokaryotes, all proteins are synthesized in the cytoplasm; in eukaryotes they are synthesized either in the cytoplasm or on ribosomes attached to the membrane of the endoplasmic reticulum. (A small number are also synthesized in chloroplasts or mitochondria.) Many finish up elsewhere in the cell, or outside the cell altogether. The story of how proteins are "targeted" to their different final destinations is a long and fascinating one (too long to tell here!), but it is important to realize its importance. The different locations that proteins move to can have profound effects on their properties, because of the formation of disulfide bonds, for example, or the attachment of large sugar chains (glycosylation). One of the major ways that proteins are recognized by the cellular machinery which enables them to cross membranes is by the possession of an amino acid sequence at their N terminus, called a signal sequence. Signal sequences are generally cleaved after the proteins have crossed the membrane, but if the signal sequence is not present when they are translated, or is not recognized by the host organism, the protein will not be secreted out of the cytoplasm. Fortunately, in exactly the same way as it is possible to tag proteins as described in Box 9.1, it is also possible to remove the codons for one signal sequence and replace them with codons for a different one, or even to add the codons for the signal sequence of one protein to the coding sequence of a protein which is not normally secreted. There is no single sequence of amino acids for all signal sequences, rather the amino acids in them tend to follow certain patterns, such as a cluster of positively charged amino acids at the N terminus. The signal sequence system is quite conserved throughout all organisms. In bacteria, the signal sequence directs proteins across the cytoplasmic membrane, into the periplasm in the case of Gram negative bacteria or outside the cell for Gram positives. In eukaryotes, a closely related signal targets proteins to the ribosomes bound to the endoplasmic reticulum, which is the first stage of a journey that may lead the proteins to the cell membrane or to be secreted from the cell. Generally, signal sequences work best when they come from the organism where the protein is being expressed. For example, if we wished to clone and express in *E. coli* a human protein that is normally secreted, the best approach would usually be to remove the codons for the human signal sequence and replace them with codons for the signal sequence of an *E. coli* protein which itself is usually secreted.

for a substantial proportion of the final molecular mass of the protein, and unsurprisingly, the presence of glycans can substantially affect the properties of the protein in different ways. These include having an effect on whether or not the protein can fold correctly, whether the protein is active, the way in which the protein is recognized by the immune system, and the

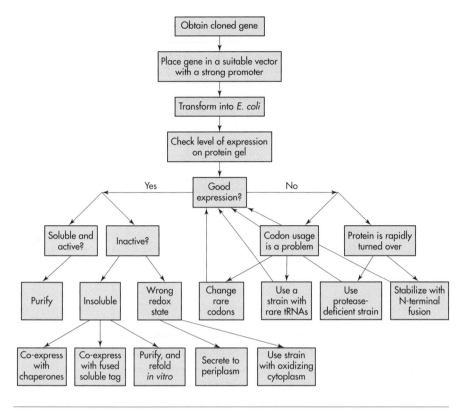

Figure 9.7 Flow diagram for trouble shooting protein expression in E. coli.

length of time that the protein circulates in the blood stream. All of these issues are obviously very important when considering proteins for pharmaceutical use. Thus it is often the case that for a protein to be studied or utilized, its glycosylation status needs to be the same as it would be in its normal cellular environment. Production in eukaryotic cells is currently the only way to achieve this. Indeed, if mammalian proteins are being studied or used, production ideally also needs to be from mammalian cells, since other organisms produce glycans which are significantly different to those of mammals.

Ideally, we would like to have a system for growing eukaryotic cells to high densities at low cost and with minimal risk of infection, these cells having been manipulated in such a way as to produce proteins whose properties are identical to those which they have in their normal host organism. In practice, it is not possible to achieve all these goals, and in deciding how to go about production of proteins in a eukaryotic system, a judgement must be made about which is the best system to use, taking into account all the issues above. In the rest of this chapter, we will look at some

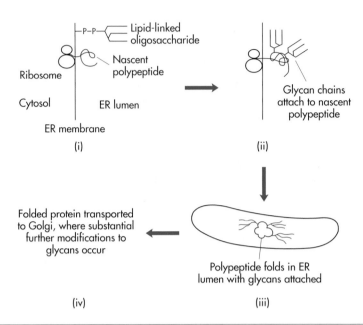

Figure 9.8 Glycosylation in a eukaryotic cell. The diagram shows the stages of glycosylation and where they occur in the eukaryotic cell.

selected examples of systems that have been devised to produce proteins in eukaryotic cells, and mention the advantages and drawbacks of each.

Expression in yeast

Yeast species have many parallels to *E. coli* when it comes to protein production. They are genetically easy to manipulate, and can be grown quickly and cheaply to high cell densities. They thus have many of the advantages of *E. coli* for protein production, with the further advantage that as eukaryotes they can be better suited for the production of eukaryotic proteins. In addition, some yeast strains are very efficient at producing large amounts of protein and secreting it into the growth medium, which simplifies the harvesting and purification of the protein.

Saccharomyces cerevisiae, the common baker's yeast, is the yeast species most widely used for protein production. A particular advantage of this strain is not scientific as such but regulatory: as humans have been eating and drinking this organism for millennia (in bread, beer and wine) it is regarded as safe and so there are fewer hurdles to overcome when trying to establish the safety of a new product purified from it. Vectors for protein expression are widely available, most being based on a small multi-copy plasmid (called the 2μ plasmid) which is naturally occurring in this organism. Such vectors are called YEps (for yeast episomal plasmids), and an example of one of these is shown in Figure 9.9. Selection for the presence

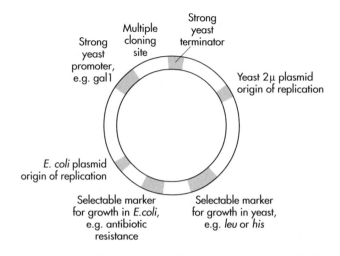

Figure 9.9 YEp for protein expression in yeast.

of the vector is generally done by using an auxotrophic mutant as the host strain (that is, one which requires a particular nutrient to be added to the growth medium in order to grow). A gene for synthesizing the essential nutrient is placed onto the YEp, and the host is then grown in a medium lacking the nutrient concerned. Only those cells which contain the plasmid can then grow. For example, the host may engineered to be Leu⁻, and hence unable to grow in a medium lacking leucine. A YEp with a functional *leu* gene on it will enable growth of this host in the absence of leucine in the growth medium, but if the plasmid is lost (and YEps do tend to be unstable when yeast cells are grown to high density) then the cell that has lost the plasmid will no longer be able to grow. In this way, a culture can be obtained where the growing cells all contain the YEp.

Plasmids engineered for protein expression in yeast (and indeed all eukaryotic hosts) are so-called shuttle plasmids, which are able to grow both in the host species and in *E. coli*, so that the gene cloning and manipulations required for high expression can be done in *E. coli* before the plasmid is transferred to the eukaryotic host. This is achieved by engineering in a region from a suitable *E. coli* plasmid that possesses an origin of replication and a selectable marker (usually an antibiotic resistance gene), enabling maintenance of the plasmid in *E. coli* (Figure 9.9). To get high expression of the cloned gene, a variety of strong yeast promoters have been identified. These include strong constitutive promoters such as the triose phosphate isomerase (TPI) promoter and the alcohol dehydrogenase I (ADHI) promoter, as well as promoters which can be induced, such as the *gal*-1 promoter, which is induced by the addition of galactose to the medium.

Other yeast species that are able to produce higher levels of protein than *S. cerevisiae* are becoming more widely used. These include the yeast *Pichia pastoris*, which has a number of advantages over *S. cerevisiae*. First, it can be grown to higher cell densities in culture and in fermentors, potentially increasing protein yield. Second, it can secrete proteins to much higher levels than *S. cerevisiae*, with protein concentration sometimes reaching several grams per liter of culture medium. Third, it has a very good regulated promoter system available. This is the promoter of the alcohol oxidase gene AOX1, which is induced when methanol is added to the culture, but strongly repressed in its absence. This enables the expression in *P. pastoris* of proteins which are deleterious, by growth to relatively highly cell densities followed by the addition of methanol.

A problem that exists with using yeast strains to express mammalian proteins is that although yeasts do glycosylate their proteins, the nature of the glycans used is different (significantly so in some cases, including that of *S. cerevisiae*) from those which are attached to proteins in mammalian systems. As we have already discussed, this can be very undesirable, leading to proteins which are rapidly removed from the host, or are inactive or active in an inappropriate way. Recently, significant progress has been made in "humanizing" the glycosylation pathways of *P. pastoris*, by deleting the yeast genes involved in modification of glycans and replacing these with the relevant genes from humans. This is a complex and challenging process, not yet complete, but we may yet see the development of strains of yeast which can produce heterologous proteins with highly defined and mammalian type glycans attached to them.

Expression in mammalian cell systems

At the start of Section 9.3, the statement was made that *E. coli* was the commonest organism used for protein expression. While this remains true, it is interesting to note that the majority of proteins produced for pharmaceutical purposes that have been licensed for clinical use are actually produced using cultured mammalian cells. The major reason for this is that many of the proteins with pharmaceutical uses are large, complex proteins with many disulfide bonds and with glycans attached. Neither *E. coli* nor yeast is yet able to produce these proteins in a state suitable for use. So despite the fact that there are some drawbacks with using cultured mammalian cells, this remains a highly important method for producing recombinant proteins.

Cells from rodent species have the interesting property that when grown in culture, most eventually become senescent and die but a small proportion continue to grow indefinitely: these are referred to as being immortalized and are widely used for production of recombinant proteins. Several different "lines" of cells are used, which have been established from different rodent tissues, the commonest being a cell line established from the ovary of a Chinese hamster, called CHO cells. These cells can be grown in appropriate culture medium to a relatively high density (though much lower than yeast or bacterial cells), and they can also take up and stably maintain foreign DNA, a pre-requisite for using them for heterologous protein production.

As with all expression systems, the nature of the expression vector is all-important. Essentially two types of vectors are used in mammalian cell culture. Mammalian cells do not possess naturally occurring plasmids, but they are susceptible to infection by viruses, some of which replicate as free circles in the cell nucleus, and vectors based on such viruses are often used. However, they tend to be unstable, and for this reason the use of vectors which allow the stable integration of DNA onto the chromosomes, which are hence passed to daughter cells on cell division in the same way as any other gene, is more favored. The drawback with these vectors is that the level of expression obtained from them can vary enormously depending on exactly where in the genome of the host cell the vector integrates. Integration is mostly by recombination between vector and host DNA that does not require DNA homology – in other words, it is relatively random rather than targeted at specific sites in the DNA. The influence that the site of integration has on expression is referred to as the position effect, and arises principally from the fact that different parts of the genome are transcribed to very different levels, with some areas being effectively transcriptionally silent. Genes inserted into these areas, even if they are under the control of strong promoters, rapidly lose their high levels of expression, due to modifications that take place in the chromatin (i.e. the DNA–protein complex that constitutes the chromosomes). The simplest way of surmounting the problem of the position effect is to produce a large number of individually transformed cell lines and screen them each for levels of expression of the gene of interest. This is laborious and time-consuming, and much effort has gone into identifying sequences of DNA which can be added to the vector to improve expression of the gene irrespective of its position of insertion in the genome, and into deliberately targeting the vector into regions of the genome which are transcriptionally active.

A variety of strong promoters have been used to drive expression of introduced genes in mammalian cells. Many of these are derived from viral promoters, since these have often evolved to be very active irrespective of the state of the cell (the priority of many viruses being to maximize transcription of their own genes). The viruses SV40 and cytomegalovirus are two examples of viruses whose promoters have been widely used for expression purposes. Eukaryotic promoters are also used, either for strong constitutive expression (for example, the promoter for the gene ubiquitin) or for regulated expression for cases where prolonged expression of a foreign protein is deleterious. Examples of regulated promoters which have been used for protein expression include promoters for heat shock genes (turned on by an increase in temperature) and for ferritin genes (turned on by the presence of iron ions in the culture medium). As you know, eukaryotic genes mostly contain introns which, if expressed in a eukaryotic cell, will be spliced out of the message before the message is transferred to the cytoplasm for translation. Most heterologous proteins are expressed from cDNA clones which of course lack introns, but as it has been found that more efficient transport of mRNA from the nucleus is seen with transcripts that have been through a splicing event, one feature that can be added to improve overall expression is an intron upstream of the cDNA coding sequence but within the mRNA that is produced from the vector.

The requirements for obtaining a mammalian cell line where every cell contains vector DNA expressing the gene for a particular protein are the same as for any other transformation – there needs to be a way to get the DNA into the cell, and a way to select for those cells that have successfully taken up the DNA. Several different methods have been developed to get naked DNA into mammalian cells, including the following:

- Incubating the cells with the transforming DNA in the presence of chemicals that encourage the uptake of DNA by the cells. Calcium phosphate was the first chemical shown to do this, but a variety of others have been shown since also to work, and many companies sell their own proprietary chemicals for this purpose.
- Electroporation: In the presence of DNA, cells can be treated with a brief electrical pulse which transiently opens pores in the membranes and enables DNA molecules to enter the cells.
- Biolistics, or particle bombardment, can be used. Here, small inert particles (typically of gold) are coated with DNA and fired into cells using a "gene gun".
- Lipofection where DNA is enclosed within artificial membranes called liposomes, which are then fused with the cells that are to be transformed.

None of these methods are particularly efficient, so it is essential to have a method for selecting those few cells which have become transformed from the large majority which have not. Several different chemicals are used for selecting transformed mammalian cells. Among the most widely used are:

- Methotrexate (MTX), a compound which is lethal to cells that do not contain a functional dihydrofolate reductase (DHFR) gene. CHO and other mammalian cell lines lacking DHFR are available.
- Methionine sulphoxamine (MSX). Again, this is a compound which is toxic to mammalian cells. Unlike the situation with MTX, no special cell lines are needed for this toxicity to be shown. However, cells can

become resistant to the compound if they express high levels of the enzyme glutamine synthetase, so as above if the gene for this enzyme is incorporated on the plasmid and high levels of MSX resistance selected for, transformants can be obtained where the GS gene – and the gene for the heterologous protein, present on the same piece of DNA – are expressed at high levels.

- G-418, an antibiotic which inhibits protein translation. Resistance to it can be obtained by expressing a gene which encodes a phosphotransferase enzyme which transfers a phosphate from ATP to the compound, inactivating it.
- Blastcidin-S, an antibiotic which inhibits DNA synthesis in mammalian cells. A resistance gene is available which removes an amine group from this compound, rendering it inactive.

Q9.6. How do you suppose MTX selection might work, given that cell lines lacking the endogenous gene for DHFR are available?

Other methods

In this chapter we have considered three types of organism for expression of heterologous proteins, and have flagged up advantages and drawbacks to each. These are not the only hosts in which heterologous proteins are expressed. Another system which is often used is the baculovirus system, where cultured insect cells are infected with vectors derived from the insect-specific baculovirus. High yields of protein can be obtained in such systems, although the cells do not become stably transformed and so the transformation procedure has to be repeated each time protein is needed, and the glycosylation events are not identical to those in mammalian cells. Another route which is developing more for protein production is the generation and use of transgenic animals and plants, where whole organisms are essentially used as bioreactors, an application that will be considered in more detail in Chapter 12.

An enormous amount of research has been done over the last two decades into improving the various systems described above to give better or more tightly regulated expression of proteins, modified in the right way when needed. But such developments need to continue, as none of the systems described above is perfect for all cases (nor is any one method ever likely to provide a universal solution to the problems of protein expression). For bacterial systems, the major issue that can still cause difficulty is proteins which fail to fold correctly inside the bacterial cytoplasm, and work continues on methods to get round this problem. The inability of microbiological hosts, whether prokaryotic or eukaryotic, to glycosylate proteins in the form in which they are found in mammals remains a serious issue particularly for pharmaceutical proteins.

9.6 A Final Word About Protein Purification

Expression of a gene and efficient translation of the expressed mRNA into protein is only the start of the story when it comes to working with proteins. Once high level amounts of the protein have been produced inside cells, the protein must then be purified before it can be studied in detail. This means that the protein of interest has to be separated from all the other components of the cell, including all the other proteins. Protein purification is a complex process requiring a high degree of skill and experience, and protocols for purifying proteins often have to be laboriously devised and tested. Most protocols rely on finding particular properties of the protein in question which can be used to gradually enrich the protein away from all the other proteins and components in the cell. For example, proteins can be separated by size on columns containing gels of the right sort, or by charge by binding them to a column and then gradually washing them off in series of solutions of increasing salt concentration. Often, several of these procedures have to be done one after the other until the protein obtained is highly pure.

More recently, a new range of techniques have come into general use which can speed up the process by cutting down the number of steps necessary, and which are broadly applicable to a range of proteins with very different properties. These techniques all involve adding extra amino acids to the protein that give it a new property, which is to bind very tightly to a particular substance. This binding can then be used to isolate the protein away from all the other materials in the cell which do not bind this substance. The method, called affinity chromatography, is best explained with a specific example.

A common example is the use of "his-tags" to help purify a protein. A his-tag is a string of (usually six) histidine residues which is added to the N or C terminus by adding the nucleotides for these extra his residues when the gene is cloned. Commercial plasmid vectors are available that already contain the codons for this his-tag, making the production of such proteins relatively straightforward. The presence of the extra his-tag on the protein often does not interfere with the ability of the protein to fold correctly and carry out its normal function, but it does make the protein bind very tightly to metal ions such as Ni^{2+}. A commercial resin is available that has Ni^{2+} ions attached to it, so if this resin is placed in a column, and a crude extract of *E. coli* cells that contain the his-tagged protein is passed down the column, the *E. coli* proteins will pass through the column but the protein with the attached his-tag will be bound to the column. If a buffer containing a salt such as imidazole is subsequently passed down the column, the bound his-tagged protein is now released from the nickel ions, and can be collected in the buffer as it flows out of the column. This process is shown diagrammatically, with an example of a protein purified on such a column, in Figure 9.10.

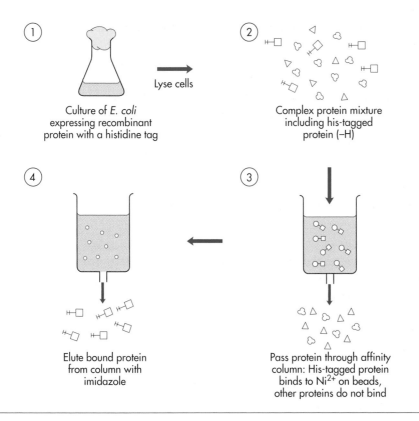

Figure 9.10 Purification of his-tagged protein on an affinity column.

Several different examples exist of the use of "tags" that can be expressed as parts of proteins in this way to help aid protein purification. Obviously such an approach could only be used once the techniques of recombinant DNA technology, coupled with the manipulation of ways of expressing proteins in cells as described in this chapter, became available.

Questions and Answers

Q9.1. Why do you think it is generally best to study proteins in pure form?

A9.1. *Proteins are typically studied using assays which detect some property of the protein, such as catalyzing a particular step in a metabolic pathway, or breaking down ATP. They can also be studied using structural methods, which use various techniques to yield information about the actual structure of the protein. If the protein is not pure, then it is not possible to be sure that the activity seen in the assay is indeed a property of the protein of interest – there may for example be other contaminating proteins in the mix that have a similar activity. If a protein is being studied structurally, and is not pure, the impurities in the mix will make it hard or impossible to carry out the* *study, as they will provide a background signal that will distort or abolish the signal from the protein of interest.*

Q9.2. What do you think would happen to the expression of a gene under the control of the λ P_L promoter if the plasmid it was on was transformed into a cell that did not contain a gene for the λ repressor protein cI?

A9.2. *The gene would be constitutively expressed, irrespective of the temperature used. The λ repressor cI has to be present to repress expression from the λ P_L promoter.*

Q9.3. List the key features of a plasmid used for expression.

A9.3. *Plasmids require an origin of replication, a selectable marker, and a unique restriction enzyme recognition site. Plasmids used specifically for expression will generally contain a strong promoter that can be regulated by the experimenter, and the cloning sites will be downstream from this promoter. A ribosome binding site may be included to improve translation of the mRNA produced from the cloned gene. A terminator will often be present downstream from the cloning sites to prevent read-through transcription from the strong promoter into other genes on the plasmid.*

Q9.4. Would either of these methods alter the amino acid sequence of the final protein produced?

A 9.4. *No, because of the degeneracy of the amino acid code. Several different codons can code for the same amino acid, so changing a codon for a given amino acid from one which is rarely used in a given organism to one which is frequently used, would not change the amino acid which is actually inserted in the polypeptide chain.*

Q9.5. If you wanted to obtain a protein identical in sequence to its naturally occurring form, and had to do this by expressing it initially as a fusion protein with a factor Xa cleavage site separating it from its cleavage partner, would the fusion partner need to be at the N-terminal or the C-terminal end of the protein?

A9.5. *The fusion partner would have to be at the N-terminal end. As factor Xa cleaves at the C-terminal end of the four amino acid sequence, if the fusion partner were at the C-terminal end of the protein then the four amino acid recognition site would remain part of the protein of interest after cleavage.*

Q9.6. How do you suppose MTX selection might work, given that cell lines lacking the endogenous gene for DHFR are available?

A9.6. *To use MTX as a selectable marker, the DNA which is being used for transformation needs to have on it a functional DHFR gene, so that once cells have taken up the new DNA and (assuming it is an integration vector)* *recombined it onto their chromosomes, they now become resistant to the effect of MTX. The degree of resistance seen increases with the level of expression of the DHFR gene, so this can be used to select for lines where the transforming DNA has been integrated into a highly transcribed region in the chromosome.*

Further Reading

Many of the approaches described in this chapter for maximizing protein expression, and overcoming the problems associated with trying to express high levels of protein, have been commercialized. This means that the product literature of the companies selling the particular solution is often an excellent place to start to read. Examples of such literature are provided below in addition to references to normal research papers. We'd better add that their inclusion here does not imply endorsement of these particular products!

The tac Promoter: A Functional Hybrid Derived from the trp and lac Promoters (1983) De Boer HA, Comstock LJ and Vasser M, Proceedings of the National Academy of Sciences of the United States of America, Volume 80 Pages 21–25.
An early and straightforward paper describing the construction of a hybrid strong promoter for use in E. coli.

Tight regulation, modulation, and high-level expression by vectors containing the arabinose pBAD promoter (1995) Guzman LM, Belin D, Carson MJ and Beckwith J, Journal of Bacteriology, Volume 177 Pages 4121–4130.
The first description of the use of the tightly regulated pBAD promoter for protein expression in E. coli.

http://www.emdbiosciences.com/docs/NDIS/inno01-000.pdf
A summary of the system which uses T7 RNA polymerase for tightly controlled expression of proteins in E. coli.

Construction and characterization of a set of *Escherichia coli* strains deficient in all known loci affecting the proteolytic stability of secreted recombinant proteins (1994) Meerman HJ and Georgiou G, Biotechnology, Volume 12 Pages 1107–1110.
A description of the engineering and testing of multiply protease-deficient E. coli *strains, used here for the specific case of secreted recombinant proteins.*

http://www.emdbiosciences.com/docs/NDIS/inno03-003.pdf
One particular system for helping to get soluble protein, by fusion to a soluble partner.

http://www.invitrogen.com/downloads/PichiaBrochureP40.pdf and
http://www.invitrogen.com/content/sfs/brochures/710_021524_pyes_bro.pdf
Two brochures on yeast expression in S. cerevisiae *and* P. pastoris.

http://www.promega.com/vectors/mammalian_express_vectors.htm
Surf to this page for a look at maps and descriptions of the many different types of mammalian expression system sold by one company.

10 Gene Cloning in the Functional Analysis of Proteins

Learning outcomes:

By the end of this chapter you will have an understanding of:

- *why a genetic approach is needed to test predictions derived from sequence analysis*
- *how it can be shown that a given gene is actually transcribed into mRNA*
- *how analysis using techniques such as gene knockouts or RNAi can give insights into the role of a given gene*
- *how properties of a particular protein such as its location in the cell, its topology in a membrane, or the protein partners that it interacts with can be probed in a variety of ways using gene cloning techniques*
- *the technique of site-directed mutagenesis and its application in the analysis of gene and protein function*

10.1 Introduction

Bioinformatics and sequence analysis allow you to develop hypotheses about the presence of a gene, and the nature and role of the protein that the gene encodes, but how do you test these hypotheses? This chapter will focus on the use of cloning methods in analyzing the roles of genes and determining the function, location, and mechanism of action of the proteins that they encode.

The identification of a gene, whether by a genetic screen or through a genome sequencing project, is only the first step in the long journey to determining the role of that gene in the cell. Many different experimental methods can be used to get from detecting the presence of an open reading frame (ORF) in a piece of DNA to an understanding of what the protein encoded by that gene actually does. In this chapter, we will focus on the approaches that involve gene cloning methods, but you should bear in mind throughout that these are only a part of the tool box available to biological scientists.

The first question to be asked about an ORF is, does it actually encode a protein? Once we have answered that question, we may want to look and see whether the protein is important or essential for the cell. We then may want to know more about the protein itself in the context of the cell: where in the cell it finishes up after synthesis, and what other proteins it interacts with. We may also be interested in the detailed mechanism of the protein, down to the level of understanding the role of individual amino acids in the activity of the protein. In the sections that follow we will see how these issues can all be addressed through gene cloning techniques. We only have space to describe a few selected examples of the many methods that are now available, more of which are being added virtually daily.

As more organisms have their complete genome sequence determined, so it is becoming more possible to ask molecular questions of entire genomes, rather than of individual genes. Several of the examples in this chapter will illustrate this point, with descriptions of high throughput projects where large numbers of genes or proteins are studied.

10.2 Analyzing the Expression and Role of Unknown Genes
Searching for expression
The fact that a stretch of DNA has the potential to encode a protein does not mean that it necessarily does so. ORFs can arise in random sequence purely by chance, and genes that were once functional can lose their expression signals and become "relics" in the genome.

> **Q10.1.** DNA in different organisms varies in the relative proportion of A+T vs. G+C base pairs it contains. Some organisms are referred to as being "GC-rich", which simply means that G and C base pairs are found more frequently in their genomes than would be expected by random chance; others are "AT-rich" for the converse reason. If you were to compare the genomes of an AT-rich and a GC-rich organism, which would you predict would contain the higher number of long ORFs that do not encode proteins but have simply arisen by chance, and why?

How can we tell if a given region of DNA is actually making RNA, and if it is, how do we know that the RNA is being translated into a protein? Transcription can be detected by using RNA-based methods that will be described in Chapter 11. These vary in sensitivity and in the nature of the information that they give you, so the absence of a detectable transcript need not necessarily mean that the gene is never transcribed (it may only be transcribed under particular conditions which you have not yet defined, for example). Moreover, the presence of a signal for RNA may not, depending on the method used, tell you that the whole ORF is transcribed. (You will be able to understand why after studying the RNA-based methods in Chapter 11.)

Even if we can show that a given ORF is transcribed into mRNA, this does not tell us that the mRNA is translated into a protein. In fact, this can in some cases be a surprisingly difficult thing to prove. Bioinformatics analysis of DNA sequences as discussed in Section 8.1 can help us make a good estimate of whether or not a given ORF is likely to be an active gene producing a protein, but ultimately this has to be proven by experiment. What follows is a short list of some of the methods that we might use to determine whether a protein is produced from an RNA transcript or not. The method chosen in a particular case will depend on the nature of the protein and the precise question being asked about it.

First, we can make a mutation in the gene, such as deleting it. Methods for doing this will be described in the next section and in Chapter 12. If the protein is essential for growth of the organism, we will be unable to make a deletion of the gene that encodes it (or both copies, in the case of a diploid organism), as the organism would not be viable. This will prove that the ORF does indeed represent a functional gene. In the case of a protein which is non-essential, we can compare the total protein profile, as detected by one or two-dimensional gel electrophoresis (Box 5.1), of the organism where the gene has been deleted, with the parent which still contains the gene. If we are fortunate, the protein will be sufficiently abundant to see following electrophoresis, and we will then see a band or spot disappear in the deletion mutant. However, many proteins are not this abundant, or may be masked by other proteins.

Second, we may be able to obtain an antibody which recognizes the protein. This is usually done by comparing the predicted protein to other proteins in a database. It is often the case that antibodies against one protein recognize other proteins which are homologues, so if there are proteins similar in sequence to our protein, and if antibodies exist against these, the antibodies have a good chance of cross-reacting with our protein. Western blotting (Box 5.1) can then be used to see whether the protein is present, and whether it disappears when we delete the gene which we think encodes it. This method can be used in combination with the first method, and as it is more sensitive it is useful for less well expressed proteins. However, it still may not be sufficiently sensitive if the protein is poorly expressed and, of course, we may not be able to find an antibody that recognizes the protein we are looking for.

Q10.2. Imagine the following scenario. You have detected an ORF in the genome of an organism in which you are interested, and obtained an antibody which has been raised against a protein with a high degree of sequence similarity to the one encoded by this ORF. A western blot with the antibody does reveal a cross-reacting band. You construct a deletion of the ORF from the genome, but when you do another western blot against the organism carrying this deletion, the same band still appears. What could be going wrong?

Third, we can make a translational fusion between the ORF for the protein, and for another protein which is easier to detect, such as the green fluorescent protein GFP. (See Box 9.1 for a discussion of translational fusions.) This approach can make proteins which are present at low level more "detectable", and, as we shall see later, it can also be a powerful route to determining the cellular location of the protein.

The construction of gene deletions

One of the first questions that is often asked of a newly identified gene is, what is its role in the organism? This is a question which is being asked more frequently in the age of large-scale sequencing projects, since often these identify large numbers of genes which either do not resemble genes in other organisms, or which are found in other organisms but for which no function is known. To answer this question, we can use genetic means to inactivate the gene, and then study the changes (if any) that result in the organism. In this section we will consider some ways in which this can be done, focusing on simple organisms such as bacteria and yeast. Gene deletions or disruptions can also be made in complex organisms and may be very informative; methods for doing this together with examples will be discussed in Chapter 12. In the following section, we will look at methods for reducing expression of genes without actually deleting the genes themselves.

There are several approaches to disrupting the function of (often referred to as "knocking out") a gene. The most complete is to delete the whole gene from the genome of the organism. Another is to insert a large piece of "junk" DNA into the gene, which will prevent the mRNA being translated into the normal protein, analogous to the way that a chunk of text pasted into a sentence can disrupt the meaning of the sentence. We could also make a more subtle change in the sequence of the gene, for example by introducing an early stop codon into the gene, or by making a single base pair addition or deletion so as to alter the reading frame. In general, the first two methods are preferred, since they are unable to revert back to the wild-type.

Homologous recombination is always used to make either a targeted deletion or disruption. For the simplest example of this, consider the situation shown in Figure 10.1. Here, the regions on either side of the gene of interest (which is shown in blue) have been cloned onto a plasmid which has then been linearized and introduced back into the cell. A single homologous recombination event between the linear DNA and the chromosome will produce a broken chromosome, as shown. (If the chromosome itself is linear, this will convert it into two linears; if it is circular, this will linearize it. In either event, the chromosome will now be lost, either because it cannot replicate or because it cannot successfully segregate to the daughter cells when the cell divides.) However, a double recombination event as shown would essentially remove the gene from the chromosome, and it would then be lost as it would

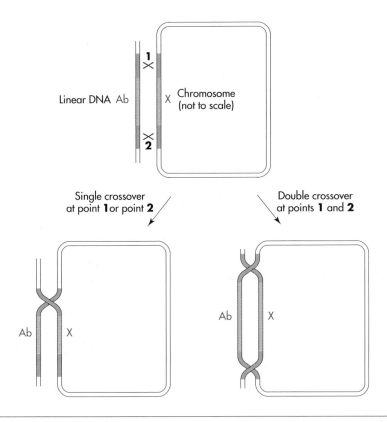

Figure 10.1 Deletion of a gene by a double recombination event between linearized DNA and chromosome DNA. The plasmid has had both the upstream and downstream flanking regions which surround the gene which we wish to delete (labeled X) cloned into it, on either side of a selectable marker (labeled Ab). A single cross-over between one of the two flanking regions leads to breakage of the chromosome. A double cross-over involving both flanking regions will swap the selectable gene Ab onto the chromosome, exchanging it with the gene that we wish to delete.

have no way of replicating itself. Such events do indeed occur, but they are rare, and it would be impossible to detect them without very laborious screening. However, the need for this screening can easily be removed by including a selectable marker (shown as Ab in the figure) between the flanking sequences on the linear DNA. As can be seen, a double recombination event leads to integration of the selectable marker at the position where the gene formerly was, and thus the rare cases where this has occurred can be selected for by growing the cells under the appropriate conditions. The marker is often (particularly in bacteria) for antibiotic resistance. However, this kind of experiment can also be done using a strain which is auxotrophic for a particular amino acid due to mutation in a gene which encodes an enzyme involved in the synthesis of that amino acid. A copy of this gene is placed between the flanking sequences for the gene that is to be deleted, and plating the cells out on a medium lacking that amino acid after transformation with the linear DNA will select for homologous gene replacement events.

This method is unlikely to yield 100% authentic gene replacements, for several reasons. First, the gene which is being used for selection may itself have a homologue on the chromosome. For example, antibiotic resistance genes are frequently modified versions of existing bacterial genes and, in the case of using an auxotrophic mutant, the gene which is mutated in the auxotroph may still be present on the chromosome. Thus recombination may take place between the selectable marker and its homologue on the chromosome rather than between the flanking regions of the gene which is being deleted. Second, when linear DNA is present in cells, there can be quite a high level of non-homologous recombination, where the linear DNA becomes inserted at a region with little or no homology to itself. In both these cases, the target gene will remain untouched.

Q10.3. In the light of the above point, what steps would you take to ensure that the gene of interest had indeed been deleted following transformation with linear DNA containing the flanking regions around an appropriate selectable marker?

Gene deletion by transformation with linear DNA is a powerful and rapid method, and for this reason was used in a genome-wide survey of gene function in the baker's yeast *Saccharomyces cerevisiae*. In this experiment, investigators used a two-step PCR procedure to generate linear fragments for every yeast gene (Figure 10.2). First, they used pairs of primers that contained 18 bases of either the immediately upstream or downstream sequence of every gene including either the start or stop codons, followed by a short synthetic "tag" sequence, followed by 18 bases of DNA homologous to either the 5' or the 3' end of a kanamycin resistance gene. PCR of the kanamycin cassette in a plasmid with these primers thus gave, for every gene in the genome, a product with the flanking region of the gene surrounding the tag sequence and the kanamycin resistance gene. In a second round of PCR, primers with 45 bases overlapping with the initial 18 bases of up- or downstream sequence were used to generate a product with more flanking DNA present at both ends. Each product of the second PCR reaction (i.e. one for each gene) was then used to transform yeast, selecting for kanamycin resistance. Of 6131 ORFs targeted, 5916 had one copy successfully deleted in a diploid state. The genes could then be tested to see whether they were essential by sporulation of the yeast cells and recovery of haploid strains. In total, this large experiment showed that nearly a fifth of the yeast genes are essential for growth in rich medium, of which only about a half had previously been identified by other methods.

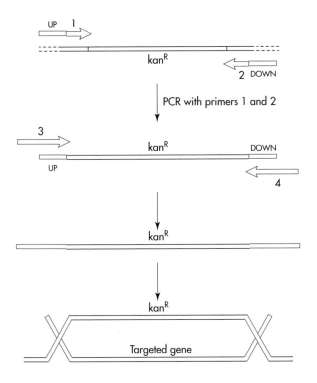

Figure 10.2 Construction of gene deletions for entire S. cerevisiae genome. Two rounds of PCR led to the generation of a linear DNA fragment with a kan^R selective marker flanked by 45 base pairs of DNA homologous to the upstream and downstream regions of the gene which is to be targeted for deletion. Transformation of this DNA into S. cerevisiae followed by growth on plates containing kanamycin leads to the selection of cells which have undergone a double recombination event, which replaces the targeted gene with the kan^R marker as shown.

Linear DNA transformation is often less successful with bacteria, which contain many nucleases that rapidly degrade linear DNA before it can recombine with homologous regions on the genome. A more widely used method is a two-step method, sometimes referred to as "pop-in, pop-out", where a plasmid carrying the disrupted gene is used rather than linear DNA.

The method is illustrated in Figure 10.3. The plasmid has the flanking regions of the gene of interest ligated into it, surrounding a selectable marker. Also present on the plasmid is another gene which can be selected against (referred to as a counter-selectable marker). A commonly used counter-selectable marker is the gene *sacB*; this encodes a protein called levansucrase, and cells which express it are killed by growth on sucrose. The reason for the presence of this gene will become clear in a moment.

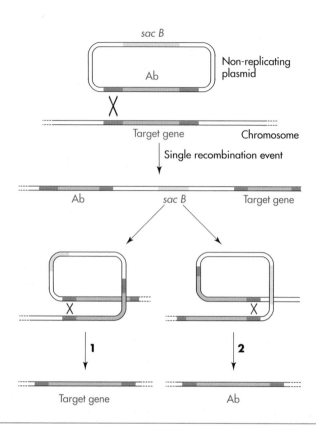

Figure 10.3 Pop-in, pop-out in bacteria. The first recombination event leads to insertion of the entire plasmid into the chromosome. Cells that undergo this event will become antibiotic resistant. Selection against the presence of the plasmid sequences (by growth on sucrose, which is lethal in the presence of the *sacB* gene) will lead to colonies that have lost the plasmid sequence by a second recombination event. Depending on which of the flanking sequences is involved in this second event, it will either regenerate the original sequence (1) or lead to the swapping of the target gene with the antibiotic resistance marker (2), which is the desired outcome.

A key feature of this method is that the plasmid cannot replicate in the organism into which it is introduced, and is hence referred to as a "suicide vector". The plasmid is transformed into the organism, and selection is made for the resistance marker on the plasmid. As the plasmid cannot replicate, the only way that the cells can become resistant to the marker and give rise to a colony is for recombination to occur between the plasmid and the chromosome. The most common event will be a single cross-over, which as shown in Figure 10.3 will integrate the entire plasmid onto the chromosome. Once this has happened, the target gene is still present. The recombination event can now be reversed by selecting against the presence of the plasmid, by growing the cells on sucrose. Cells that still have the plasmid on their chromosome will die when plated onto sucrose, due to the presence of the *sacB* gene, but if they lose the plasmid by excision (and the plasmid will then be lost from the cell as it cannot replicate) they will be able to survive. Plasmid excision can occur in one of two ways, as shown in Figure 10.3. If it is simply the reverse of the integration event, the selectable marker will also be lost. However, if the recombination event takes place on the other side of the selectable marker gene, the marker will be retained, but the wild-type gene will now be lost, giving rise to the desired gene deletion.

Many modifications of this basic protocol exist in bacteria and other organisms, for example to enable the selectable marker itself to be subsequently removed so that multiple mutations can be made in the same strain. However, the same basic principle is used in constructing gene knockouts from *E. coli* to complex mammals such as mice (as discussed in Section 12.5) and these are now fundamental tools in genetic research.

Inhibition of transcription: the use of RNAi

Although targeted gene deletion methods are very powerful for assessing the role of genes in organismal function, they are not always the method of choice. For example, in some organisms they are challenging or lengthy to do, particularly in diploids where both gene copies need to be removed before any conclusions can be drawn about their roles. Recently, another method for assessing gene function has emerged which is proving to be very valuable for "high throughput" experiments where a large proportion of the genes in a genome are investigated. This method can also be used in working with systems where it can be tricky technically to produce targeted knockouts, such as mammalian cells. It relies on a phenomenon discovered accidentally in plants, which is that the presence of double-stranded RNA, homologous to a particular gene, often leads to inhibition of expression of that gene. This method was first called co-suppression, but is now more usually referred to as RNAi (RNA-mediated interference).

The biological reason for the existence of RNAi is still not fully understood, although recent results strongly support the hypothesis that it may have evolved in cells as an anti-viral defense mechanism. What has been shown is that, in certain cells, if a double-stranded RNA molecule (dsRNA) is present, it is cleaved into small (20–25 nucleotides) lengths of dsRNA by an enzyme called "dicer". These small dsRNAs, referred to as siRNA (for "small interfering RNA") then assemble together with particular proteins into a complex called RISC, for "RNA-induced silencing complex". In these complexes, the dsRNAs are separated to form the single-stranded state, and these now anneal with the mRNA to which they are complementary and the RISC complex breaks down this mRNA (see Figure 10.4). The net effect is that the gene for which the dsRNA was first introduced is "silenced", as the mRNA that it produced is cleaved, and this can happen

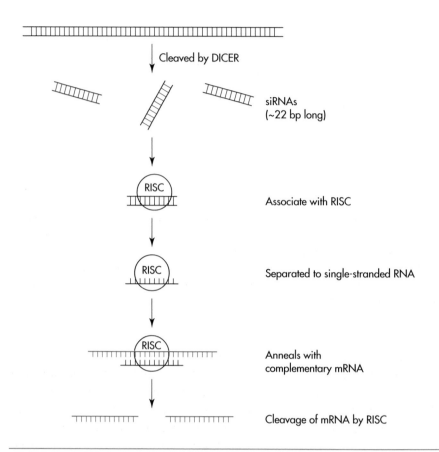

Figure 10.4 Steps in RNA interference (RNAi). Double-stranded RNA becomes associated with the DICER enzyme complex, which cleaves the RNA into short (about 22 base pairs) sequences. These in turn associate with the RISC complex. The RNA is separated to give bound single-stranded RNA, which anneals with the RISC complex to the complementary mRNA molecules. RISC then cleaves these mRNAs, which hence prevents their translation into protein.

very effectively as the RISC complexes are catalysts that can cleave many mRNA molecules.

Fortunately for scientists, three of the organisms where this method works well are three of their favorite experimental model organisms: the fruit fly *Drosophila melanogaster,* the nematode *Caenorhabditis elegans,* and the thale cress *Arabidopsis thaliana* (see Box 2.3 for a discussion of model organisms and their uses). This has made genome-wide screens of gene function in these organisms plausible, along the same lines as the genome-wide gene knockout study on baker's yeast described above. The first organism to be studied in this way was *C. elegans,* which is a relatively simple multi-cellular organism and a very popular model for studying many aspects of cell biology and development. This work was helped by a

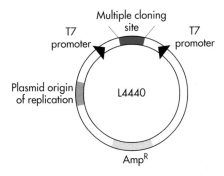

Figure 10.5 Plasmid vector for production of dsRNA in E. coli. The insert corresponding to the gene of interest is cloned into the multiple cloning site between the two T7 promoters, and the plasmid is transformed into a strain which encodes T7 RNA polymerase. This will transcribe the insert in both directions, producing both a sense and an antisense strand of mRNA which will anneal with each other.

remarkable finding: that dsRNA can be introduced to *C. elegans* cells simply by feeding them with *E. coli* (their favorite food) that are expressing the dsRNA. Experiments thus involved constructing a PCR-based library of 16,757 *C. elegans* genes in an *E. coli* vector where the insert could be simultaneously transcribed in both directions, by virtue of having a T7 promoter at each end of the site where the insert was cloned (Figure 10.5). The plasmids were transformed into a strain of *E. coli* with a gene for T7 RNA polymerase on the chromosome under the control of a *lac* promoter (see Figure 9.2). Induction with IPTG (Box 3.4) thus gave rise to mRNA from both strands of the cloned PCR fragment, which annealed in the *E. coli* cytoplasm to yield a dsRNA corresponding, in each individual clone, to a single gene in *C. elegans*. Feeding each of these clones to *C. elegans* showed that RNAi of over 10% of the genes induced a phenotype such as sterility, failure to develop at either the embryonic or larval stage, or a growth or developmental defect in the post-embryonic stage. Many subsequent studies have looked for other phenotypes in *C. elegans* using RNAi, and as *C. elegans* shares a large number of its genes with humans there is every reason to believe that insights into phenomena such as ageing and neurodegeneration in humans will emerge from these types of studies.

This method of using RNAi does not work in mammalian cells, as the long dsRNAs are degraded before they can elicit any gene silencing. However, it has been found that if siRNAs are synthesized and introduced directly into mammalian cells, they are not degraded and they do cause suppression of the genes against which they are targeted. This has caused much excitement as a potential way of developing novel therapeutics and indeed, phase II clinical trials of the first siRNA product, which has been shown in animal studies to inhibit the processes associated with macular degeneration (an eye condition that leads to blindness) should start very soon. It also provides a powerful tool to investigate gene function in tissue cultured cells, since specific genes can be turned down or silenced altogether by this method and the consequences for the cell monitored.

This can be particularly useful in cases where there are several copies of a gene present in cells, and we wish to discover whether they have distinct functions. For example, cells contain several members of the Hsp70 gene family, which encode similar proteins which are all made in higher levels after the cells are heat shocked at a temperature higher than normal growth temperature but not high enough to be lethal. siRNA experiments have recently shown that two of these are required for tumor cell growth but not for normal cell growth, making these genes obvious targets for siRNAs in cancer therapy.

10.3 Determining the Cellular Location of Proteins

One of the questions that we often want to know the answer to is, whereabouts in the cell does a particular protein do its job? Some proteins function in the cytoplasm, some function in one of the cellular compartments, and some function in membranes (i.e. any of the membranes of the organelles, or the plasma membrane that surrounds the cell). In many cases they are made where they function: cytoplasmic proteins are made in the cytoplasm, and some proteins inside mitochondria and chloroplasts are made from DNA which is transcribed and translated in those organelles. But many proteins have to move from the location where they are made to the location where they are required for their function. For example, many proteins such as histones, polymerases and transcription factors obviously need to function inside the nucleus, but all are synthesized by the ribosomes in the cytosol. Many other proteins are embedded in the plasma membrane where they can act as pumps or as sensors which bind to and detect particular molecules. These proteins are made (in eukaryotes) on the ribosomes on the rough endoplasmic reticulum, but still have to be transported through the endoplasmic reticulum and the Golgi body before being targeted to the membrane. Other proteins which are found in mitochondria or other organelles such as lysosomes and peroxisomes are synthesized in the cytoplasm and then transported to their final destinations.

How can the location of individual proteins be determined? The cell can be fractionated using appropriate techniques into reasonably pure preparations of its individual components (e.g. nuclei, mitochondria, membrane vesicles derived from the endoplasmic reticulum), so if the protein is sufficiently abundant to be easily visible following gel electrophoresis, or if it has an enzyme or other activity that can be used to determine its presence, then cell fractionation followed by electrophoresis or assay is sufficient. If an antibody to a protein is available, the sensitivity of this method can be increased by using western blotting (Box 5.1). But for many newly discovered proteins, where all that is known is that there is a gene encoding the protein, these methods are unlikely to be suitable.

Some predictions can be made about the eventual location of proteins by looking at the predicted amino acid sequence, derived from the DNA sequence. Many proteins which are sorted to different cellular targets have characteristic signals, which are stretches of amino acids of particular types that direct the protein to different cellular locations, and computer programs are available that have quite a high success rate at detecting these. However, these alone cannot prove the location of a protein, but only provide hypotheses which need to be tested by a direct experiment.

Gene cloning methods introduce a new approach to this problem, which relies upon the construction of translational fusions (Box 9.1). As you may recall, these are proteins that are made by fusing different ORFs or parts of ORFs, so that a novel protein is created which has properties derived from both of the ORFs concerned. For protein localization studies, a powerful method is to fuse ORFs for proteins of interest with the gene encoding the fluorescent protein, GFP (green fluorescent protein).

How does this help determine the location of the protein of interest? First, GFP is a protein that is easy to image *in vivo*, using a technique called fluorescence microscopy. It also requires no external co-factors, and so the cells expressing GFP or a fused derivative of it do not need any particular treatment to make the protein visible. When fused to other ORFs and translated, the GFP usually retains its full fluorescent properties. The presence of a GFP fused to a protein often has no detrimental effect on the protein itself, which can continue to function in its normal way. Most importantly, once GFP has been translationally fused to an ORF, it generally then becomes subject to the cellular localization signal encoded by that ORF. Thus if a fusion is made between GFP and a protein which is normally found in the nucleus, the whole protein will still be transported to the nucleus, which will now be fluorescent when looked at with a fluorescence microscope.

This technique has been used to study many proteins in many different organisms, particularly model organisms such as *D. melanogaster* and *C. elegans*. It can also be used in tissue culture to study protein localization in human or other mammalian cells. Generally, this is done with genes for individual proteins. But perhaps the most ambitious use of the method is in a high throughput study, where an attempt was made to determine the localization of every protein encoded by the yeast *S. cerevisiae*. It is worth looking in some detail at how this was done, to see how such methods are now gaining ground with the availability of complete genome sequences. Several other examples will be considered in the sections that follow.

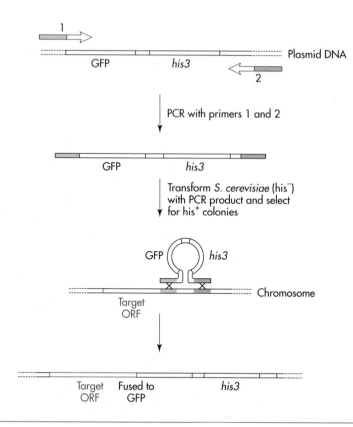

Figure 10.6 Construction of a GFP fusion to any ORF in the yeast genome. The ORF is shown in blue. PCR with primers 1 and 2 yields a PCR product with ends homologous to the 3′ end of the ORF and a region downstream from the ORF (shown in gray). Double recombination between this linear piece of DNA and the yeast genome produces a fusion between the ORF and GFP, with the selectable marker *his3* immediately downstream.

Q10.4. Why is it important that the addition to a protein of amino acids encoding GFP does not usually interfere with the function of the protein?

The steps involved are shown in Figure 10.6. The first step was to obtain fusions of all the ORFs in the yeast genome with GFP. This was done by producing a plasmid that contained a gene encoding GFP and also the *his3* gene, which is required for biosynthesis of the amino acid histidine. A PCR primer was then designed for each of the 6234 ORFs in the yeast genome. The primers had one end complementary to the 3′ end of the ORF, and the other complementary to the 5′ end of the GFP. All the primers, of course, had to be designed so that the ORF and the gene for GFP would have the same reading frame when the gene was transcribed.

Q10.5. Why was it necessary in designing the primers to ensure that they would put the GFP in the same reading frame as the ORF?

A second primer was designed that was homologous to the region down-stream from the *his3* gene in the plasmid, followed by a stretch of DNA homologous to the sequence downstream from the ORF in the genome. PCR reactions were then done for each ORF, and each PCR product was individually transformed into a haploid yeast strain that was auxotrophic for histidine. The transformants were plated out on minimal medium lacking histidine. Colonies that arose on these plates are candidates for ones where a homologous recombination event has taken place between the PCR product and the ORF, since this will insert the *his3* gene onto the chromosome, enabling the growth of the transformed strain in the absence of histidine. PCR on individual transformants was then used to check that the PCR product had indeed integrated at the correct locus. A total of 6029 transformants were obtained. The others presumably represent cases where the presence of the GFP is deleterious to the normal function of the protein, and where the protein is essential for cell growth.

All transformants were screened for fluorescence, and approximately two thirds showed an increase. The remainder probably represent fusions with genes which were not expressed, or were expressed at very low levels, under the conditions of the experiment. Those colonies that showed fluorescence were then studied individually using fluorescence microscopy, and categorized into different groups according to where the fluorescence signal was seen to localize (nucleus, nuclear periphery, actin cytoskeleton, mitochondrion, plasma membrane, cytoplasm, etc.). Careful analysis enabled 22 different cellular locations to be defined. It was found that 44% of the proteins studied were found in locations other than the two major ones (nucleus and cytoplasm), and in many cases these were assignments that had not been made before.

10.4 Mapping of Membrane Proteins

One of the most interesting and challenging areas of cell biology is the study of membrane proteins. These proteins are of great importance to the cell, acting as pumps and channels that maintain internal homeostasis, and detectors that enable the cell to respond to conditions and materials outside the cell even when they do not penetrate the membrane. They are of considerable interest to structural biologists since they have to interact with two very different environments (the internal hydrophobic region of the membrane and the hydrophilic aqueous milieu on both the inside and outside of the cell). They are also difficult to study, particularly from a structural point of view, because they can only maintain their structure and their function when embedded in the membrane, and so cannot be studied as totally pure proteins.

Membrane proteins by definition have at least one part of the protein embedded in the membrane, but in many cases the polypeptide chain of a membrane protein will cross the membrane several times. These are referred to as polytopic proteins. Regions of the protein which cross the membrane often form amphipathic alpha-helices. These are alpha-helices where residues on one side of the helix tend to have hydrophilic side chains, and those on the other side, hydrophobic side chains. In a typical polytopic protein, the alpha-helices will form a bundle where the hydrophobic side chains point outwards, and interact with the hydrophobic interior of the membrane, while the hydrophilic residues face inwards and interact with each other, or form a channel through which charged or polar substances can pass. Bioinformatic analysis of putative membrane proteins can be helpful in detecting the regions of the protein which are likely to form the membrane-spanning segments, but to prove the topology of a given protein in a membrane requires other methods.

Consider the protein shown in Figure 10.7. This is a purely diagrammatic representation of the protein showing the regions of the protein which are crossing the membrane opened out in a straight line for convenience: in practice, these would be bundled together. Regions between the membrane spanning helices form domains which are either on the "inside" or the "outside" of the membrane. How can we determine whether this sort of model, which could be predicted from a computer analysis of the protein

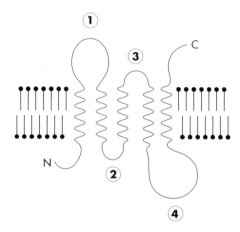

Figure 10.7 General topology of a membrane spanning protein (shown in blue) embedded in a lipid bilayer. N and C show the N and C termini of the protein. Domains 1 and 3 are clearly on the opposite side of the membrane to domains 2 and 4.

sequence, is likely to be correct? Gene cloning and the use of translational fusions provide one useful way of helping to do this.

Two particular genes are traditionally used in gene fusion studies to map the topology of membrane proteins. One of these is β-galactosidase, encoded by the *lacZ* gene, which we have encountered before as a reporter protein. This protein is active only when it is in the cytoplasm of *E. coli*; attempts to export it from the cytoplasm to the periplasm (the space between the inner and outer membranes of *E. coli*) lead to an inactive protein. The other is alkaline phosphatase, encoded by the *phoA* gene. For activity, this protein requires the formation of disulfide bonds, and these cannot form in the cytoplasm of *E. coli*. For this reason, alkaline phosphatase is only active when in the periplasm. The properties of these proteins can thus be used to map the domains of membrane proteins which are between the membrane spanning domains, which in turn can be used to check whether the topology and orientation of the protein as predicted is correct.

This is done by constructing a series of in-frame translational fusions between the gene for the membrane protein, and either the *lacZ* or the *phoA* gene. These fusions will thus contain all the signals for membrane protein insertion up to the point where the join with the reporter protein occurs. They can then be introduced into an *E. coli* cell that is *lacZ*⁻ or *phoA*⁻ (to remove any background activity), and the amount of activity determined. If the *lacZ* fusion occurs in a cytoplasmic domain of the protein, it would be predicted to be active, whereas if it occurs in a periplasmic domain, it would be inactive. The reverse is true for a *phoA* fusion. Thus by measuring the level of activity of translational fusions with *lacZ* and *phoA* at multiple points throughout the gene of interest, a map of the topology of the membrane protein can be built.

This is illustrated in Figure 10.8, which shows a membrane-spanning protein called ProW that was mapped in this way. As can be seen, at every position where a *lacZ* fusion was made, a *phoA* fusion was also made. In every case, one of the two enzymes showed high activity while the other was low, and this enabled the predicted domains of the membrane protein to be assigned to either the cytoplasm or the periplasm. One point worth noting is that in the cases where the fusion position was in one of the membrane spanning helices, the activity seen was that of the domain immediately before this helix. The reason for this is that the complete helix is needed for the protein to cross the membrane. If only a part of the helix is present before the fusion to the reporter protein, that part of the protein will remain either inside or outside the membrane.

Q10.6. In Figure 10.8, no fusions have been shown with the large periplasmic N-terminal domain. Had they occurred, what would you have predicted about their activities?

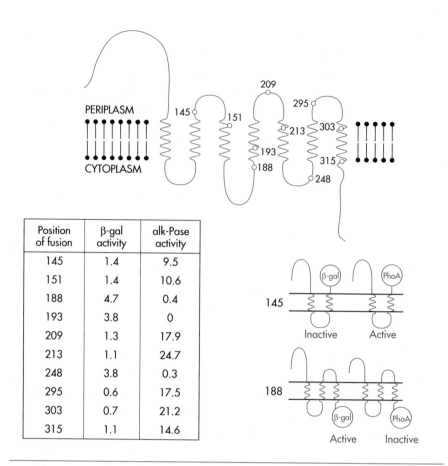

Position of fusion	β-gal activity	alk-Pase activity
145	1.4	9.5
151	1.4	10.6
188	4.7	0.4
193	3.8	0
209	1.3	17.9
213	1.1	24.7
248	3.8	0.3
295	0.6	17.5
303	0.7	21.2
315	1.1	14.6

Figure 10.8 Mapping of membrane topology using *lacZ* and *phoA* fusions. The example of ProW mapping is shown, with fusion positions marked on the diagram and the activities of the *lacZ* and *phoA* fusions at each position shown in the table. The smaller figures show diagrammatically the appearance of fusions at positions 145 and 188.

How are the fusions made? One approach is to simply decide, on the basis of where the domains are predicted to be, exactly where the fusions should be placed. PCR primers can then be designed to construct the fusions between the gene of interest and the fusion partner, ensuring of course that the two reading frames are in frame with each other. Another approach which has been quite widely used is to generate random fusions between the target gene and the fusion partner. This has been done by using transposons, which carry on them close to their 5′ end the genes either for β-galactosidase or for the periplasmic form of alkaline phosphatase lacking an initiator methionine and with no stop codons upstream before the start of the transposon. The transposon carries a kanamycin-resistance marker downstream from the fusion partner. If the gene of interest is now cloned onto a plasmid and this plasmid is transformed into a strain containing one of these transposons, insertion of the transposon into the plasmid will lead to the plasmid acquiring a kanamycin-resisitant marker, which can then be selected for. Genes that have received an insert that is both in-frame and in a domain where the fusion gene is active can then be directly detected, by plating out on a medium where the colonies turn blue if active β-galactosidase or active alkaline phosphatase are present.

This method is particularly useful for the study of membrane proteins from higher eukaryotes. It has been shown that generally these insert into the membrane of *E. coli* with the same topology that they use in the organism from which they originate. Thus, by isolating the gene for the membrane protein of interest, expressing it in *E. coli*, and constructing fusions either by direct cloning or by using one of the transposons described above, a collection of translational fusions can be generated which can be used to map the positions of the domains of this protein.

10.5 Detecting Interacting Proteins

As the amount of information available about proteins in cells grows, so it becomes more important to consider not only what individual proteins do but also what other proteins they interact with. The majority of proteins in the cell act not in isolation but in contact with other proteins, and in some cases a given protein may interact with a large number of partners. How can these interactions be demonstrated and studied? Is it possible to define, for any given protein in the cell, the set of all "partner" proteins with which that protein interacts? In the section that follows, we will look at two very different methods that can be used to detect interactions between proteins, both of which rely on our ability to manipulate the genes that encode those proteins.

Two-hybrid methods

The two-hybrid methods are a set of ingenious methods that enable us to use relatively simple genetic selections to study the interaction between two proteins, or to identify the partners for a particular protein. They rely on an understanding of how activation of gene expression occurs, as shown in Figure 10.9. This simplified diagram shows how expression of the yeast GAL genes, expression of which is required for the yeast to grow using galactose, is regulated by the Gal4 protein. This protein binds to a sequence upstream of the GAL gene promoter (known as a UAS or upstream activator sequence) and contacts the transcription complex bound at the promoter, thus activating transcription of the genes.

Analysis of the Gal4 protein showed that the regions involved in binding of the protein to the UAS, and in activation by contact with the transcription complex on the promoter, were distinct parts of the protein. This led

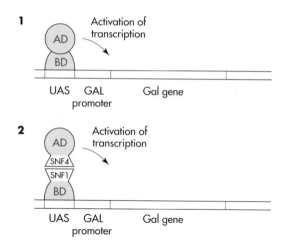

Figure 10.9 The two-hybrid assay in yeast. 1) shows the two domains of the Gal4 protein. BD, the binding domain, positions the Gal4 upstream of the GAL promoter. AD, the activation domain, activates transcription of the GAL gene. 2) Shows that the BD and AD of Gal4 can be separated. If they are fused to two proteins which themselves interact – here, SNF1 and SNF4 – activation of transcription of the downstream gene will still occur.

researchers to test an idea which at first sight seemed far-fetched: would the process of gene activation still work if the two parts of the Gal4 were actually on separate proteins, but these two proteins themselves contacted each other?

To test this idea, the following experiment was done. A fusion was made between a GAL promoter and a *lacZ* gene, thus placing *lacZ* under the control of the GAL promoter. This was done in a host strain where the endogenous *gal4* gene had been removed, so the cells were unable to respond to galactose in the medium. Two translational fusions were then cloned onto two separate plasmids. In one, the part of the *gal4* gene encoding the N-terminal end of the Gal4 protein, which is the part of the protein that binds to the UAS sequence, was fused to the 5′ end of a gene called SNF1, which encodes a protein kinase. In the other, the part of the *gal4* gene encoding the activator domain of Gal4 was fused to the 3′ end of a gene encoding SNF4, which was known from other studies to physically interact with SNF1. The two constructs were introduced into the yeast strain described above, and the cells were plated on galactose.

What did the investigators hope to see? The hope was that SNF1 and SNF4 would interact, and in doing so would effectively make a complex that was capable of activating expression from the GAL promoters, since it now contained both the UAS binding region and the activator region of the Gal4 protein, linked by the SNF1/SNF4 interaction. This should cause expression of the *lacZ* gene, which as you recall had been placed under the control of the GAL promoter, and it should also enable the strains to now grow on galactose, as the new complex could now activate expression from the GAL promoters within the cell (including the endogenous promoter driving the GAL gene). Both these predictions were met: even though the two essential domains of the Gal4 protein were now on different proteins, the fact that these two proteins interacted enabled them to come together to activate gene expression.

This remarkable system allows a number of different experiments to be done. First, it can be used to define the regions of two proteins that interact with each other. In the first experiment described above, complete coding regions for the two interacting partners were used. However, by expressing smaller parts of the proteins and checking for *lacZ* activity, it would be possible to map the domains on the two proteins that are required for them to interact. Such experiments have indeed been done.

It is important to realize that the two-hybrid system is not foolproof and can give rise to both "false positives" (i.e. an interaction is apparently seen even though it does not exist or is not significant *in vivo*) and "false negatives" (an interaction which does occur is not seen using this system), and any results from the yeast system need to be verified using other methods, but the overall principle is a very useful one.

Another application of the yeast two-hybrid system is to look for partners to a particular protein, rather than investigating known partners in more depth. In such a system, the protein for which partners are being sought is referred to as the "bait" protein, and it is the gene for this protein which is fused to the DNA sequence encoding the DNA binding domain of the Gal4 protein, exactly as in the method described above. Next, a cDNA library is made in such a way that the cDNAs are inserted downstream from the activator domain, such that a fusion protein will be produced. The library is then transformed into a yeast strain containing the "bait" plasmid. Plasmids in the library that contain a protein or protein domain that interacts with the bait protein will lead to the formation of an active Gal4, which will hence activate expression from a GAL promoter. Frequently in this kind of experiment, a GAL promoter is placed upstream of a *his* gene, and the experiment is done in a *his* auxotroph, enabling selection of only those clones that contain a protein that will bind to the bait. The entire procedure is shown in Figure 10.10.

Q10.7. With such a system, what medium should be used to select for the clones of interest?

Q10.8. Can you think of a modification to this method that would enable you to identify all the proteins that bind to a particular DNA sequence?

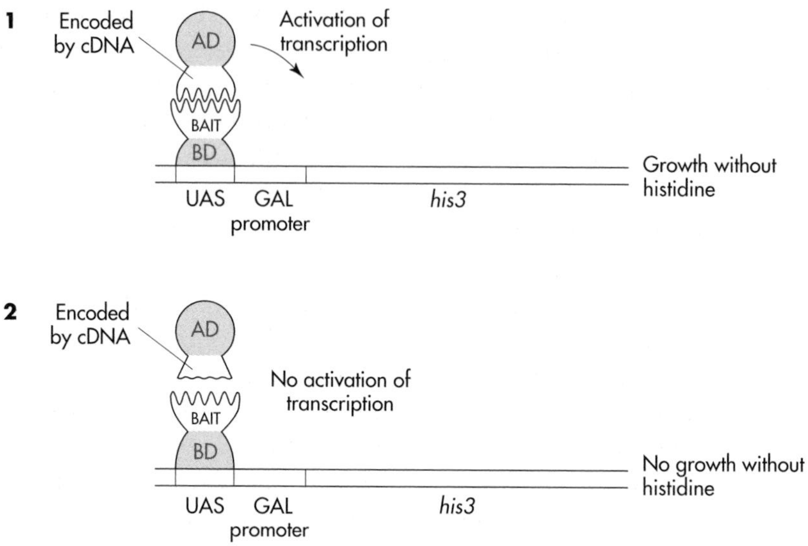

Figure 10.10 Use of a two-hybrid system with a bait to identify interacting partners. A his⁻strain containing the bait protein fused with the binding domain (BD) of Gal4 is transformed with an expression library of random cDNA clones fused to the activator domain of Gal4. If (as shown in 1) a particular cDNA clone encodes a protein which interacts with the bait, the downstream gene (his3) will be transcribed, and this can be selected for by growth without histidine. If (as shown in 2) the protein encoded by the cDNA does not interact with the bait, the BD and AD of Gal4 will not interact and the cell will not grow without histidine. This method enables very large numbers of clones to be screened for those that encode proteins that interact with the bait protein.

This method can be adapted to look for protein interactions at a whole genome level. It is not an experiment for the faint-hearted, but, in principle, every ORF in a genome could be fused separately with both the Gal4 activator domain and with the Gal4 UAS domain, and by introducing every possible pair of ORFs into a suitable yeast strain, potential interactions between every pair of proteins could be tested. A moment's thought will hopefully convince you that some short-cuts are needed for such an experiment to be successful; yeast itself, for example, has around 6000 ORFs so to check every possible protein–protein interaction would require not only a huge effort in cloning all the genes but 36,000,000 (i.e. 6000 × 6000) separate transformations. In practice, a number of approaches have been used to make this method more workable. First, special vectors were used for the cloning of the ORFs such that the same PCR fragment could be used both for cloning downstream of the UAS binding domain and upstream of the Gal4 activator domain. Second, use was made of the mating behavior of yeast. Yeast exists in two haploid mating types, a and α, which can easily be crossed to produce a diploid containing material from both types, including, if plasmids are present and appropriate selection is used, both plasmids. Thus for the large scale screen, plasmids with the UAS binding domain fusions were in cells of the α type, and those with the Gal4 activator domain fusions were in cells of the a type. Third, all the cells containing Gal4 activator domain fusions were pooled to form an activator domain library, and it was this entire library that was mated with each of the individual clones carrying the UAS binding fusions. The progeny of the mating were selected first (under conditions where neither of the parents could grow) and then screened for evidence of interactions. Obviously, any one protein present as a UAS binding domain fusion might interact with many proteins present in the activator domain library, and for this reason 12 independent clones from each mating were sequenced to identify potential partners. Not all interacting partners would be identified in this way (since some proteins may interact with more than 12 partners), but the sheer scale of the work (72,000 sequencing reactions if all matings produced at least 12 colonies) required does impose limitations despite the use of automation. In fact, larger studies have subsequently been undertaken using the same approach but with organisms with bigger genomes, including one on *D. melanogaster* which identified 10,021 different protein–protein interactions.

Co-immunoprecipitation and TAP tagging

The two-hybrid approach is only one of several that can be used to detect proteins that act with other proteins and, as stated above, it does suffer from a rather high rate of false positives and negatives. Ideally, any experimental deduction should always be backed up by results from a totally different method, and this is particularly the case with a method that has a rather high failure rate. Several other methods do indeed exist for detecting protein–protein interactions, which rely on a more biochemical approach than the two-hybrid system, where the power of yeast genetics is used to study interactions *in vivo*. For example, a method often used is called co-immunoprecipitation (usually shortened to co-IP). This relies on the use of antibodies to detect not only a given protein, but also the proteins that interact with it. The approach is very simple. Most antibodies bind very tightly to a protein called protein A, derived from the bacterium *Staphylococcus aureus*. Protein A can be coupled to a high molecular mass substance such as agarose or sepharose, and if mixed with a solution of the antibody and centrifuged at high speed, this will pellet the antibody and anything to which it may be bound.

Imagine a complex mixture of proteins, as shown in Figure 10.11, where an antibody is available against one of the proteins, and where this protein is itself interacting with several other proteins. Adding the antibody to this mixture (with the antibody in excess) will lead to the antibody binding to its

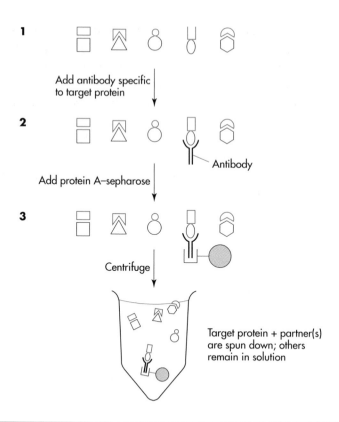

Figure 10.11 The principle of co-immunoprecipitation. The objective is to identify partners that interact with a particular protein in a complex mixture. To do this, an antibody specific to that protein is added to the mixture, followed by protein A conjugated to a large structure such as a Sepharose bead. The protein A binds the antibody which has bound the protein, and when centrifuged the entire complex is sedimented. The other proteins remain in solution. The protein in the pellet can now be studied to see what other proteins have sedimented with it. These will be candidates for partners that interact with the protein.

target protein, and if protein A–sepharose is now also added, the antibody in turn will be bound by the protein A. When the solution is now spun, not only will the target protein be spun down ("immunoprecipitated") but also any other proteins with which it is interacting. These proteins can in their turn now be identified by methods such as N-terminal sequencing or mass spectrometry.

Although this is a good method and widely used, it relies on having an antibody available for the protein which is highly specific so that there is negligible background. As high throughput methods of the type described above become more widely used, so generic methods are needed which do not rely on having antibodies and which can be used for most proteins.

One such method is known as TAP tagging, for tandem affinity purification, and as with many of the approaches described in this chapter, it relies on the construction of translational fusions. In this method, a fusion partner is added to the gene encoding the protein of interest, at either the N- or the C-terminal end of the protein. The fusion partner encodes the domain from protein A that binds to IgG antibodies, and a domain that binds strongly to the protein calmodulin. These two domains are separated by a stretch of amino acids that contain a site recognized by a protease derived from tobacco etch virus (TEV protease). The fusion partners have been designed in such a way that the protein A part is always at the extreme end (see Figure 10.12).

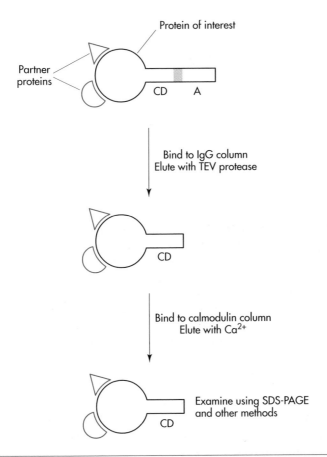

Figure 10.12 The basics of the TAP tagging method for identifying protein interactions. The gene for the target protein is engineered so that the protein is expressed with a calmodulin binding domain, followed by an IgG-binding domain derived from protein A, at either the N or C terminus. The two are separated with a TEV protease cleavage site shown in pale blue. A cell extract containing this protein (and its interacting partners) is passed down an IgG column. The bound protein is eluted by cleavage with TEV protease, and further purified by passage down a calmodulin column. Bound protein can be eluted with Ca^{2+}, and the interacting partners which copurify can be studied.

In a TAP tagging experiment, the gene for the protein of interest is fused to the TAP partner, and the translational fusion is then expressed back in the organism of interest. An extract is then made from this organism under conditions where interactions between proteins are minimally disrupted. The extract is then passed down a column to which IgG antibody has been attached. The protein A part of the TAP tag will bind tightly and specifically to this antibody, thus immobilizing the protein to which it has been fused and any proteins interacting with this protein, but other proteins in the cell extract will pass through the column. To elute the bound protein from the column, the column is treated with the TEV protease which cleaves at the site next to the protein A region. The eluted proteins are then passed down a second column containing bound calmodulin. Again, the fusion protein will bind tightly to this column, and it can subsequently be eluted by treatment of the column with calcium ions. The presence of other bound proteins can now be detected, and these can be identified using one of the biophysical methods described above.

Q10.9. Why do you think two columns are used in this procedure, rather than just one?

Q10.10. Even given the high degree of purity that this method can achieve, can you think of any good controls that would make it likely that the only proteins detected are the fusion protein and the proteins that it interacts with?

These methods and others like them that rely on protein-protein interactions are referred to as "pull down" assays. As with the two-hybrid system, they can give both false positive and false negative results, so the results need to be interpreted with caution and, ideally, backed up with other approaches. Also, as with the two-hybrid system, this approach lends itself to high throughput methods, particularly with yeast where it is relatively straightforward to insert a TAP tag upstream or downstream of any ORF. The TAP tagging approach is particularly useful for detecting large complexes of interacting proteins, whereas the two-hybrid approach can only find interactions one pair at a time.

10.6 Site-Directed Mutagenesis for Detailed Probing of Gene and Protein Function

Mutations are, of course, the basic tool of any kind of genetic analysis. Classically, mutations had to be selected for after doing random mutagenesis, but with the development of recombinant DNA methods, it became possible to construct specific mutations in any sequence of interest. Up to now we have mainly considered examples of large mutations such as deletions and gene knockouts. In this section, we will look at how cloning methods can be used to produce mutations at a much finer level than this. In fact, as you will see, it is possible to change any base in any DNA sequence to any other base, which means that we can also change any amino acid in any protein to any other amino acid. This method is known as site-directed mutagenesis, since the mutation is engineered at a precisely predefined site in the DNA.

The methodology of site-directed mutagenesis

The principle behind site-directed mutagenesis is simple, although (as we shall see) technical fixes are required for it to work at high efficiency. To explain this principle, we will start by describing the original method; refinements will be discussed subsequently. Figure 10.13 shows a gene,

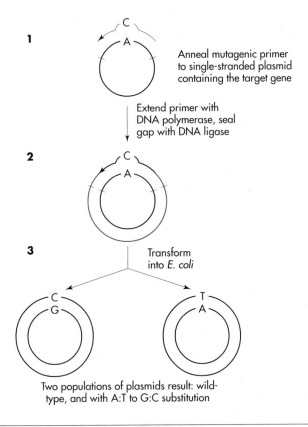

Figure 10.13 The basics of site-directed mutagenesis. An oligonucleotide, complementary to the target gene except for the altered base at the desired position of the mutation, is annealed to a single-stranded plasmid containing the gene that we wish to mutate 1). DNA synthesis (catalyzed by DNA polymerase) of the second strand is primed by the oligonucleotide, and the new strand is sealed to 3′ end of the primer by DNA ligase 2). When transformed into E. coli, two plasmids will result as the plasmid replicates: one wild-type and one mutated at the desired position 3). The plasmids can be distinguished by sequencing.

cloned in a plasmid, that we wish to mutate by changing a particular A to a G. To do this, we design an oligonucleotide which is exactly complementary to the DNA that flanks the base we wish to change, but which has a mismatch at the position we wish to change, such that instead of having a T it now has a C. We now produce the plasmid in single-stranded form. (This traditionally was done by using for the cloning vector a plasmid that could be produced either as a double-stranded or single-stranded plasmid; as this method is rarely used these days the details of how this was done will not be considered further.) The oligonucleotide is annealed to the plasmid, and will bind only at the position to which it is complementary. The single-stranded plasmid with the annealed oligonucleotide is now incubated with DNA polymerase, all four dNTPs, and DNA ligase. DNA polymerase will use the annealed oligonucleotide as a primer, and will synthesize a strand of DNA which is exactly complementary in sequence to that of the single-stranded plasmid. When the DNA polymerase has extended all the way round the plasmid, the DNA ligase can now seal the gap by joining the newly synthesized DNA strand to the 5′ end of the annealed oligonucleotide. We now have a double-stranded plasmid where one strand carries the wild-type sequence, and the other the mutated sequence.

If this plasmid is now transformed into *E. coli*, two populations of plasmid will result as the DNA replicates. One half will have the wild-type sequence. The other half will have the mutated sequence, as when the first round of DNA synthesis is done inside the cell, the new base that was part of the oligonucleotide will be copied as though it were the correct base. Thus we have produced a defined mutation, and can now go on to look at the consequences of this.

Q10.11. Looking at the gene sequence and its translation below, design two different oligonucleotides that could be used to change the asparagine (N) residue to a lysine (K) residue. In both cases design the oligonucleotide such that there are 10 bases of flanking DNA on either side of the base that you choose to change. *Hint.* Refer to the genetic code in Table 8.1.

```
5′–CGATCGCTAATCGCGATCGATAACGATATCGCGATCCTCGGCTCGACT
     R   S   L   I   A   I   D   N   D   I   A   I   L   G   S   T
```

The technical challenge with site-directed mutagenesis is to make it as efficient as possible. Theoretically, from the description above, it should be clear that half the plasmids arising from the mutagenesis procedure will carry the mutation. In practice, the number is very much less than this, for a variety of reasons. The main reason is that *E. coli* contains very efficient systems to spot and correct mutations, and is able to distinguish newly synthesized DNA, so when the plasmid carrying the mutation is

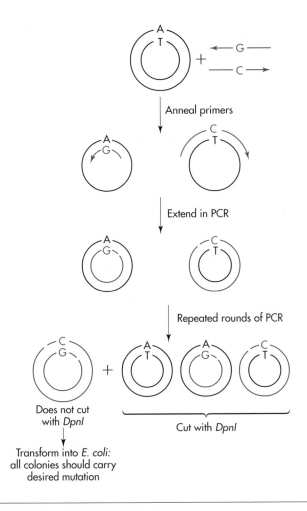

Figure 10.14 One method for high efficiency site-directed mutagenesis.
Two primers are produced which are complementary to each other and to the
target gene, but containing the desired mutation. In the first round of a PCR
reaction using these two primers, plasmids with a single mismatch and a single
nicked strand (as no DNA ligase is present) will be produced. Subsequent PCR
reactions will lead predominantly to a double-stranded plasmid with nicks on
both strands and the new base present on both strands. Digestion with *Dpn*I
before transformation removes plasmid molecules containing non-mutated bases.

transformed into *E. coli*, most of the mutated strands will be converted
back to wild type. Over the years that site-directed mutagenesis has been
in use as a technique, there have been many methods devised to get round
this problem. Below, and in Figure 10.14, we describe one of them which
is currently widely used.

In this method, two primers are used rather than one. The two primers
are complementary to one another and also to the DNA to be mutated,
except that they contain the desired mutation. These primers are mixed

with a plasmid which contains the target DNA which we wish to mutate, and then used in a PCR reaction with a thermostable DNA polymerase with both high processivity and high fidelity. (Processivity simply means that the polymerase is capable of synthesizing long strands of DNA without dropping off the template; fidelity means that it makes very few errors in this process.) Using a PCR-based approach means that there is no need to have the DNA in single-stranded form, as it is denatured at the first step in the PCR cycle. Both primers are thus extended around the plasmid until the polymerase reaches the 5′ end of the primer, at which point it goes no further. As there is no DNA ligase in the mix, the new DNA strand is not covalently joined to the primer, but leaves a "nick", i.e. all the bases have been copied, but the phosphodiester backbone of the DNA is not sealed. Thus after several cycles of the PCR reaction, there will be a mixture of the original parental plasmid, and nicked plasmids which carry the mutation on both strands. To remove the parental non-mutated plasmid, the entire mix is now digested with the enzyme *Dpn*I, which only cuts DNA which is methylated on one or both strands. DNA isolated from most *E. coli* strains is methylated, and hence is cut by *Dpn*I, but the newly synthesized DNA from the PCR reactions is unmethylated and will not be cut. *E. coli* can repair nicked DNA, as long as the nick is opposite an intact strand of DNA, so when the nicked mutated plasmids are transformed into *E. coli*, they are repaired and replicate as normal linearized DNA. This method is very successful and mutations can be obtained with a frequency greater than 80%.

In the examples considered above, we looked at altering a single base. However, site-directed mutagenesis is not limited to single bases. Several bases can be changed at once, by using a primer or primers with several mismatches to the wild-type sequence. In this case, longer oligonucleotide primers are used, to ensure that they anneal to the template strand despite the mismatches. It is also possible to create deletions using oligonucleotide primers. To do this, a primer is designed with sequences complementary to the sequences that flank the desired deletion. When the primer anneals to the plasmid, the bases between these regions are "looped out" of the DNA, as they do not anneal to the primer, and so they are lost from the extended PCR product.

Uses of site-directed mutagenesis

Site-directed mutagenesis is a very widely used technique. It can be used to probe both DNA function and protein function. For an example of the former, imagine that we are studying a stretch of DNA that is known to bind a particular protein, and we wish to find out exactly which bases interact with the protein to enable this binding. We can map the position of the protein on the DNA using a variety of methods that will be discussed in Section 11.4. Having determined where on the DNA the protein binds, individual bases in this region can now be mutated to any other desired base using site-directed mutagenesis, and the mutated DNA checked for the extent to which protein will still bind to it.

Site-directed mutagenesis is perhaps most widely used, however, to probe protein function. By changing the DNA sequence of the gene that encodes a protein, we can change any amino acid to any other amino acid, and this is used to test hypotheses about the roles of particular amino acids in proteins. These hypotheses are often generated by examining the structure of a protein, either on its own or in a complex with other proteins or ligands. For example, it is known that prior to infection of human cells by the HIV virus, the virus has to bind to the surface of the cells. This binding is due to an interaction between the viral coat protein gp120 and a cell surface protein present on some human cells (all parts of the immune system) called CD4. Early structural studies on CD4 identified a number of amino acids which were good candidates for being important in the interaction with gp120, including an exposed hydrophobic residue, phenylalanine, as position 43 in the chain. This residue was mutated to alanine, and it was shown that when this was done the affinity of CD4 for gp120 dropped by about 500-fold, confirming the importance of this residue for gp120 binding. When a structure was obtained (some 5 years later) of the two proteins in a complex together, it was confirmed that the side chain of this phenylalanine protruded into a pocket on the gp120 protein, explaining in terms of structure why this residue is so important in binding.

Similar experiments have been done to probe the importance of individual amino acid residues in enzyme mechanisms, protein structure, interactions with other proteins and DNA, post-translational modification of proteins, localization of proteins, and many more issues.

Questions and Answers

Q10.1. DNA in different organisms varies in the relative proportion of A+T vs. G+C base pairs it contains. Some organisms are referred to as being "GC-rich", which simply means that G and C base pairs are found more frequently in their genomes than would be expected by random chance; others are "AT-rich" for the converse reason. If you were to compare the genomes of an AT-rich and a GC-rich organism, which would you predict would contain the higher number of long ORFs that do not encode proteins but have simply arisen by chance, and why?

A10.1. *GC-rich DNA has a higher frequency of long ORFs. This is because the stop codons (TAA, TAG, and TGA) are relatively AT-rich – only two of the nine bases are not A or T. On a purely random basis, then, these stop codons will be relatively rarer in DNA that has fewer As and Ts, and so there will be more longer ORFs in this DNA.*

Q10.2. Imagine the following scenario. You have detected an ORF in the genome of an organism in which you are interested, and obtained an antibody which has been raised against a protein with a high degree of sequence similarity to the one encoded by this ORF. A western blot with the antibody does reveal a cross-reacting band. You construct a deletion of the ORF from the genome, but when you do another western blot against the organism carrying this deletion, the same band still appears. What could be going wrong?

A10.2. *The most likely scenario is that the organism of interest contains multiple copies of the gene you are looking at, and one or more of these are more highly expressed than the gene which you have deleted. This is in fact not an uncommon finding.*

Q10.3. In the light of the above point, what steps would you take to ensure that the gene of interest had indeed been deleted following transformation with linear DNA containing the flanking regions around an appropriate selectable marker?

A10.3. *The simplest way to check that a particular gene has disappeared from the chromosome is to use PCR with primers internal to that gene, or primers from the flanking sequences. In the first case, the PCR product seen in the wild-type will disappear completely in the mutant; in the second case, a new band will be seen the size of which will depend on the size of the selectable marker gene which has been introduced. Southern blotting and probing with the target gene is an equally valid method for confirming loss of the target gene from the chromosome. In the case of diploid organisms, one copy of the gene is likely still to be present. This situation is discussed in the next chapter.*

Q10.4. Why is it important that the addition to a protein of amino acids encoding GFP does not usually interfere with the function of the protein?

A10.4. *If the protein lost its function, it would be harder to investigate the localization of essential proteins in the cell, since the cell would die when the GFP-fusion was expressed. It is of course often the essential proteins that we are most interested in! There are ways round this, such as expressing the gene for the fusion protein from a plasmid while the normal protein is still present on the genome, but this can cause problems with altered levels of the fusion protein, which may hence not finish up in its correct location, or interference between the fusion protein and the wild-type protein, which may cause loss of function.*

Q10.5. Why was it necessary in designing the primers to ensure that they would put the GFP in the same reading frame as the ORF?

A10.5. *The aim of a translational fusion is, as the name implies, to produce an mRNA which when translated gives rise to a single polypeptide chain possessing the properties of both the fusion partners. This can only be done if the genes for the two partners are fused so that the reading frame does not change between the first and the second partner. If it does, then the chain is likely to terminate prematurely as new stop codons may now be present, and in any event will not have the correct amino acids present for the second partner if the reading frame changes between the two.*

Q10.6. In Figure 10.8, no fusions have been shown with the large periplasmic N-terminal domain. Had they occurred, what would you have predicted about their activities?

A10.6. *Your immediate answer might be that the* phoA *fusions would be active, and the* lacZ *fusions not, as this domain is periplasmic. However, bear in mind that a fusion to this domain would produce a protein with no membrane spanning helices, since the fusion occurs before the first helix begins. Thus, there would be no information in the protein to get it into and across the membrane, and so this domain would remain cytoplasmic, giving high activity for the* lacZ *fusions and low for the* phoA *fusions.*

Q10.7. With such a system, what medium should be used to select for the clones of interest?

A10.7. *The medium will need to select for both plasmids, and also will need to select for the activation of the* his *gene under the control of the GAL promoter. Thus a medium lacking in histidine should be used. The strain containing the bait plasmid alone, or the strains containing the bait and prey plasmids but with no interaction between them, will be unable to grow as they cannot manufacture the histidine that they need, whereas a successful bait–prey interaction will lead to activation of the* his *gene and successful growth on this medium.*

Q10.8. Can you think of a modification to this method that would enable you to identify all the proteins that bind to a particular DNA sequence?

A10.8. *Suppose you create a plasmid where the DNA sequence you are interested in is placed upstream from a promoter that controls a selectable marker (e.g. a* his *gene) or a detectable marker (e.g. a* lacZ *gene). You now create a library where cDNAs are cloned upstream of only the Gal4 activator domain. No Gal4 UAS binding domain is used in this experiment. If one of your clones expresses a protein which recognizes the DNA sequence, the protein will bind to the DNA sequence, and (via the Gal4 activation domain to which it is fused) will activate expression of the selectable or screenable marker. If you answered this question correctly: congratulations! You have independently invented what is known as the "one-hybrid" system. This will be further discussed in the following chapter (Section 11.5).*

Q10.9. Why do you think two columns are used in this procedure, rather than just one?

A10.9. *This makes the procedure much more specific. A small number of proteins in cell extracts will bind to either the protein A column or to the calmodulin column, but it is very unlikely that any protein, other than the fusion protein of interest, will bind to both columns. Thus a very high degree of purification can be achieved by the use of two columns.*

Q10.10. Even given the high degree of purity that this method can achieve, can you think of any good controls that would make it likely that the only proteins detected are the fusion protein and the proteins that it interacts with?

A10.10. *An obvious control is to do the same experiment but with the protein of interest present in its normal state, no longer as a fusion with the TAP tag. In this case, it would be predicted that no proteins will be found eluting from the second column.*

Q10.11. Looking at the gene sequence and its translation below, design two different oligonucleotides that could be used to change the asparagine (N) residue to a lysine (K) residue. In both cases design the oligonucleotide such that there are ten bases of flanking DNA on either side of the base that you choose to change. *Hint.* Refer to the genetic code Table 8.1

```
5′–CGATCGCTAATCGCGATCGATAACGATATCGCGATCCTCGGCTCGACT
     R   S   L   I   A   I   D   N   D   I   A   I   L   G   S   T
```

A10.11. *Lysine can be coded for by either AAA or AAG, while in this example asparagine is encoded by AAC. Thus we could either change the last base in the triplet to an A or a G. To do this, any of the four oligonucletotides below would be suitable. Two will anneal to the strand shown in the figure; the other two to the opposite strand. In each case the mutated base is shown in bold.*

(1) *5′-CGATCGATAA**A**GATATCGCGA*
(2) *5′-CGATCGATAA**G**GATATCGCGA*
(3) *5′-TCGCGATATC**T**TTATCGATCG*
(4) *5′-TCGCGATATC**C**TTATCGATCG*

Further Reading

Functional profiling of the *Saccharomyces* genome (2002) Glaever G *et al.*, Nature, Volume 418 Pages 387–391.
A monumental piece of work where deletions in nearly all yeast genes were constructed and analyzed. The methods used are described in the Supplementary Materials, available online and on http://yeastdeletion.stanford.edu

http://www.ambion.com/techlib/resources/RNAi/
A page of links to excellent resources on all aspects of RNA interference, hosted by a company that has turned making tools for RNAi into a successful business

"Where's Waldo in yeast?" (2003) Wohlschlegel JA and Yates JR, Nature, Volume 425 Pages 671–672.
A News and Views review of papers which used GFP and TAP tags to study location of yeast proteins

Use of *phoA* and *lacZ* fusions to study the membrane topology of ProW, a component of the osmoregulated ProU transport system of *Escherichia coli* (1996) Haardt M and Bremer E, Journal of Bacteriology, Volume 178 Pages 5370–5381.
A paper describing in detail the mapping of a membrane protein using translational fusions, as discussed in Section 10.4 and illustrated in Figure 10.8

A novel genetic system to detect protein–protein interactions (1989) Fields S and Song O, Nature, Volume 340 Pages 245–246.
The paper that first demonstrated use of the yeast two-hybrid system

11 The Analysis of the Regulation of Gene Expression

Learning outcomes:

By the end of this chapter you will have an understanding of:

- *the various methods that can be used to map the transcription start site for a gene*
- *how regulation of gene expression can be investigated*
- *the different approaches that can be used to identify the DNA sequences required for proper regulation of gene expression*
- *the strategies that can be used to identify and then clone the gene for a transcription factor*
- *the techniques that can be used to study the transcriptome and proteome*

11.1 Introduction

The regulation of gene transcription is an essential process in all living organisms and viruses. Even for bacteria and viruses not all genes are expressed at any one time. At its simplest, regulation of gene expression is required for the response to a changing environment, so that nutrients can be efficiently utilized and so that biomolecules not present can be synthesized when required. Complex gene regulation patterns are required for the growth and development of multicellular organisms from a single cell, the fertilized egg. An example of the importance of gene regulation in development is the observation that mice and men have virtually the same "set of genes" and it is therefore how the expression of these genes is regulated that is responsible for correct development, rather than the genes themselves. Regulation of gene transcription is therefore a very active field of research and many techniques involving gene cloning and analysis have been developed. In this chapter you will learn about a range of techniques that can be used to characterize the regulation of expression of a gene. Initially most techniques that will be discussed are used to study regulation of expression of a single gene, but in the latter part of the chapter we will discuss how techniques have been developed to help us exploit the genomic sequences that are available and allow us to monitor gene transcription and translation on a genomic scale. For most of the questions that you might want to ask about gene regulation the experimental techniques are essentially the same for prokaryotic and eukaryotic organisms with only minor modifications.

The co-ordinated expression of a gene involves DNA elements in the vicinity of the gene and many protein factors. The level of complexity is somewhat different between prokaryotes and eukaryotes. In bacteria, all the DNA sequences required for correct control of transcription are located within a few hundred base pairs of the transcription start (see Figure 8.1). Immediately upstream from the transcription start is the RNA polymerase binding site and then upstream from this there may be activator binding sites. Repressor binding sites may be located at different positions within the promoter; they may overlap, be upstream of or downstream of the RNA polymerase binding site. Figure 11.1a shows the sequence elements required for the regulation of the *lac* operon from *E. coli*.

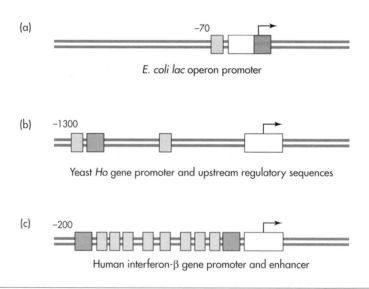

(a)

−70

E. coli lac operon promoter

(b)

−1300

Yeast *Ho* gene promoter and upstream regulatory sequences

(c)

−200

Human interferon-β gene promoter and enhancer

Figure 11.1 Schematic representations of the regulatory regions for three genes. The upstream limit of regulatory sequences is indicated. Regions responsible for binding RNA polymerase and general transcription factors are shown as white boxes, activator and repressor binding sites are shown as gray and blue boxes respectively. a) The *E. coli lac* operon contains the binding sites for RNA polymerase, for one activator protein and for one repressor protein. b) The yeast *HO* gene contains the core promoter responsible for recruiting the general transcription machinery, two sites that bind two different activator proteins and a repressor binding site. c) The human interferon-β gene contains the core promoter responsible for recruiting the general transcription machinery, several binding sites for activator proteins and two sites that bind proteins involved in repression.

In eukaryotes the core promoter, where the transcription machinery assembles, is normally upstream from the first exon. Some genes have several core promoters that drive transcription from different start points, maybe even from within an intron. Immediately upstream from the core promoter is the proximal promoter, which typically contains binding sites for several transcription factors. In addition most genes have associated enhancer elements. Enhancers are regions of DNA that contain binding sites for transcription factors, they do not contain binding sites for RNA polymerase or the general transcription factors and therefore function by activating transcription from a core promoter. Enhancers can be several kilobases upstream or downstream from the core promoter. A gene may have several enhancers that can be utilized at different times, by different signals or in different cell types. Examples of two eukaryotic promoters are shown in Figure 11.1b and c. It is beyond the scope of this book to discuss the complexities of transcription, but you should be aware that in eukaryotes these DNA elements drive the assembly of more than fifty protein components including RNA polymerase, and that many activators work by not only interacting with the transcription machinery but by altering the chromatin environment to make it more suitable for transcription. Despite the differences in the mechanisms for control of transcription in prokaryotes and eukaryotes the experimental techniques used to investigate these processes in the two groups of organisms are essentially the same with only minor modifications.

This chapter is organized into five sections, each of which deals with one aspect of gene expression, each one answering a particular question regarding the transcription process. The types of questions you may have when you first start to investigate a particular gene are:

- Where does transcription start?
- Is the gene subject to regulation?
- When is the gene switched on and when is it off?
- Where are the important regulatory elements in the DNA sequence?
- What protein factors bind the DNA elements identified?
- How do these protein factors interact with each other and with other proteins involved in transcription?

Normally, the cloned gene is the starting point for any study of the regulation of a gene's expression. In prokaryotes, most regulatory elements are within the first 200 bp upstream from the transcription start; therefore if studying a bacterial promoter you would require a clone spanning the 5′ end of the coding region and approximately 200–300 bp of sequence upstream from the translation start codon. In eukaryotes, since enhancer sequences can be several kilobases away from the transcription start you would ideally have a genomic clone that, in addition to containing at least the first exon of the transcribed region, would contain several kb of upstream sequence. Even so, this may not encompass all the elements required for proper regulation. The DNAse I hypersensitivity assay can be used to help identify and determine the position of distant enhancers (see Section 11.4).

11.2 Determining the Transcription Start of a Gene

One of the first questions that needs to be answered when studying the transcriptional regulation of a gene is the location of the transcription start, i.e the point on the DNA template at which mRNA synthesis begins. One of the reasons that this is important is that once you have located the transcription start it is often possible to deduce possible promoter elements based on their location with respect to the start point. For example, most genes in *E. coli* are transcribed by RNA polymerase that is directed to the transcription start by a transcription factor, called σ^{70}, which recognizes and binds sequence elements positioned 10 and 35 bases upstream from the transcription start. In eukaryotes the core promoter may contain conserved sequences overlapping the transcription start, 25 bases upstream or 30 bases downstream.

There are several techniques that can be used to map the transcription start point. We will discuss three approaches: primer extension, nuclease protection, and rapid amplification of cDNA ends (RACE). All three approaches require that you have some sequence information within the 5′ end of the transcript.

Primer extension

Primer extension is a relatively easy way to determine the transcription start point of a gene (Figure 11.2). All that is required is a source of cells in which the gene is actively transcribed and a radiolabeled oligonucleotide primer complementary to a region within the 5′ end of the target transcript, reverse transcriptase and nucleoside triphosphates. How to radiolabel the 5′ end of an oligonucleotide is discussed in Box 11.1.

The technique can use either total RNA or mRNA, which is purified from cells isolated in conditions when the gene is expressed. How you purify mRNA away from the abundant rRNA and tRNA was discussed in Chapter 4 (Section 4.18 and Figure 4.12). The RNA population is used as template for a reverse transcription reaction primed by the radiolabeled oligonucleotide. The oligonucleotide, which is specific to the gene of interest, should hybridize only to the mRNA of interest. Reverse transcriptase (see Box 4.2) can then be used to synthesize a DNA copy of the mRNA using, as a primer, the labeled oligonucleotide. DNA synthesis will continue until reverse transcriptase reaches the 5′ end of the transcript when it will "fall off" the mRNA template. Assuming that all the original transcripts started from the same point, i.e. had the same 5′ end, the product of this reaction will be a labeled

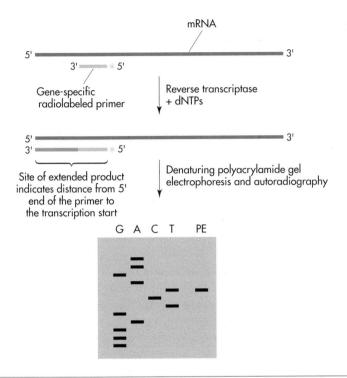

Figure 11.2 Transcription start site mapping: primer extension. RNA is purified, a gene-specific radiolabeled (*) primer is hybridized to the RNA for the gene of interest and the primer extended by reverse transcriptase in the presence of dNTPs. The size of the extended product (PE) is determined by denaturing polyacrylamide gel electrophoresis. A sequencing reaction (G, A, C and T), using the same primer and the cloned gene as template, is used to calibrate the gel and determine the length of the extended product.

DNA fragment of one specific size. The size of the fragment, relative to the position where the primer hybridized, allows you to map the transcription start point. You should note that if you are mapping the start point for a eukaryotic gene, your primer should anneal within exon 1.

Q11.1. You are mapping the transcription start for a eukaryotic gene using primer extension and the intron/exon boundaries are not known. What effect would using a primer that anneals to a sequence in exon 2 have on the predicted transcription start?

Nuclease protection
An alternative approach to primer extension that is more robust, but technically slower to complete, is nuclease protection mapping (Figure 11.3). Two alternative nucleases can be used (S1 nuclease or RNAse), but the

Box 11.1 End Labeling

In end labeling, an enzyme such as T4 polynucleotide kinase is used to add a radioactive phosphate on to the 5′ end of a DNA molecule. This technique can be used to label both oligonucleotides and double-stranded DNA fragments.

Preparing a DNA fragment that is radiolabeled at one 5′ end

For some assays, such as primer extension and footprinting, it is important that one specific 5′ end is labeled. In (a) in the figure below, the DNA is cleaved with a restriction enzyme that cuts adjacent to the region to be analyzed. The cut DNA is then treated with the DNA modifying enzyme alkaline phosphatase (Section 3.8). Alkaline phosphatase removes the 5′ phosphate that is present on the ends of the restriction fragments. The enzyme is inactivated and the DNA cleaved with a second restriction enzyme downstream from the region to be sequenced. You do not treat with alkaline phosphatase after the second restriction enzyme digest and therefore the second cut site will still have the 5′ phosphate. So at this point you have a DNA fragment with the 5′ phosphate removed from one end but not the other. This allows you to add a radiolabeled phosphate to the dephosphorylated end by treating with polynucleotide kinase, in the presence of ATP that contains ^{32}P at the gamma position, $[\gamma\text{-}^{32}P]$ATP (Box 5.3). The product of this procedure is a DNA fragment with a ^{32}P-labeled phosphate on only one of the 5′ ends.

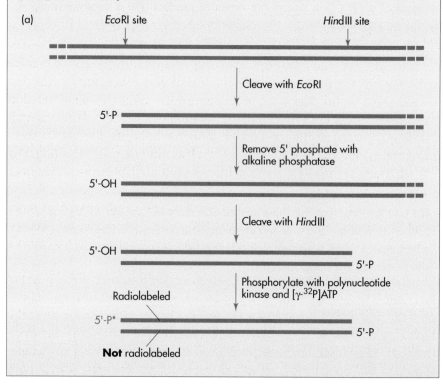

In (b), the strand that is radiolabeled is now cleaved and the DNA is denatured, this generates three single-stranded DNA molecules, one full length and the other two from the strand that was cleaved. The one fragment upstream from the cleavage site will be radiolabeled, whereas, the fragment downstream from the cleavage point and the full length complementary fragment will not be. The important outcome of this procedure is that only the radiolabeled DNA is visualized.

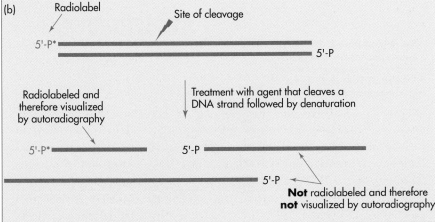

principles of the two approaches are essentially the same. A synthetic DNA or RNA probe is hybridized to the 5′ end of the mRNA. The region of the probe that is hybridized to the mRNA is protected from the action of enzymes which cleave single-stranded nucleic acid molecules, the size of the heteroduplex is determined and the transcription start mapped relative to the 5′ end of the probe. The S1 nuclease mapping procedure was the first to be developed, but has since been superseded by the RNAse protection assay. Both approaches rely on the generation of a probe of known length covering the transcription start and complementary to the mRNA transcript, this is labeled using a technique such as random prime labeling (Box 5.4). S1 nuclease mapping requires a single-stranded DNA probe, whereas RNAse mapping requires a single-stranded RNA probe. Since the RNAse protection assay is used more routinely this will be discussed in more detail. How a single-stranded labeled RNA probe is generated is discussed in Box 11.2. The labeled RNA probe is hybridized to the mRNA to form a double-stranded RNA–RNA duplex. The two ends of the duplex are determined by the 5′ end of the mRNA (the transcription start) and the 5′ end of the labeled probe. Since the probe is generated specifically for this experiment the position of the 5′ end is known. By determining the length of the duplex you will know how far the 5′ end of the probe is from the transcription start. The length of the duplex is determined by digesting any single-stranded RNA probe by RNAse. RNAse will not digest the RNA–RNA duplex.

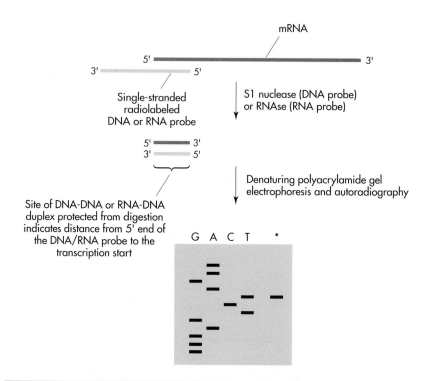

Figure 11.3 Transcription start site mapping: Nuclease protection. RNA is purified and a labeled RNA probe complementary to the 5′ end of the mRNA hybridized. The probe extends from within the gene near the 5′ end to past the likely transcription start site. A duplex forms that extends from the 5′ end of the probe to the transcription start. The size of the duplex is determined by digesting the single-stranded part of the probe with S1 nuclease or RNAse. The size of the protected DNA/RNA probe (*) is then determined by denaturing polyacrylamide gel electrophoresis. A sequencing reaction (G, A, C and T) is used to calibrate the gel.

Box 11.2. Production of Labeled RNA Probes for Use in the RNAse Protection Assay

It is possible to purchase a number of different plasmids which allow you to make RNA probes from a cloned DNA template. One example is the pGEM5Zf(+) plasmid. This plasmid contains a multiple cloning site that is flanked by an SP6 promoter and a T7 promoter on either side. The SP6 and T7 promoters are recognized by RNA polymerase encoded by the phages SP6 and T7, respectively. For an RNAse protection assay a fragment covering the transcription start of your gene is cloned into the multiple cloning site of pGEM5Zf(+). Depending on the orientation of cloning, either SP6 RNA polymerase or T7 RNA polymerase is used to generate an RNA probe that is complementary to the mRNA for the gene and overlaps the transcription start. Plasmid DNA can be purified and used as template by one of the bacteriophage RNA polymerases in the presence of nucleoside triphosphates (GTP, CTP, ATP and UTP) and one labeled nucleoside triphosphate (typically [α-^{32}P]UTP). This labeled RNA probe can then be used in an RNAse protection assay.

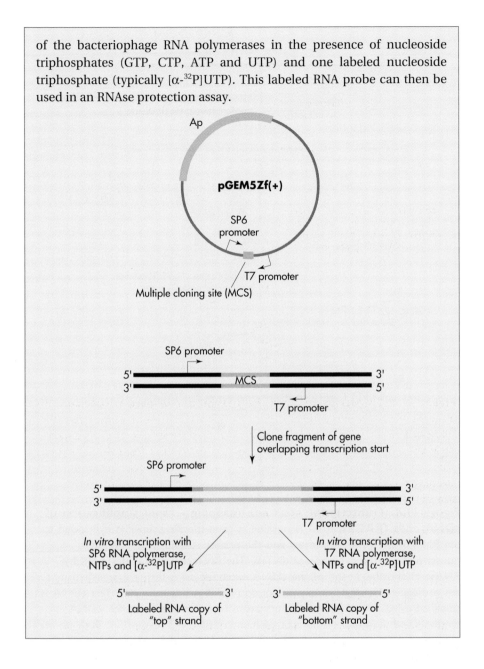

Following digestion the length of the labeled probe protected from digestion is determined by analysis on a denaturing polyacrylamide sequencing gel.

Rapid amplification of cDNA ends

Rapid amplification of cDNA ends (RACE) is a technique that is designed to facilitate the cloning of the 5′ end of a cDNA which should include the region where transcription starts (Figure 11.4). In addition to being a

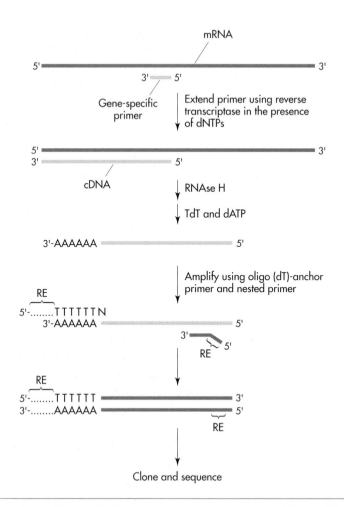

Figure 11.4. Transcription start site mapping – rapid amplification of cDNA ends (RACE). RNA is purified, a gene-specific primer is hybridised to the RNA of the gene of interest and the primer extended by reverse transcriptase in the presence of dNTPs. The RNA template is degraded by RNAse H and the 3′ end of the cDNA extended using terminal transferase in the presence of a single deoxynucleotide (dATP in this example). The cDNA is amplified using a "nested" primer upstream to the primer used for cDNA synthesis and a second primer complementary to the 3′ extension. Both primers contain sites for a restriction enzyme at their 5′ end (RE). In this example an oligo (dT)-anchor primer is used, the "primer" is in fact a mixture of four primers which, in addition to having a run of Ts, have either G, A, C or T (shown as N in the diagram) at their 3′ end (the anchor), this ensures that the primer anneals to the cDNA at the junction of the polyA tract and the original extension product. The amplification product is cloned and sequenced. The transcription start site is adjacent to the poly T tract in the sequence.

technique often employed to generate full-length cDNA clones, it can also be used to map transcription start points. The technique has some advantages over primer extension and nuclease mapping. Firstly, it involves a PCR step and is therefore very sensitive, requiring only a small amount of mRNA from the gene of interest. Secondly, the RACE product is normally sequenced and therefore the actual sequence of the 5′ end of the transcript is confirmed, rather than obtaining a relative distance from a fixed point within the gene. This can therefore avoid problems that can arise due to the presence of unknown intron sequences (see Question 11.1).

The initial step in the RACE protocol is a primer extension but in the absence of label. This yields a single-stranded cDNA where the 3′ end represents the transcription start. The next part of the procedure is to use PCR to amplify the cDNA. PCR requires two primers flanking the region to be amplified. One primer could be the same as used for primer extension, but rather than using the same primer, one just internal to the original is used to help reduce non-specific amplification. The sequence at the 3′ end of the cDNA is not known and so a specific primer cannot be used. Instead, a sequence is added to the 3′ end to which a primer can hybridized. Two methods are generally used to do this. One method is to extend the 3′ end by the addition of a short single-stranded oligonucleotide using RNA ligase, an enzyme that can be used to ligate single-stranded nucleic acids. Alternatively, a homopolymer can be added to the 3′ end using a terminal deoxynucleotidyl transferase (TdT) and one deoxynucleoside triphosphate, e.g. dATP. In this case a polyA tail will be added to the 3′ end and therefore a polyT primer can be used in the PCR reaction. PCR is then used to amplify a DNA fragment containing the 5′ end of the transcript and this is cloned into a plasmid vector. Several independent clones are then sequenced to determine the transcription start point.

Q11.2. In the final PCR step of the RACE protocol, rather than use the same primer that was used for the original primer extension step, one just internal ("upstream") is used. Why does this help reduce non-specific amplification?

You may wonder which of the three techniques discussed above would be best for determining the transcription start. Each approach has different advantages and disadvantages and best practice would be to use at least two different techniques so that the transcription start site is mapped with more confidence. Primer extension is the easiest technique to do, but is often prone to artefacts. Primer extensions also require more RNA template because the product of the extension reaction is only labeled once, at the 5′ end. RNAse protection is more sensitive because the RNA probe is labeled several times during synthesis, but generating the RNA probe requires a cloning step. RACE is very sensitive because it includes a PCR amplification step, although reverse transcriptase is prone to pausing before the end of the RNA template is reached and can give rise to a truncated cDNA product, which results in the transcription start being mapped incorrectly.

11.3 Determining the Level of Gene Expression

Gene regulation is of course not an all-or-nothing phenomenon, with genes either being "on" or "off"; rather, individual genes can display a huge range of levels of expression relative to one another, and the level of expression of a single gene can also vary considerably under different conditions or at different developmental phases. One of the most important pieces of data that you require if you wish to study the transcriptional regulation of a gene is to know under what conditions the gene is transcribed, and the amount of transcription which takes place. In many cases you will determine the level of gene regulation before you map the transcription start. The techniques described below for determining the level of gene regulation do not require the precise transcription start to be known, but they do require sequence information for at least part of the gene. Three of the techniques that will be discussed below (northern analysis and dot blots, QRT-PCR and nuclear run-ons) are used predominantly to determine the level of the wild-type promoter in its normal genetic context. However, when you want to characterize regulatory elements, for instance by mutagenesis, you would normally use reporter gene assays. In addition to the techniques described in this section it is also possible to use primer extension and nuclease protection assays to measure not only the transcription start but also the level of transcription if the products of the primer extension or nuclease protection assay are quantified. However, these techniques are not as robust as northern blots and QRT-PCR.

Northern analysis and dot blots

Northern analysis and dot blots can be used to determine the level of a particular RNA species in a population of RNA that has been purified from either cells in culture or from a tissue sample. The technique is very similar to Southern analysis which was described in Chapter 6 (Box 5.2). When using Southern analysis a labeled DNA probe is hybridized to its complementary DNA sequence, which allows the detection of this "target" sequence, following agarose gel electrophoresis and transfer of the DNA to a membrane. Northern analysis (also called northern blotting) is essentially the same, with the main difference being that a labeled DNA probe is hybridized to RNA that has been separated by gel electrophoresis, rather than DNA.

Northern analysis involves first purifying RNA from tissue samples or cells growing in culture. It is important that the sample contains no DNA contamination (see Question 11.3). Once the RNA has been purified the

RNA molecules are separated according to size on an agarose gel before being transferred to a nitrocellulose or nylon membrane. The membrane is than placed in a sealed container with a labeled probe complementary to the gene of interest. The probe will hybridize to complementary mRNA. The probe that has not hybridized to the target is washed from the membrane and hybridization detected by autoradiography or phosphor-imager analysis. Separating the RNA population by gel electrophoresis means that you obtain information about the length of the mRNA as well as its abundance; for example, if the mRNA is cleaved or transcription starts from a different start point then the size of the transcript detected will differ from that predicted.

However, often the mRNA has been previously characterized and you simply need to determine the amount of mRNA for a particular gene. In this case a modified version of northern analysis, called dot blots, can be used. The dot blot procedure is shown in Figure 11.5. For dot blots, rather than separating the RNA sample on an agarose gel, the sample is applied directly as a single spot to the membrane. Many RNA samples can be applied to the same membrane; these samples may be from different cell types, from cells growing under different conditions or cells at different stages of development. The membrane with the RNA samples bound is placed in a sealed container with a labeled probe complementary to the gene of interest. If a spot on the membrane contains mRNA for the gene of interest the probe will hybridize. Probe that has not hybridized to its target mRNA is washed off the membrane. The presence of hybridized probe is then detected by autoradiography or phosphor-imager analysis. A spot on your autoradiograph or phosphor-image tells you that the transcript from the gene was present in the RNA samples and therefore the gene was being transcribed in the cell from which the RNA was purified. If, as is normally the case, the labeled probe is in excess, the amount of probe that hybridizes to the RNA sample will be proportional to the amount of RNA for that gene in the sample. Since the probe is labeled we can quantify the amount of probe that has hybridized which gives a readout for the amount of RNA in the starting sample.

Q11.3. Why is it essential to remove all DNA from the RNA sample before northern analysis?

Quantitative real-time reverse transcriptase–polymerase chain reaction
Quantitative real-time reverse transcriptase–polymerase chain reaction (QRT–PCR) is now a common method for determining the amount of a particular mRNA (Figure 11.6). The protocol has essentially two steps. In the first, a single-stranded cDNA is made from the mRNA using reverse transcriptase (Section 4.18) and then the amount of this cDNA is determined using quantitative real-time PCR (see Section 3.18).

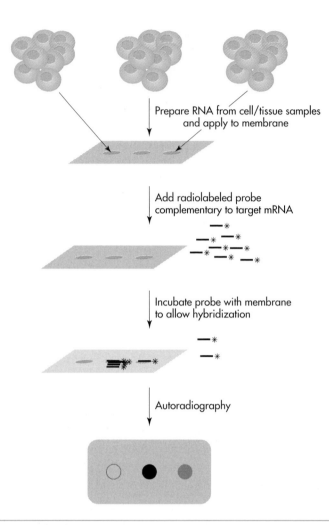

Figure 11.5 Northern analysis. RNA is prepared from different test cells or tissues; it can either be separated by electrophoresis or, as in this example, spotted onto a nylon or nitrocellulose membrane. A labeled (*) probe specific for the gene of interest is incubated with the membrane and will hybridize to its complementary RNA. Hybridization can be detected by autoradiography if using a radiolabeled probe or chemiluminescence if the probe is labeled with biotin. The intensities of the spots can be quantified to give an estimate of the amount of probe hybridization to each spot, which in turn indicates the amount of the corresponding mRNA in each of the starting samples.

The cDNA synthesis step can use a primer that is specific for a single gene, if so reverse transcriptase will only copy the gene of interest into cDNA. Alternatively, random hexanucleotides can be used to prime the reverse transcription reaction. Random hexanucleotides are a population of oligonucleotides representing all possible six base sequences. Using

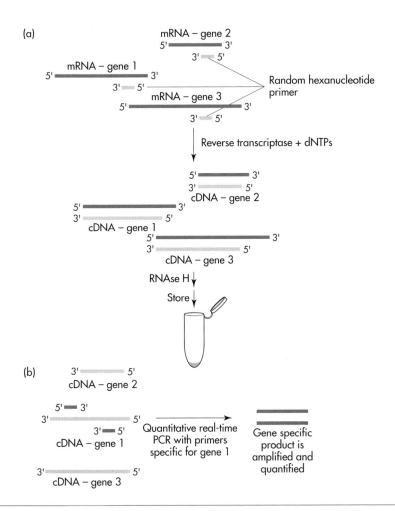

Figure 11.6 Quantitative real-time reverse transcriptase-PCR. a) RNA is purified and random hexanucleotides are used to prime the synthesis of cDNAs by reverse transcriptase for all RNAs present in the starting sample. RNA is removed by RNAse H and the samples stored. b) The amount of any one particular cDNA in the sample can be quantified by quantitative real-time PCR using gene-specific primers.

random hexanucleotides means that all RNA molecules should be reverse-transcribed into cDNA. For the subsequent QRT–PCR reaction gene-specific primers are used. These primers direct the amplification of a short segment of the cDNA. How QRT–PCR is used to quantify the amount of DNA, in this case cDNA, is described in Section 3.18.

Q11.4. The QRT–PCR protocol utilizes random hexanucleotide primers to generate the cDNA prior to QRT–PCR. What is the advantage of using random hexanucleotides rather that a gene-specific primer?

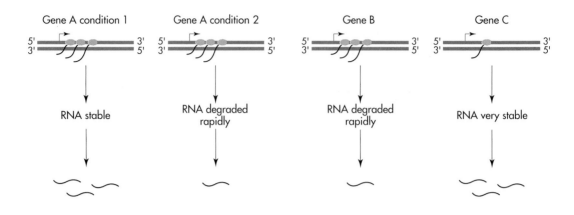

Figure 11.7 RNA stability as well as the rate of transcription initiation can affect the steady-state levels of a mRNA. Genes A (condition 1) and B are transcribed at the same rate, however, the mRNA for gene B is rapidly degraded leading to a difference in the steady-state mRNA levels of the two genes. Gene C is transcribed at low levels, but its mRNA is very stable and therefore it has the same steady-state mRNA levels as gene A. Altering the rate of mRNA degradation can alter the steady-state mRNA levels (gene A conditions 1 and 2).

Nuclear run-on

Although the expression of many, if not most, genes is controlled at the level of transcription initiation, many are regulated at later steps such as elongation, termination and RNA stability, and in eukaryotes by splicing, polyadenylation and mRNA transport (e.g. Figure 11.7). Using both northern blots and QRT–PCR you can determine the steady-state level of an RNA transcript. However, these techniques will not provide any information as to whether it is transcription initiation or one of the other steps that is being regulated. One technique that does provide some evidence that the level of mRNA is regulated at the point of transcription initiation is the nuclear run-on (Figure 11.8). This procedure is normally used when studying regulation of a eukaryotic gene. The nuclear run-on assay indicates the number of RNA polymerase molecules transcribing a gene at the time of the assay. It can therefore be used to differentiate between regulation due to changes in transcription initiation or changes in another step in gene expression. Changes in the rate of initiation will result in an altered nuclear run-on readout, whereas regulation of a subsequent step, for example control due to changes in RNA stability, will not result in a difference in the nuclear run-on readout.

The nuclear run-on assay is shown in Figure 11.8. Cells are isolated at different times or under different conditions when levels of expression of the gene of interest are thought to change. The cells are cooled on ice, the cells lysed and nuclei isolated by centrifugation. The nuclei are then warmed to 37°C in the presence of all four nucleoside triphosphates

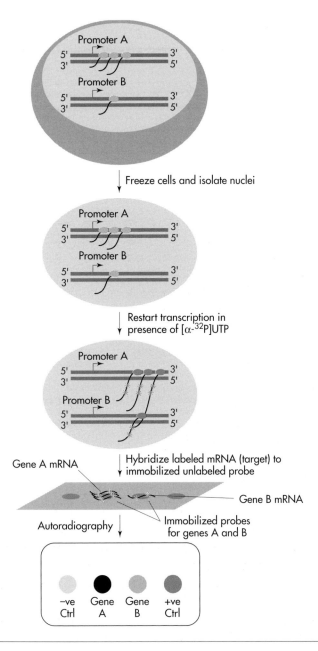

Figure 11.8 The nuclear run-on assay. Cells are isolated and cooled on ice to pause transcription and nuclei are isolated. The cells are warmed in the presence of $[\alpha\text{-}^{32}P]UTP$ (*) and transcription of genes that were being transcribed when the cells were cooled resumes; there is no new initiation of transcription. The radiolabeled mRNA produced is incubated with a membrane containing an immobilized probe. The probe is a single-stranded DNA complementary to the mRNA for the gene of interest. Several probes for different genes can be spotted onto the membrane. Following hybridization and autoradiography the intensities of each spot can be quantified to give the amount of labeled target hybridized to each probe, which indicates the level of transcription of each gene when the cells were isolated.

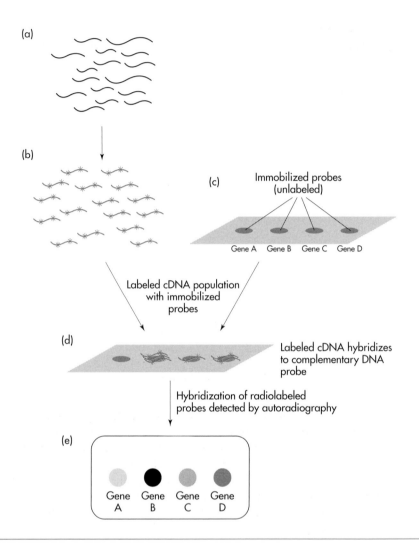

Figure 11.9 Reverse northern dot-blot analysis. a) RNA is purified and random hexanucleotides used to prime cDNA synthesis by reverse transcriptase in the presence of dNTPs and a labeled deoxynucleoside triphosphate (*) to give labeled cDNAs b). Unlabeled DNA probe(s) complementary to the gene(s) of interest are immobilized on a membrane c). The labeled cDNA is incubated with the membrane d) and hybridization quantified by autoradiography e).

including radiolabeled [α-^{32}P]UTP. The paused RNA polymerase molecules that were transcribing the gene when the cells were isolated and cooled now resume transcription and generate a radiolabeled product. The labeled product for the gene of interest is quantified by using "reverse" northern analysis (Figure 11.9). In this method, a DNA probe complementary to the gene of interest is immobilized on a nylon or nitrocellulose membrane (Figure 11.9c) and the labeled products from the nuclear run-on (Figure 11.9b) are hybridized to the immobilized probe. Since the

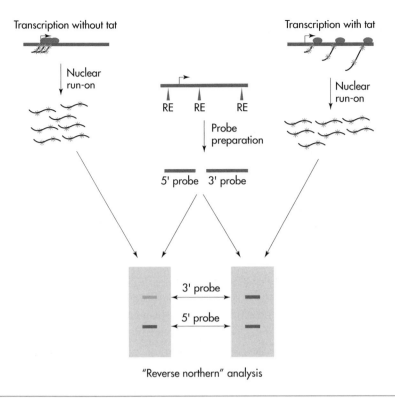

Transcription without tat

Nuclear run-on

RE   RE   RE

Probe preparation

5' probe   3' probe

Transcription with tat

Nuclear run-on

3' probe

5' probe

"Reverse northern" analysis

Figure 11.10 Using the nuclear run-on assay to study regulation of gene expression by the HIV-1 activator, tat. A plasmid containing the HIV-1 DNA was cut with two restriction enzymes, the fragments (probes) separated by gel electrophoresis and transferred to a nylon membrane. One probe was complementary to the 5' end of the transcript including the TAR sequence and one probe was complementary to the 3' end. Labeled (*) RNA was hybridized to the immobilized probes on the nylon membrane and the level of hybridization quantified by autoradiography. (See Kao *et al.* 1987 for further details.)

mRNA is labeled we can quantify the amount of mRNA that has hybridized to the immobilized probe, which gives a readout for the level of transcription of the gene at the time the cells were isolated.

The nuclear run-on assay can often provide information about which step, post-initiation, is being regulated. The nuclear run-on assay has been used to study activation of HIV-1 gene expression by the virally encoded activator, tat (Figure 11.10). Tat increases the steady-state levels of full length mRNA, not by increasing the rate of transcription, but by stopping RNA polymerase from terminating transcription prematurely while transcribing the gene. Following termination, any truncated transcripts are degraded to a 59-base RNA species called TAR. A slightly modified nuclear run-on protocol to that described above was used. The experiments were completed in tissue culture using two strains, one producing the activator tat and one in which tat was not produced. The assays were started by addition of radiolabeled [α-^{32}P]UTP to permeabilized cells and the amount of labeled RNA produced determined following a 15-min incubation. The RNA was purified and quantified by "reverse" northern analysis. A plasmid containing the HIV-1 DNA was cut with two restriction enzymes which generated several DNA fragments; these were separated by gel electrophoresis and transferred to a nylon membrane, as discussed for Southern analysis. Two of the restriction fragments (immobilized DNA probes) covered the tat regulated transcript. One probe was complementary to the 5′ end of the transcript including the TAR sequence and one probe was complementary to the 3′ end. The labeled RNA was hybridized to the immobilized probes on the nylon membrane and the level of hybridization quantified by autoradiography. The results showed that the amount of RNA complementary to the 5′ end of the HIV-1 transcript was the same in the presence or absence of tat, so the level of transcription initiation is the same in both cases. The amount of RNA complementary to the 3′ end of the transcript was increased in the presence of tat. This indicates that transcription can initiate at high levels in the presence or absence of tat, but in the absence of tat transcription terminates before a full-length message is produced.

> **Q11.5.** Which technique, standard northern or reverse northern analysis would be best to study (a) the levels of mRNA for the same gene in several RNA samples (e.g. from different cell types) or (b) the levels of mRNA for several genes in the same RNA sample?

Reporter gene assays

If you want to determine the level of expression of a particular gene you might think it possible to simply measure the level of the protein product directly. However, most gene products are difficult or impossible to assay. Even more importantly, the promoter of interest may be one for an essential gene, so if you want to subsequently study the promoter by introducing mutations and studying the effect on transcription any loss of activity would lead to the death of the cell. By cloning the promoter alone, fusing it to a reporter gene, and introducing the fusion into the organism, the regulation of the promoter can be studied without any effect on the endogenous gene from which the promoter was originally isolated.

Promoter probes are specialized vectors that contain a promoter-less gene, upstream from which is a cloning site into which potential promoters can be cloned. Promoter probes rely on a simple concept, which is that a promoter can be fused to any other gene, and that gene will now be regulated by that promoter. This is an enormously useful idea, since it means that any promoter can be isolated and fused to a gene which produces a protein which is simple to assay (Figure 11.11). Promoters can then be

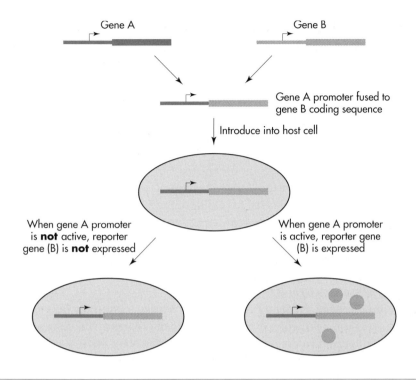

Figure 11.11 Reporter gene assays. Gene A is subject to regulation in different cell types or under different environmental or developmental signals. Regulation is due to the properties of the promoter. Gene B is a gene which expresses a protein with an easily detectable or assayable property. Fusing the coding sequence of gene B with the promoter for gene A produces a reporter gene construct in which the regulation of the gene A promoter can be studied by measuring or detecting the gene B product.

examined to see how their activity varies in specific tissues or cell types, over time as an organism grows and develops, or in response to external effects such as hormones or stresses. Also, mutations can be made in the promoter (still fused to the gene which produces the easily assayable protein) and the effects of these mutations on levels of the protein can be studied. Mutations which alter DNA bases which are essential for promoter function will result in the loss of the protein, whereas mutations in non-essential bases will have no effect. Thus by using such assays, a picture of the entire promoter can be built up. The proteins which are assayed are referred to as reporter proteins, and the genes which encode them are called reporter genes.

An effective reporter protein needs a number of properties, listed below. It is probably true to say that no one reporter protein has all of these properties fully optimized, and the choice of reporter used depends on the particular organism under study and the particular study being done.

Box 11.3 Examples of Reporter Proteins and their Assays

β-galactosidase

This protein, originally from *E. coli*, is encoded by the *lacZ* gene; there are a variety of assays used for this protein. The presence of β-galactosidase can be revealed using the dye X-gal, which is cleaved to form a blue product when β-galactosidase is present. This has already been discussed in the context of blue–white selection (Chapter 2) but can also be used for staining tissue sections of transgenic organisms. This gives a quick indication of where the gene is expressed, and can be used very accurately to look at expression in individual cells. In addition, the substrate ONPG is cleaved by β-galactosidase to give a yellow product that can be measured in a spectrophotometer, and this is widely used to quantitate the level of β-galactosidase.

Green fluorescent protein (GFP)

This protein from the jelly fish *Aequorea victoria* is naturally fluorescent, emitting green light when excited with blue or UV light. The fluorescence can be measured to determine the amount of protein expressed. A great advantage of this reporter is that it is active in cells without any need for fixing or adding co-factors or substrates, and so can be used to monitor gene expression in living tissue. This includes monitoring gene expression in "real time"; for example, following the change in expression from a given promoter during tumor development. Variants have been also produced which fluoresce at different wavelengths, enabling two different promoters to be monitored in the same cell at the same time.

Luciferase

The luciferase protein, from the firefly *Photinus pyralis*, is luminescent; it converts chemical energy to light energy (unlike GFP, which absorbs light energy in one wavelength and emits it in a different one). When supplied with its substrate (luciferin) in the presence of oxygen and ATP, luciferase emits yellow–green light, which can be detected in a variety of ways. If the luciferase is in a crude cell-free extract, light is measured using a photomultiplier to measure the intensity of the light. If the luciferase is in a whole organism, a video camera (which has to be fitted with appropriate devices to detect low light intensity) can be used to detect the luminescent signal.

A reporter protein should:

- Be easy to assay. Ideally the assay will be fully quantitative, so that a precise measure of the level of the protein (and hence by inference the extent to which the promoter is active) can be obtained.

- Be non-toxic to the cell. There is no point in inducing the expression of a protein which will kill the cells that you wish to study!
- Be stable under the conditions of the experiment. You need to be sure that any variation in level of the protein that you see is due to changes in expression from the promoter, rather than the reporter protein itself varying in stability under different conditions.
- Not be present in the organism under study. It is important that there is no background activity produced by the organism, so that the level of activity assayed is due solely to that produced from the promoter of interest.

Many reporter genes producing a variety of proteins are now in common use, some of these are listed in Box 11.3.

An example of the use of a reporter gene assay is a study completed on the *nirB* promoter from *E. coli*. This promoter is controlled in response to anaerobiosis and nitrite. This can be demonstrated by cloning the *nirB* promoter upstream from the *lacZ* gene in a promoter probe vector; transcription from the *nirB* promoter will lead to the expression of the *lacZ* gene product, β-galactosidase. *E. coli* was transformed with the recombinant plasmid and the transformant grown in four conditions: aerobically, aerobically with nitrite, anaerobically, and anaerobically with nitrite. The promoter was shown to be turned on only during anaerobiosis, and that nitrite could drive increased rates of transcription during anaerobic growth (Figure 11.12).

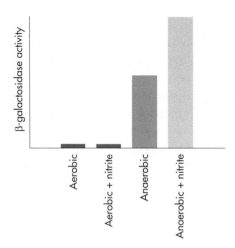

Figure 11.12 Analysis of the *E. coli nirB* promoter using the reporter gene assay. The levels of β-galactosidase expressed under each growth condition is shown in the graph. (See Jayaraman *et al.* 1988 for further details.)

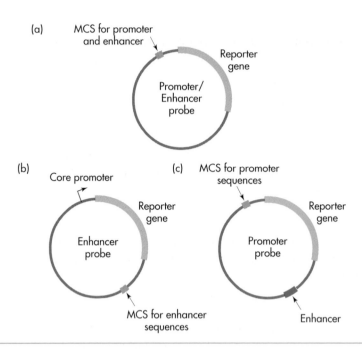

Figure 11.13 Vectors for reporter gene assays. a) Basic promoter probe vector; promoter regions, with associated enhancer regions when present, can be cloned upstream from the reporter gene using the multiple cloning site (MCS). b) Enhancer probe vector contains a core promoter upstream from the reporter gene; downstream from the reporter gene is a MCS into which potential enhancers can be cloned and studied. c) Promoter probe vector contains a generic enhancer downstream from the reporter gene and a MCS upstream into which a potential core promoter can be cloned and studied.

When studying regulation of a eukaryotic gene, a family of promoter and/or enhancer probes can be exploited (Figure 11.13). An enhancer probe vector contains a "generic" core promoter and a multiple cloning site into which potential enhancer can be cloned, so that enhancer dependent activation of the core promoter can be studied. A promoter probe vector contains a "generic" enhancer and a multiple cloning site immediately upstream from the reporter gene into which potential core promoters can be cloned. A third vector that contains neither a promoter nor enhancer can be used to study genome fragments that encompass all the regulatory elements, promoter and enhancers, required for proper regulation. Reporter gene assays in eukaryotes require a method for introducing the recombinant constructs into cells in culture (Box 11.4), using protocols analogous to transformation of bacteria cells.

11.4 Identifying the Important Regulatory Regions

Once you have determined the transcription start site for a gene and shown that changes in gene expression are due to regulation of transcription

---

Box 11.4 Methods for Introducing DNA into Eukaryotic Cells

Introduction of DNA into eukaryotic cells is termed transfection. There are several transfection procedures.

- Calcium phosphate, DEAE-dextran and polycations

The DNA is mixed with either calcium phosphate, DEAE-dextran or polycations and then applied to eukaryotic cells resulting in the uptake of the DNA via endocytosis.

- Electroporation

In a similar procedure to transformation of bacterial cells, the application of a high voltage to cells in the presence of DNA results in the uptake of the DNA by the cells.

- Liposome-mediated transfer (lipofection)

The DNA is treated with a cationic lipid which binds the DNA. This complex then fuses with the cell membrane allowing entry of the DNA into the cell.

---

initiation, the next task is to identify the sequences within the DNA that are responsible for this control. In prokaryotes these sequences are normally the RNA polymerase binding sites and sites for activator and repressor sequences. The RNA polymerase binding site is within 35 base pairs of the transcription start and in most cases activator and repressors will normally bind within a few hundred base pairs upstream and/or downstream of the transcription start. It is therefore relatively straightforward to identify these sites using a combination of approaches including sequence analysis (see Chapter 8), deletion analysis, site-directed mutagenesis (Section 10.6) and footprinting (see below). In eukaryotes, although the core promoter and proximal promoter element are within a few hundred base pairs of the transcription start site, identification of all the sequence elements required for the correct regulation of gene transcription is much more difficult because enhancer elements can be many thousands of base pairs away. Added complexity arises because in many cases a single gene may be controlled by multiple enhancers and there are often other function elements such as locus control regions (LCRs) and matrix attachment regions (MARs). One approach to identify regulatory elements would be to clone a large (about 100 kb) genomic fragment containing the gene of interest and create a reporter gene construct which can be used to create transgenic mice or be introduced into a cell line by transfection. Once proper regulation had been demonstrated deletion analysis could be used to map important regulatory sequences. This approach would be very time consuming and involve a large amount of cloning in order to characterize such a large fragment. An alternative approach is to use the DNAse I hypersensitivity assay.

DNAse I hypersensitivity assay

In eukaryotes the DNA is tightly packaged into chromatin and therefore protected from cleavage by endonucleases such as DNAse I. However, *in vivo* promoter and enhancer elements tend to be either nucleosome-free or have altered nucleosome structure. These regions may be more accessible to nucleases and are therefore cleaved when the chromatin is treated with DNAse I, i.e. they are DNAse I hypersensitive. The DNAse I hypersensitivity assay if often used to identify possible enhancer elements (Figure 11.14). The assay is essentially a modified Southern blot, in which nuclei are prepared and treated with DNAse I prior to purification of the genomic DNA and Southern analysis. The important point is that the chromatin structure should not be disrupted prior to the DNAse I treatment so that the nuclease only cleaves DNA that would be accessible in the normal living cell. Cells should also be isolated when the gene of interest is being expressed.

If you were completing a standard Southern analysis you would expect your DNA probe to hybridize to a genomic fragment of precise size as determined by the distance between the sites of cleavage of the restriction enzyme used. If the DNA is cleaved by DNAse I, however, the fragment that is bound by the labeled probe will be shorter than expected. This difference in size between DNAse I treated and an untreated control can be used to map the DNAse I hypersensitive site and therefore the possible location of potential enhancers and other regulatory elements. The DNA is digested with different restriction enzymes and several different probes used so that the whole region in the vicinity of the gene can be analyzed and the location of each hypersensitive site mapped. If the genome of interest has been sequenced it is relatively simple to design PCR primers to amplify the DNAse I hypersensitive region and hence clone the potential enhancer or promoter region. If sequence information is not available and the enhancer is upstream or downstream of your cloned gene you can use inverse PCR to clone the DNA sequences flanking your gene, which should include the region identified by the DNAse I hypersensitivity assay. Once cloned the region can be further characterized e.g. using EMSAs (see below) and reporter gene assays.

Electromobility shift assay

The electromobility shift assay (EMSA) can be used to identify important regions of a promoter and/or enhancer as discussed below. The assay involves observing the migration of a DNA fragment during polyacrylamide gel electrophoresis (PAGE) in the presence or absence of one or

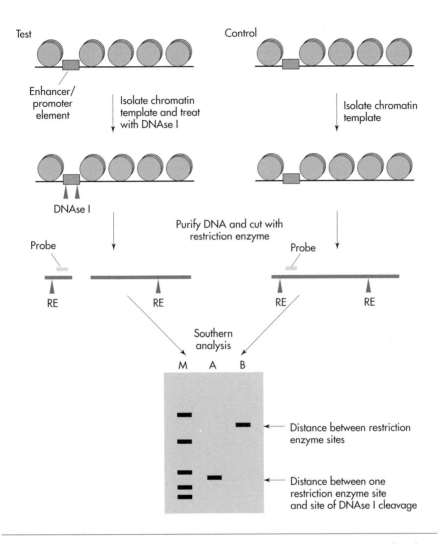

Figure 11.14 DNAse I hypersensitivity assay. Intact chromatin is isolated from cells expressing the gene of interest and treated with DNAse I, which cleaves the DNA where there are no nucleosomes or where the nucleosome structure is altered. The DNA is purified and analyzed by Southern analysis using DNA that has not been treated with DNAse I as a control (lane B). DNAse I cleavage results in detection of a shorter fragment in the Southern blot (lane A), the difference in size can be used to map the position of the DNAse I hypersensitive site relative to the position of the restriction enzyme recognition sites. DNA standards are used to calibrate the gel (M).

more transcription factors (Figure 11.15a). As you will be aware, the migration of a linear DNA fragment during gel electrophoresis is mainly dependent on the length of the DNA fragment. It is possible to analyze DNA–protein complexes using gel electrophoresis and, since these complexes have a higher mass than the free DNA, they will migrate at a slower rate

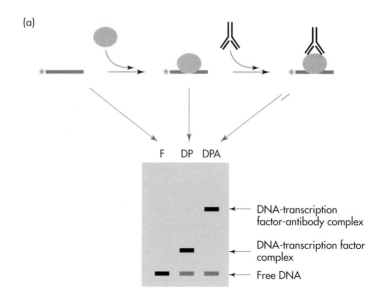

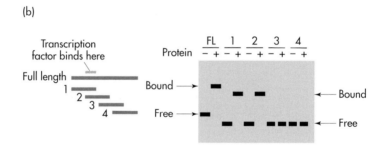

Figure 11.15 The electromobility shift assay (EMSA). a) A labeled (*) DNA fragment migrates according to its size during gel electrophoresis (lane F). When bound by a protein the DNA fragment's mobility during electophoresis is reduced and therefore protein binding to DNA can be detected as a shift in mobility (lane DP). Increasing the mass of the complex further by binding of a second protein, in this case an antibody, results in another shift, often called a supershift. Binding of the antibody to the complex is used to help confirm the identity of the protein present in the original complex. b) The full length fragment binds the transcription factor (lane FL). Four shorter overlapping fragments are used to map the position of the transcription factor binding site. The transcription factor binds to fragments 1 and 2, but not 3 and 4. The binding site on the DNA must be within the common sequence between fragments 1 and 2.

compared with the free DNA. When visualized the position of the DNA–protein complex will have shifted compared with the free DNA hence the term electromobility shift assay. The assay normally requires the formation of the DNA–protein complex *in vitro* with a radiolabeled DNA fragment.

Once separated on a polyacrylamide gel, the free DNA and DNA–protein complex are detected by autoradiography. You may wonder why you cannot visualize the DNA using standard staining such as ethidium bromide. The reason is that the concentration of the DNA used in most EMSAs is too low to be detected by conventional staining; low DNA concentrations are required because the amount of protein is often limiting because you are either using purified proteins and do not have large quantities or you are using a cell extract in which the protein will be present at low concentrations. Although the EMSA will tell you if the DNA fragment contains a binding site for a transcription factor, it will not tell you where on the fragment the protein binds. To map where proteins bind to a large DNA fragment a set of smaller overlapping fragments can be used (Figure 11.15b).

The assay requires a source of the transcription factors thought to bind the DNA. These can be either purified proteins or cell extracts from cells where the gene is thought to be regulated. When using cell extracts, the identity of the transcription factor may not be known, in which case the EMSA may be the starting point for the purification of the factor and then the cloning of its gene.

In other cases once a potential transcription factor binding site has been identified, sequence analysis may provide insight into which transcription factor binds. In this case the EMSA can again be used to help confirm which transcription factor binds the promoter/enhancer region. To do this, an antibody specific for the transcription factor suspected to bind the DNA can be added to the DNA–protein complex, forming a tertiary DNA–protein–antibody complex. This tertiary complex will have a larger mass than the transcription factor–DNA complex and therefore will migrate even more slowly during gel electrophoresis. The antibody is said to "supershift" the complex. So a supershift after addition of an antibody specific to a transcription factor helps confirm that the factor is present in the DNA–protein complex and therefore binds the promoter and/or enhancer fragment. Antibodies specific for a large number of transcription factors can be bought from commercial suppliers.

Q11.6. Figure 11.15b shows how a series of overlapping fragments can be used to map where a transcription factor binds within a large DNA fragment. Why would you use overlapping fragments rather than a series of non-overlapping fragments?

Footprinting

The EMSA is a relatively easy technique that can be used to study DNA–protein interactions and, although it will confirm that a protein binds to a DNA fragment, it gives no direct information about the position on the DNA where the protein binds. As described above, this can be deduced

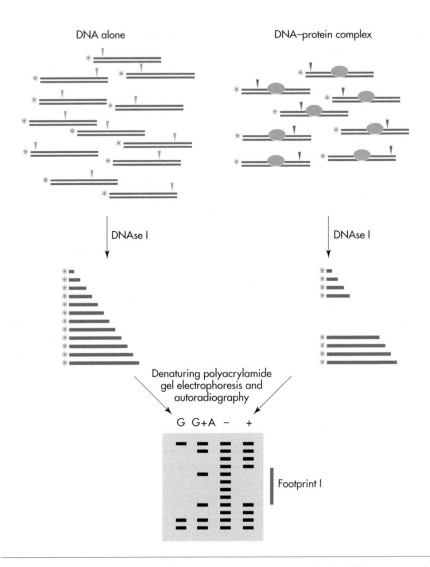

Figure 11.16 DNAse I footprinting. A DNA fragment labeled (*) at one end of one strand is treated with a limiting amount of DNAse I, such that each molecule is cleaved once, the position of cleavage being random (▼). When analyzed by denaturing polyacrylamide gel electrophoresis followed by autoradiography, only the labeled DNA fragments are visualized (one from each cleavage reaction). This produces a ladder (lane −). If a DNA–protein complex is treated with DNAse I, the DNA will be protected from cleavage where the protein is bound and a population of fragments will be missing (lane +). Mapping where the DNA is protected from cleavage indicates where the protein is bound. Sequencing reactions are used to calibrate the gel.

using the EMSA by characterizing several overlapping fragments. A more direct way to determine where a protein binds on a DNA fragment is to use a footprinting protocol. In a footprinting experiment we exploit the ability of a protein bound to DNA to protect the DNA from cleavage. There are several enzymes and chemical reactions that can be used to cleave DNA at random points. Determining which region of the DNA is protected from cleavage tells us where the protein was bound (Figure 11.16). We will discuss footprinting using DNAse I as an example.

All footprinting protocols require an end-labeled DNA fragment labeled only at one end of one strand (Box 11.1). This end-labeled fragment is then subject to cleavage by DNAse I at a very low concentration of DNAse I such that each molecule is cleaved just once. This will generate a population of cleavage products which in theory contains DNA fragments representing cleavage at every base in the sequence. When these are separated by denaturing polyacrylamide electrophoresis and analyzed by autoradiography a ladder will be produced. If the same DNAse I cleavage reaction is done in the presence of a DNA–protein complex, the DNA will be protected from cleavage where the protein binds the DNA, and therefore fragments of a certain length will be missing. Mapping where the DNAse I fails to cleave the DNA will tell us where the transcription factor was bound to the DNA. In addition, the binding of the protein to the DNA may change the conformation of the DNA which results in certain positions within the sequence being more readily cleaved by DNAse I than in the unprotected fragment, and therefore giving more intense, hypersensitive, bands in the footprint. These too can thus be used as a way of mapping the binding site of the protein. An example of a DNAse I footprint is shown in Figure 11.17.

Because DNAse I is a large protein it generates a low resolution footprint; all the DNA in the region that is bound by the protein is protected rather than individual nucleotides that are in close contact with the protein. An alternative to DNAse I footprinting which has better resolution is hydroxyl radical footprinting. This is a chemical based approach, in which a transcription factor–DNA complex is formed and then a chemical reaction started that will generate hydroxyl radicals. Hydroxyl radicals cleave DNA, but because they are very small molecules, you only see protection from cleavage where the protein comes into close proximity with the DNA. Hydroxyl radical footprinting, therefore, generates a high resolution footprint.

Footprinting experiments not only tell you whether a protein binds to a DNA fragment but they also tell you where they bind. It would therefore appear that they would be the preferred choice over EMSAs for studying DNA–protein interactions. However, EMSAs are more straightforward and more robust, and they can be done using crude cell extracts or partially purified proteins, whereas footprinting normally requires purified proteins. Another advantage of EMSAs is discussed in Question 11.8.

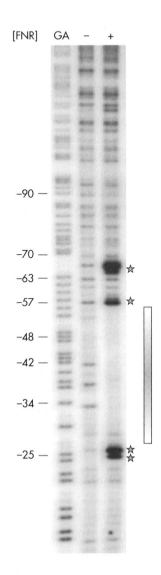

[FNR]   GA   –   +

-90 —

-70 —
-63 —
-57 —

-48 —

-42 —

-34 —

-25 —

Figure 11.17 DNAse I footprint of the transcription factor FNR bound to the *E. coli nirB* promoter. The DNAse I cleavage pattern in the absence (–) and presence (+) of FNR is shown. The gray bar indicates the region of the promoter protected from cleavage by FNR and stars indicate the positions of enhanced cleavage. Enhanced cleavage is due to FNR bending the DNA and therefore giving DNAse I better assess. A G+A sequencing reaction was used to calibrate the gel. (Courtesy of Dr D. Browning, University of Birmingham.)

Q11.7. You have a limiting amount of protein or cell extract which you use in an EMSA or footprinting experiment which results in only 10% of the fragments being bound by the protein. Would you be able to detect the DNA–protein complex using either the EMSA or footprinting protocol?

Deletion analysis and site-directed mutagenesis

So far, we have discussed using reporter gene assays, EMSAs and footprinting to characterize DNA containing promoters and/or enhancer regions. Footprinting allows us to map the precise position of binding sites for transcription factors within a DNA fragment, but it is not a robust technique and normally requires purified protein. If you do not have a purified system, you need another strategy to try and map the important elements within a larger genomic fragment. Deletion analysis and site-directed mutagenesis combined with reporter gene assays and EMSAs provide a relatively robust strategy to further characterize a promoter or enhancer region.

As an example of deletion analysis we will look at how you might analyze the promoter region of a prokaryotic gene. The first task is to clone the promoter region into a promoter probe vector to generate a reporter gene construct as discussed in Section 11.3 and to determine the conditions in which proper transcriptional regulation is observed in the reporter gene assay. You would then construct a series of derivatives in which sequences are deleted from the promoter region and the activity of these promoter derivatives would then be determined. You could use a reporter gene assay and/or EMSAs to study the different promoter derivatives. For example, if you wished to determine the upstream limit of your promoter, you would create a set of nested deletions where the 5′ limit of your promoter is successively closer to the transcription start (Figure 11.18a) and analyze the deletions using a reporter gene assay. Deletion of sequences that are not important for regulation would not affect promoter activity, but deletion of an important region will lead to deregulation. The deletions can either be made by exploiting sites for restriction enzymes or more typically by PCR. If using PCR you would use a common 3′ primer in all PCR reactions and then use a set of 5′ primers to generate the series of nested deletions. The primers would incorporate a recognition site for a restriction enzyme to facilitate cloning (see Section 3.17).

In addition to making nested deletions you can use site-directed mutagenesis to create specific deletions or change specific base sequences within a promoter region (Figure 11.18b and c). This would allow you to map more precisely any important sites within the DNA. These defined deletions or sequence changes can be characterized in a reporter gene assay or EMSAs.

Deletion analysis was used to investigate the *nirB* promoter discussed above (Figure 11.19). Six promoter constructs were cloned in which successively larger regions of the upstream promoter sequence was deleted. A promoter that extends to 208 base pairs upstream from the transcription start (−208) has normal regulation, as does a promoter that extends to −149. Promoters that extend to −87 or −73 have slightly higher activity under anaerobic conditions but can no longer respond to nitrite. A promoter that only extends to −47 shows limited activation, and one that extends to −19 is

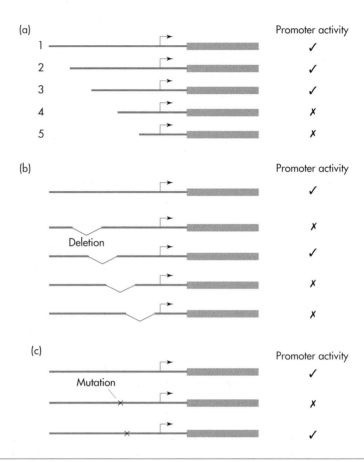

Figure 11.18 Deletion and site-directed mutagenesis analysis of promoter regions. a) A set of nested deletions where the 5′ limit of your promoter is successively closer to the transcription start. Construct 3 contains all sequences required for activity, whereas construct 4 is missing sequences essential for promoter activity. b) Deleting different regions and checking promoter activity helps identify DNA sequences important for activity. c) Changing individual bases helps confirm which bases are important for promoter activity.

inactive. This information was used to develop a model for the regulation of this promoter. The sequence between −87 and −149 contains elements vital for an activator that responds to nitrite and also contains a site for a repressor that reduces the anaerobic activity of the promoter in the absence of nitrite. Sequences between −47 and −73 contain elements that bind an activator that turns on transcription in response to anaerobiosis. Having identified important regions, site-directed mutagenesis was used to identify which individual bases are important for regulation.

Creating a large number of deletion and site-directed mutants and cloning them into a promoter probe vector is a major undertaking. To then

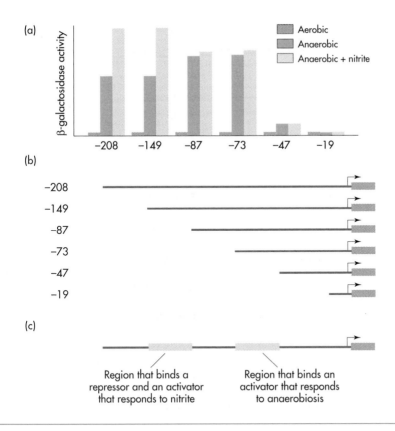

Figure 11.19 Deletion analysis of the E. coli nirB promoter. The graph a) shows the levels of β-galactosidase, during various growth conditions, expressed from each of the deletion constructs shown in b). This allowed the identification of important regulatory regions c). (See Jayaraman *et al.* 1988 for further details.)

isolate and label each derivative for EMSA analysis also requires a large amount of effort. However, the effect of mutations on protein binding can be characterized using the EMSA, without the need to prepare a labeled fragment for each mutant, by looking at how well they compete for binding with the wild-type fragment. In an EMSA, autoradiography is used to detect the DNA fragments following electrophoresis and therefore you only visualize the radiolabeled fragment. If you add a large excess of unlabeled fragment this will bind the transcription factor and stop it binding the labeled copies of the fragment and so the shift will be lost. The ability of a large excess of unlabeled fragment to specifically compete for binding can be exploited to analyze promoter mutants. Rather than having to prepare labeled fragments for each mutant, you simply prepare a labeled fragment carrying the wild-type promoter. This labeled "wild-type" fragment is used in EMSAs in which unlabeled, "cold", fragments carrying mutations are added. If the mutations do not affect binding the mutant fragment will still bind the transcription factor and no shift will be observed. If, however, the mutation stops the transcription factor binding, the unlabeled fragment will not be able to compete and you will get a shift of the labeled "wild-type" fragment.

> **Q11.8.** Are laboratory-based techniques, such as deletion analysis, the only approach to identify possible binding sites for transcription factors?

11.5 Identifying Protein Factors

Techniques such as reporter gene assays and EMSAs in combination with deletion analysis and site-directed mutagenesis can identify sites within promoter/enhancer regions that are important for proper regulation of gene transcription. However, often you do not know the identity of the protein factor(s) that bind these sites. In order to fully characterize the promoter it is vital that the transcription factors responsible for regulation are identified. In the following section we will discuss three techniques – affinity purification, the yeast one-hybrid screen and expression library screening – that can be used to identify the transcription factors that bind a specific sequence.

Affinity purification

One strategy that can be used to purify and identify a transcription factor is to use its ability to bind specifically to a DNA sequence as an aid to protein purification (Figure 11.20). Once the site on the DNA which is bound has been defined using deletion or site-directed mutagenesis as described above, this DNA can be synthesized as an oligonucleotide which is then attached to a matrix that is packed into a chromatography column. If a crude cell extract prepared from cells that are known to contain the active transcription factor is now passed down the column, the target factor will bind to its DNA site and will therefore be retained on the column. The column can be flushed with buffer to remove all proteins that are interacting non-specifically before a different buffer is applied that results in the transcription factor being washed from the column and collected. The transcription factor is thus purified based on its affinity for its target DNA sequence. The purification process can be monitored using EMSAs and SDS PAGE, a technique designed to analyze proteins (Box 5.1) which provides information about the subunit composition of the transcription factor, i.e. whether it contains more than one different protein component, and can also be used to measure the size of the protein components. Affinity purification can be used in conjunction with more conventional protein purification techniques such as size exclusion chromatography and ion exchange chromatography. Once any protein factors have been purified you can determine the amino acid sequence of the N terminus of the protein and then design an oligonucleotide probe that can be used to

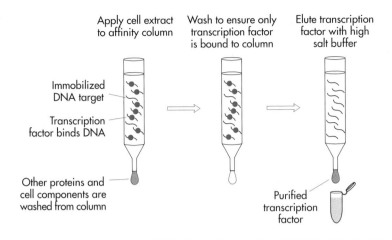

Apply cell extract to affinity column

Wash to ensure only transcription factor is bound to column

Elute transcription factor with high salt buffer

Immobilized DNA target

Transcription factor binds DNA

Other proteins and cell components are washed from column

Purified transcription factor

Figure 11.20 Affinity purification of a transcription factor. The DNA sequence that the transcription factor binds is immobilized on a matrix. A cell extract containing the factor is passed over the column; the factor binds its target DNA sequence whereas other proteins are washed from the column. The transcription factor is eluted by washing the column with a high salt buffer.

screen a cDNA library and so clone the gene(s) for the transcription factor, as discussed in more detail in Section 5.8.

Yeast one-hybrid screen
An alternative approach to affinity purification is to attempt to clone the gene for the transcription factor that recognizes a DNA site within your promoter/enhancer regions using a functional screen. The yeast one-hybrid screen (Figure 11.21 and Question 10.8) exploits the fact that the product of the gene you are trying to clone binds to a specific DNA sequence. It relies on the fact that in eukaryotes many transcription factors have a modular design with the DNA binding function present on one domain of the protein and the activation function present on another domain. Importantly, it is possible to "pick and mix" activation domains and DNA binding domains. In the yeast one-hybrid screen you first clone multiple tandem copies of your DNA binding site (target sequence) upstream of a reporter gene and then integrate the reporter gene construct into the yeast genome. Transcription of the reporter gene is dependent upon a transcriptional activator recognizing the target sequence. The integrated reporter gene can be used to screen a special cDNA library. The cDNA library is cloned into a specialized vector such that the cDNAs are cloned upstream of a sequence of DNA that codes for a strong activation domain from a yeast transcription factor. The cDNA and activation domain coding sequence are transcribed and translated into a single polypeptide, a translational fusion (see Box 9.1). Your cDNA library will therefore consist

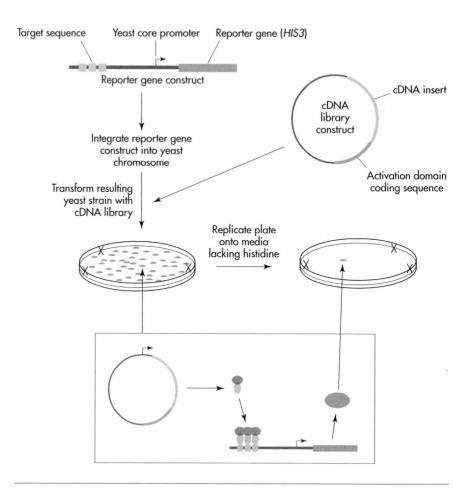

Figure 11.21 The yeast one-hybrid assay. Several copies of the DNA sequence that the transcription factor binds are cloned in tandem upstream from a yeast core promoter and reporter gene (in this example *HIS3*). The construct is integrated into the yeast genome. The resulting yeast strain is used to screen a cDNA library in which the cDNA (blue) is cloned upstream from a coding sequence for a strong activation domain from a yeast activator (gray). The cDNA and activation domain coding sequence are transcribed and translated into a single polypeptide, a translation fusion. If a cDNA clone codes for a DNA binding domain that recognizes the target sequence, this translational fusion will drive the transcription of the integrated reporter gene construct. Colonies expressing the reporter protein can be selected by their ability to grow in the absence of histidine.

of a very large collection of genes coding for translation fusions. Any cDNA clones that contain a DNA binding domain will encode an artificial yeast transcription factor, because the DNA binding domain will be fused to a yeast activation domain. If the DNA binding domain recognizes your target sequence, this artificial activator will drive the transcription of the reporter gene construct.

An example of a reporter gene that can be used in the assay is the yeast *HIS3* gene. This gene is required for the biosynthesis of histidine, so cells that lack *HIS3* cannot grow on media that lack histidine. The yeast strain used in the one-hybrid screen lacks the *HIS3* gene. Integration of the reporter gene construct has the *HIS3* gene downstream of the target DNA binding sequence The cDNA library containing the cDNA-activation domain translation fusions is used to transform the yeast strain with the integrated reporter gene construct and the transformants are plated on a medium lacking histidine. A cDNA that codes for a DNA binding domain that recognizes the target sequence will therefore generate an artificial activator that will drive transcription of the reporter gene and support growth on the medium lacking histidine. The advantage of this technique is that it enables the direct selection of the cDNA for the transcription factor of interest. There are limitations to this technique, however, that are discussed in Question 11.10.

Expression library screening
In Section 5.4 we discussed the construction of a cDNA expression library and the immunological screening of this library. This procedure can be modified to allow screening of an expression library with a labeled DNA probe rather than an antibody. The DNA probe should contain multiple tandem copies of your DNA binding site (target sequence). Rather than binding an antibody to the nitrocellulose filter to which the "plaques" have been transferred, the labeled DNA probe is hybridized. This will identify which plaque contains the cDNA clone encoding the transcription factor that binds the target sequence.

Q11.9. Which approach would you use to clone the genes for a transcription factor that you suspect contains two different protein subunits and where neither subunit can bind the DNA target on its own?

Further analysis
Once a transcription factor has been purified and/or its gene cloned a large number of techniques can be used to further characterize the transcription factor, including some strategies already discussed in this chapter (e.g. footprinting) and other chapters. For example, interacting partners can be identified using the yeast two-hybrid system or the pull-down assay, described in Section 10.5.

11.6 Global Studies of Gene Expression
Until recently most scientists interested in the regulation of gene expression would study a single gene or operon. However, with the advent of genomics and the availability of whole genome sequences, techniques

have been developed which allow us to study the regulation of expression of all, or a very large number of, genes from an organism in a single experiment. These are very powerful techniques that can provide a large amount of information including an insight into the function of genes of unknown function. We will look at techniques that allow us to monitor the level of RNA transcription and protein expression.

Transcriptomics

Transcriptomics is a term that has been coined to describe the study of global gene expression at the level of RNA. The transcriptome is the complete set of RNA transcripts produced by a cell or organism at any one time. Since different genes are being transcribed at different times and in different cell types, the transcriptome is very dynamic (unlike the genome of an organism which is always the same). We discussed northern analysis in Section 11.3 as a method to determine the level of an RNA transcript and how this technique is used to study a single gene. During northern analysis you immobilize your RNA sample on a nylon or nitrocellulose filter; you then incubate a labeled probe with the filter, and the level of probe hybridization gives you a readout of the level of RNA for that gene in the original sample. Transcriptomics exploits a technique that is essentially "reverse" northern analysis (Figure 11.9). In this technique, rather than immobilizing the RNA sample, the DNA probe is immobilized, for example applied to a nylon filter; the DNA probe is not labeled. RNA is then isolated from your sample cells and labeled cDNA prepared representing all the RNA in the sample; this is then incubated with the immobilized DNA probe. The level of probe hybridization gives a readout of the level of RNA for the gene represented by the probe. If you immobilize several probes on the same nylon filter you can determine the RNA levels for each gene, based on the level of hybridization to each probe, since the labeled RNA will only hybridize to its complementary probe on the membrane. It is possible to array thousands of probes onto a single nylon membrane just a few centimeters square, for example a commercial array that is readily available contains probes for nearly 3000 cancer-related human genes on an 8 cm × 24 cm nylon membrane. So by using this array you are in essence doing northern analysis of nearly 3000 genes in a single experiment. Rather than using nylon membranes to immobilize your probe it is possible to use glass slides. The glass slides are high quality microscope slides that have been chemically treated so that they can bind the DNA probes. The glass slides have some advantages, the main one being that they allow the detection of fluorescently labeled cDNA. This permits independent detection of two samples because each sample can be labeled with different fluorescent dyes (Box 5.3) and hybridized to the same array, facilitating direct comparison of the two samples. Arrays are printed on glass slides at a very high density with tens of thousands of probes arrayed on a single glass slide and are often termed DNA micro-arrays (nylon based arrays are often called macro-arrays). DNA micro-arrays are printed by specialized machines that deposit a single spot of each probe at a mapped position on the glass slide so that, following hybridization, you know which spot corresponds to each gene.

The DNA probes bound to the slide can be either oligonucleotides or PCR fragments. PCR fragments can either contain the whole coding sequence (ORF) or a shorter region within the gene. The PCR products need to be denatured to give single-stranded DNA when spotted onto the glass slide or nylon membrane in order to allow hybridization of the labeled cDNA. One disadvantage of using the complete ORF is that it may lead to cross-hybridization between cDNAs and probes for homologous genes leading to incorrect readout for the homologous genes. PCR primers could be designed that direct the amplification of a unique region for each gene, however this would require a large amount of bioinformatic analysis. Whether you amplify the whole ORF or a section of the gene, probe preparation requires a large number of PCR reactions; for example, a yeast array would require more than 6000 reactions. This has led to automation of many procedures with laboratory robots being used to set up reactions. In addition, if gene specific primers are being used you need to synthesize a large number of oligonucleotides; 12,000 are required to amplify the 6000 yeast genes. In many cases the DNA probes are amplified from an existing library of clones. If this is the case it may be possible to use the same primers, derived from the vector sequence, to amplify all the probes. For example, a library containing every *E. coli* gene cloned into the same vector has been constructed and so it is possible to use primers that anneal to the vector sequence on either side of the insert. However, you still need to set up thousands of PCR reactions and, as the library was originally created by using gene specific primer pairs to amplify each ORF, if starting from scratch it would not save time or money. The alternative is to use oligonucleotide probes. Oligonucleotide probes are normally 40 to 70 bases in length and are designed so that they are unique for each gene to avoid cross-hybridization with homologous genes. Gene chips are another type of array. These are created by synthesizing oligonucleotides *de novo* directly on the array using photolithographic technology, the same technology that is used to make electronic chips. The oligonucleotides are approximately 25 bases in length and normally several oligonucleotides are synthesized for each gene.

Once an array has been constructed, the basic experimental procedure for doing a transcriptomic analysis is the same whether you are using a nylon-, glass- or gene chip-based array. RNA is isolated from test and control cells. Labeled cDNA is then synthesized from this RNA and hybridized to the array, and the amount of cDNA bound to each spot is quantified. If using a nylon-based array the cDNA can be labeled with a radioisotope and detected by autoradiography or labeled with biotin and detected by chemi-luminescence (Box 5.3). When using a glass-based array or gene chip the cDNA is labeled with a fluorescent dye and detected by a specialized machine, called a micro-array scanner, which uses a laser to scan the slide. The laser excites the dyes at their respective excitation wavelengths and a detector measures the emission signal from the excited dyes as the laser passes each spot.

Figure 11.22 is a schematic representation of a glass-based array experiment. RNA is isolated from control and test cells, and the RNA is used as template for cDNA synthesis using random hexanucleotide primers in the presence of dUTP labeled with one of two fluorescent dyes, Cy3 and Cy5. Cy3 is used to label the control sample and Cy5 is used to label the test sample. Random hexanucleotides are a population of oligonucleotides six bases in length with random sequence; using random hexanucleotides means that all RNA molecules should be reverse-transcribed into cDNA. The samples are combined and the mixture of labeled cDNAs is hybridized to the glass array. The array is then scanned using a micro-array scanner. Transcript levels in the control and test samples are measured as Cy3 or Cy5 fluorescence respectively. By comparing the relative Cy3 and Cy5 fluorescence it is possible to determine if genes are up-regulated, down-regulated, constitutively expressed or not expressed under the two conditions tested. If the Cy3 signal is represented as green and the Cy5 signal as red, overlaying the fluorescent scans gives different colored spots. Spots corresponding to genes expressed in both conditions will be yellow (i.e. green + red), genes induced or upregulated in the control conditions will appear as green spots, and genes induced or upregulated in the test conditions will appear as red spots. Spots that give no signal from either dye after scanning represent genes that are not transcribed under either condition.

Q11.10. In a micro-array constructed with oligonucleotides, should the oligonucleotide probes be sense or antisense, i.e. the same sequence as the RNA or the template strand of the DNA?

Q11.11. Could an oligonucleotide array be used to analyze the labeled RNA from a nuclear run-on experiment?

Proteomics

Proteomics is a term that has been coined to describe the analysis of all proteins present in the cell. The proteome is the complete set of proteins produced by a cell or organism at any one time. The proteome, like the transcriptome, is very dynamic, since different genes are being transcribed, translated and subject to different post-translational modification at any given time in a cell. As you will learn, proteomics techniques cannot pro-

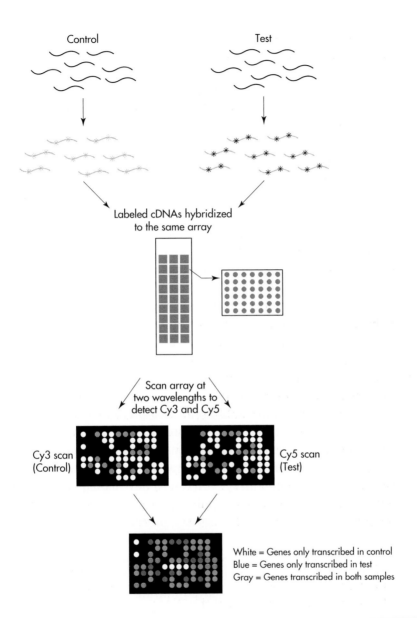

Figure 11.22. Using microarrays to analyse the transcriptome. RNA is purified and random hexanucleotides used to prime cDNA synthesis by reverse transcriptase in the presence of dNTPs and a labeled deoxynucleoside triphosphate (*). Two samples can be prepared, control and test, and labeled with the fluorescent dyes Cy3 and Cy5 respectively. The labeled DNAs are combined and hybridised to an array containing thousands of probes immobilised on a glass slide. The array is scanned with a laser which excites the dyes and fluorescent emission from each spot is detected and quantified. The figure shows the Cy3 and Cy5 scans (different levels of signal are shown as gray and white spots). Often the Cy3 output is depicted as green and the Cy5 as red, when overlaid the colour of the spot is determined by the expression pattern of the gene. Green spots represent genes only transcribed in the control sample, red spots gene only transcribed in the test sample, yellow spots represent gene transcribed in both samples.

vide a complete list of all proteins present in a cell. This therefore begs the question, why study the proteome when transcriptomics can provide what appears to be detailed information about every gene? The main reason is that gene activity is not regulated solely by changes in the amount of mRNA, especially in eukaryotes. Many genes are transcribed but not translated, and once translated many proteins are not active unless post-translationally modified. Turn-over levels for different proteins also vary dramatically and for many proteins differ depending on the conditions. Ultimately, the consequence of gene expression is not whether a mRNA is made, but rather whether an active protein is made. Transcriptomics provides an indirect readout of gene activity, whereas proteomics has the potential to give more definitive information. Transcriptomic techniques are more straightforward and can provide information about every gene, whereas proteomics gives less coverage but the information is probably more relevant to understanding the cell and the organism. Moreover, proteomic techniques not only have the potential to tell you if the protein has been translated and how much is around in the cell, but can also tell you whether it has been covalently modified.

The most common protocol used for analyzing the proteome is two-dimensional gel electrophoresis (Box 5.1) followed by mass spectrometry of the separated proteins (Figure 11.23). Current two-dimensional gel protocols allow the separation of up to 3000 proteins in a single gel but you cannot separate and detect all the proteins present in a complex sample such as a cell extract. Once the proteins have been visualized you are presented with an image containing up to 3000 spots of varying intensity, and unlike with DNA micro-arrays you do not have a map of which spot represents which protein. To identify the protein in each spot, the spot is cut out and the protein treated with a proteolytic enzyme to fragment it. The sizes of the fragments are then determined by mass spectrometry (MS) which can determine molecular masses with a very high degree of accuracy. Each protein will generate a unique proteolytic "fingerprint" because the probability of two proteins being cleaved into the same number of fragments, with each fragment having the same mass, is extremely small. The proteolytic fingerprint can be compared with a database of theoretical fingerprints generated by computer cleavage of all entries in a protein database (see Box 8.2). Various proteolytic enzymes are used. A common example is trypsin, an enzyme found in the gastrointestinal tract of animals; trypsin specifically cleaves proteins after lysines or arginines. The proteolysis products are normally analyzed by a specialized mass spectrometry method called MALDI-TOF MS (matrix-assisted laser desorption/ionization time-of-flight mass spectrometry), which can measure the mass of each fragment to a very high degree of accuracy. MALDI-TOF MS allows high throughput analysis of several hundred samples in a few hours. This approach is only of use if a database of theoretical fingerprints is available.

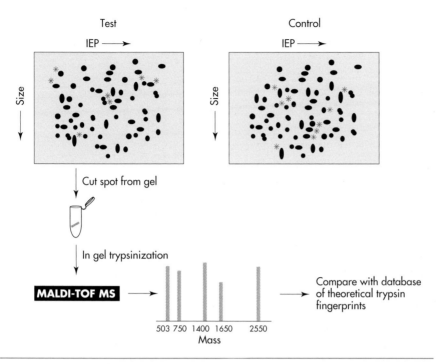

Figure 11.23 Analysis of the proteome using two-dimensional gel electrophoresis and mass spectrometry (MS). The proteins present in a cell extract are separated first by electrophoresis on an immobilized pH gradient which separates proteins according to charge (IEP) and then by traditional SDS polyacrylamide gel electrophoresis which separates proteins according to size. Differences in the proteome between two sample (test and control) are visualized as missing or additional spots, indicated by *. Individual spots can be analyzed by mass spectrometry, a spot is treated with the proteolytic enzyme trypsin and the mass of the tryptic fragments determined by MS. The MS fingerprint can be compared with a database of theoretical trypsin fingerprints.

An alternative mass spectrometry approach, tandem mass spectrometry (MS/MS), can be used to sequence a protein fragment and can therefore be used to identify proteins that cannot be identified by MALDI-TOF MS (Figure 11.24). In this approach, following proteolytic cleavage, the peptide fragments are separated by MALDI-TOF MS and then each fragment is passed through a collision chamber that breaks the fragments into a series of random smaller daughter fragments. These daughter fragments are analyzed by a second round of MS. The collision breaks each individual molecule just once; this generates two populations of fragments. In one population the fragments have a common N terminus and vary in size by one amino acid, whereas in the second they have a common C terminus and also vary in size by one amino acid. The characteristics of the N-terminal fragments are different to the C-terminal fragments and so the data from the second MS can be separated into two ladders. Each ladder starts from

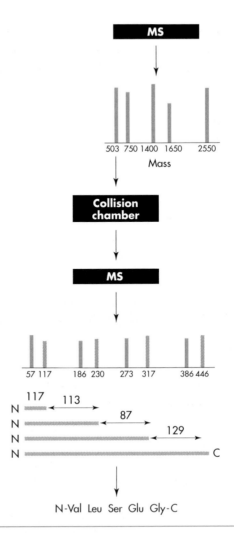

Figure 11.24 Protein sequencing by MS/MS. A protein is first digested with trypsin and the tryptic fragments separated by MS. One of the fragments is selected during the first MS and passed through a "collision" chamber which fragments the peptide into two populations of fragments. One set have a common N terminus and vary in size by one amino acid, whereas the second have a common C terminus and also vary in size by one amino acid. The mass of each fragment is determined by a second MS procedure. The N-terminal and C-terminal fragments can be differentiated. The fragments can be arranged by size giving two ladders, the N-terminal (blue) and the C-terminal (gray). The difference in size between the two adjacent fragments indicates the identity of the residue.

one end of the original proteolytic fragment, the individual "rungs" of the ladder representing the mass of one peptide fragment, and the difference in mass between two adjacent fragments is determined by which amino acid is present at that position. It is therefore relatively straightforward to determine the amino acid sequence of each proteolytic fragment. Each protein spot can be digested with at least two proteolytic enzymes, in separate experiments, to generate overlapping fragments that can be sequenced by MS/MS and assembled into a complete protein sequence.

Proteomics can be used to identify which genes are regulated in test and control conditions. Comparison of the two two-dimensional gels reveals which proteins are subject to regulation. In many cases only spots representing genes subject to regulation are picked and analyzed by MS (Figure 11.23).

> **Q11.12.** You are interested in the major differences in gene expression between cells grown in two different conditions. Which approach (transcriptomics or proteomics) would be easier to implement for an organism for which the genome has not been sequenced?

Questions and Answers

Q11.1. You are mapping the transcription start for a eukaryotic gene using primer extension and the intron/exon boundaries are not known. What effect would using a primer that anneals to a sequence in exon 2 have on the predicted transcription start?

A11.1. *It is not possible to sequence the primer extension product without using an extended protocol such as RACE (see below). If the primer hybridized to a sequence within exon 2, the intron will be missing from the mRNA but not the genome sequence, and therefore the start site will not be mapped correctly. The length of the primer extension product represents the distance from where the primer hybridizes to the 5′ end of the mRNA. If you only have the genomic sequence, and don't know where the introns are, you would map the transcription start as the distance on the genomic sequence as defined by the length of the primer extension product.*

Q11.2. In the final PCR step of the RACE protocol, rather than use the same primer that was used for the original primer extension step, one just internal ("upstream") is used. Why does this help reduce non-specific amplification?

A11.2. *If in the original primer extension step the primer annealed to RNA from somewhere else on the genome and produced a second cDNA, this cDNA would not have the same sequence as the target cDNA. If PCR was done with the same primer as was used in the initial step, this cDNA would still be amplified even though it does not correspond to a transcript. However, if a primer slightly upstream of the first one is used, it would be very unlikely to have any homology to this second cDNA which would not be amplified in the final PCR reaction.*

Q11.3. Why is it essential to remove all DNA from the RNA sample before northern analysis?

A11.3. *The labeled probe used to detect the RNA during northern analysis will also hybridize to any contaminating DNA as all genes, expressed or otherwise, will be present in the genomic DNA; this can give you a false positive for transcription of the gene.*

Q11.4. The QRT–PCR protocol utilizes random hexanucleotide primers to generate the cDNA prior to QRT–PCR. What is the advantage of using random hexanucleotides rather that a gene-specific primer?

A11.4. *Random hexanucleotides will generate a population of cDNAs reflecting the range of RNA molecules in the starting sample. It is therefore possible to subsequently use quantitative real-time PCR on the same mixture of cDNAs to determine the levels of mRNA for any gene for which the DNA sequence is known.*

Q11.5. Which technique, standard northern or reverse northern analysis would be best to study (a) the levels of mRNA for the same gene in several RNA samples (e.g. from different cell types) or (b) the levels of mRNA for several genes in the same RNA sample?

A11.5. (a) *Standard northern analysis is best suited to determining the mRNA levels for the same gene from several samples, because each RNA sample can be immobilized on the same membrane and the same labeled probe hybridized and the level of hybridization quantified. You cannot use two different probes in the same hybridization, so cannot analyze more than one gene at a time.*

(b) *Reverse northern would be best for determining the mRNA levels of several genes within the same sample because the probes are immobilized on the membrane, so several probes can be immobilized on the same membrane. The labeled cDNA will then hybridize to the immobilized probe and the level of hybridization of each mRNA can be quantified.*

Q11.6. Figure 11.15b shows how a series of overlapping fragments can be used to map where a transcription factor binds within a large DNA fragment. Why would you use overlapping fragments rather than a series of non-overlapping fragments?

A11.6. *If you used a series of non-overlapping fragments, the DNA binding site may be spilt between two adjacent fragments and therefore no binding would be detected. For example, if you have two fragments 1–40 and 40–80, the protein may require sequences from 35 to 50 to bind, and so would bind neither fragment. Overlapping fragments are used to ensure that any potential site is present on at least one fragment. For example, if you have three fragments 1–40, 20–60 and 40–80, a protein that binds from 35 to 50 could bind to the 20–60 fragment.*

Q11.7. You have a limiting amount of protein or cell extract which you use in an EMSA or footprinting experiment which results in only 10% of the fragments being bound by the protein. Would you be able to detect the protein–DNA complex using either the EMSA or footprinting protocol?

A11.7. *EMSA would still detect the small percentage of labeled DNA fragments that have bound protein because you would see the appearance of a shifted band. You would have difficulty detecting the bound protein using footprinting protocols because the 90% of unbound fragments in the "+ protein" lane would give rise to a ladder almost identical to the "– protein" lane.*

Q11.8. Are laboratory-based techniques, such as deletion analysis, the only approach to identify possible binding sites for transcription factors?

A11.8. *No, you can also use a bioinformatics approach. It is possible that your promoter or enhancer contains binding sites for known transcription factors. Therefore, you can search the DNA sequence for potential target sequences for known transcription factors. Although this approach may help you identify potential regulators for your gene, experimental work is still required to confirm that the transcription factors identified by bioinformatics really are involved in regulation.*

Q11.9. Which approach would you use to clone the genes for a transcription factor that you suspect contains two different protein subunits and where neither subunit can bind the DNA target on its own?

A11.9. *The only approach that you can use is affinity purification, because both the one-hybrid assay and an expression library screen a single clone in each colony or plaque.*

Q11.10. In a micro-array constructed with oligonucleotides, should the oligonucleotide probes be sense or antisense, i.e. the same sequence as the RNA or the template strand of the DNA?

A11.10. *The probes are sense, i.e. have the same sequence as the RNA. This is because the RNA is purified and then labeled cDNA, that is complementary to the RNA, prepared. The oligonucleotides need to hybridize to the labeled cDNA, so like the RNA, are complementary to the cDNA.*

Q11.11. Could an oligonucleotide array be used to analyze the labeled RNA from a nuclear run-on experiment?

A11.11. *No, in the nuclear run-on experiment the RNA is directly labeled and hybridized to the immobilized probe. The immobilized probe therefore needs to be complementary to the RNA (see answer to Q11.10 for more detail).*

Q11.12. You are interested in the major differences in gene expression between cells grown in two different conditions. Which approach (transcriptomics or proteomics) would be easier to implement for an organism for which the genome has not been sequenced?

A11.12. *A proteomic approach would be the easiest to implement, since it would be difficult to obtain the sequence information to generate the probes required to print the array for a transcriptomics approach. With a proteomics based approach, you would need to optimize the two-dimensional gel electrophoresis for your sample, but once this has been done you should be able to visualize differences in the protein complement of the cells grown in the two conditions, isolate individual proteins form the gels and obtain sequence information using mass spectrometry. Bioinformatics can then be used to attempt to identify the proteins due to similarity to sequences in a protein database.*

Further Reading

The *nirB* promoter of *Escherichia coli*: location of nucleotide sequences essential for regulation by oxygen, the FNR protein and nitrite (1988) Jayaraman PS, Gaston KL, Cole JA and Busby SJW, Molecular Microbiology, Volume 2 Pages 527–530.
This paper describes a study that used reporter gene assays and deletion analysis to investigate regulation of gene transcription

Anti-termination of transcription within the long terminal repeat of HIV-1 by *tat* gene product (1987) Kao SY, Calman AF, Luciw PA and Peterlin BM, Nature, Volume 330 Pages 489–493.
This paper describes a study that used the nuclear run-on assay to investigate the regulation of gene transcription

Marker proteins for gene expression (1995) Wood KV, Current Opinion in Biotechnology, Volume 6 Pages 50–58.
This is a review of reporter gene assays

Three functional classes of transcriptional activation domains (1996) Blau J, Xiao H, McCracken S, O'Hare P, Greenblatt J and Bentley D, Molecular and Cellular Biology, Volume 16 Pages 2044–2055.
This paper describes a study that used nuclease protection and reporter gene assays to investigate the regulation of gene transcription

DNAse footprinting: a simple method for the detection of protein–DNA binding specificity (1978) Galas DJ and Schmitz A, Nucleic Acids Research, Volume 5 Pages 3157–3170.
This paper describes the use of DNAse footprinting in studying protein–DNA interactions

Applications of DNA microarrays in biology (2005) Stoughton RB, Annual Review of Biochemistry, Volume 74 Pages 53–82.
This is a detailed review of DNA microarrays used in transcriptomic studies

From genomic to proteomics (2003) Tyers M and Mann M, Nature, Volume 422 Pages 193–197.
This is a review of proteomics

12 The Production and Uses of Transgenic Organisms

Learning outcomes:

By the end of this chapter you will have an understanding of:

- *what is meant by the expression "transgenic organism"*
- *why such organisms are powerful tools in biological research*
- *some of the applications of transgenic organisms*
- *some of the different methods used for constructing transgenic organisms*
- *the production and uses of knockout organisms*

12.1 What is a Transgenic Organism?

A transgenic organism is one which contains novel heritable genetic material, derived from a different species using molecular cloning techniques. Although the name "transgenic" can be applied to any organism which fits the definition, it is unusual for bacteria or single-celled eukaryotes which have been genetically modified to be referred to as being transgenic. Instead, the expression is mostly used to refer to multicellular eukaryotes. This is the only sense in which the term will be used in this chapter. Often, the term "genetically modified organism" (or "GMO") is used to refer to transgenic organisms. There is no difference in meaning between these two expressions.

The novel DNA which is introduced into cells can take many forms. In the case where it consists of one or more genes which encode novel proteins, these are often referred to as transgenes. To be expressed, the novel gene will need to be under the control of a promoter. This will determine the level to which it is expressed, and can also determine when during development and where in the organism it is expressed. For an organism to be truly transgenic, it has to contain a transgene in all of its cells (Figure 12.1). This means that the gene must be stably replicated along with the host DNA in every cell cycle for all cells (Figure 12.2). This is achieved in practice by integrating the novel DNA into one of the chromosomes of the host organism. This integration may be random or, in some cases, directed to a particular place on the chromosome.

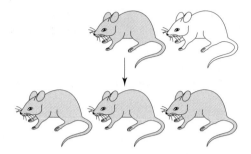

Figure 12.1 Transgenic organisms contain the transgene in all cells and transmit the transgene to their progeny. In the case shown here, a transgene is being expressed which turns the coat color of mice from white to dark. All the progeny from a cross between the transgenic mouse and a non-transgenic mouse also have the dark coat color, showing that the transgene must have been present in the cells of the transgenic parent that give rise to the gametes.

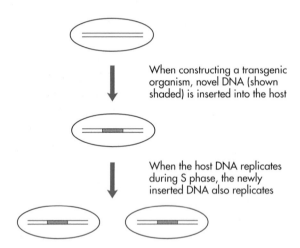

When constructing a transgenic organism, novel DNA (shown shaded) is inserted into the host

When the host DNA replicates during S phase, the newly inserted DNA also replicates

Figure 12.2 In a transgenic organism, the novel DNA becomes integrated into the DNA of the organism. It is then replicated with this DNA whenever the host DNA replicates, and passed on to both of the daughter cells which arise from cell division.

Q12.1. In Figure 12.1, all the progeny of a cross between a transgenic dark-coated parent and a non-transgenic white-coated parent have dark coats. Assuming that dark coat color is controlled by a single gene, and that individuals from the white-coated parental non-transgenic line only give rise to white progeny when crossed with each other, what does this tell us about the transgenic dark-coated parent? If the progeny of this cross were again crossed with a non-transgenic mouse, would the progeny of this second cross also all be dark-coated?

Transgenic organisms are a mainstay of modern biological research in numerous fields. They also have many applications in industry, medicine and agriculture, some of which would have seemed the stuff of science fiction only a few years ago, and which in some cases stir up considerable ethical controversy. In this chapter, we will look at some of the reasons why the use of transgenic organisms has become so widespread, both in the research laboratory and in the commercial world. We will also survey some of the many different ways in which transgenic organisms can be made, emphasizing the common principles which underlie all of them.

What kinds of organism can be made transgenic?

It is interesting at this point to simply list some of the cases where transgenic organisms have been successfully engineered. You may be surprised at the large number of organisms which are in this list. Examples of mammals which have been made transgenic include mice, rats, hamsters, rabbits, cows, sheep, goats and pigs. Amphibians such as frogs and toads have also been made transgenic, as well as birds such as chickens. Many different transgenic fish have been produced, including salmon, trout and catfish. Among invertebrates, much interest has focused on transgenic insects, including the fruit fly *Drosophila melanogaster*, but also on other insects such as moths and mosquitoes. The nematode *Caenorhabditis elegans* – a simple organism much used in developmental biology research – can be made transgenic. If we turn to plants, the list is even more impressive: all the major crop plants of the world (such as rice, maize, wheat, barley, potatoes and soy beans) have been used to develop transgenic lines, as have numerous other commercial varieties (examples being as diverse as tomatoes, tobacco, oil seed rape, petunias and roses).

The above list raises the obvious question: is it possible to make transgenic people? We will return to this at the end of the chapter.

12.2 Why Make Transgenic Organisms?

There are numerous reasons why the development of transgenic organisms has been given such a high priority in research over recent years. Some of these reasons are to do with using transgenic organisms as tools in pure research, but many are also to do with their potential in the commercial world. We will start our discussion of transgenic organisms by looking at some of these reasons.

Changing the phenotype of an organism by expressing a novel gene product

Transgenic organisms are often made where the presence of a novel gene product changes some property – that is, some aspect of the phenotype – of that organism. This was shown diagrammatically in Figure 12.1. Such changes can range from the barely detectable to the very dramatic. The

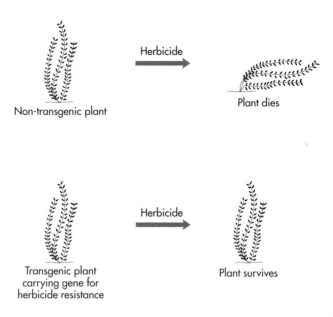

Figure 12.3 Herbicide resistance in transgenic plants. A non-transgenic plant, sprayed with a non-selective herbicide which kills all plants, will die. If a transgenic plant is constructed where the novel gene encodes herbicide resistance, the plant will survive when sprayed with herbicide.

change may be brought about for a commercial reason, or to help in the understanding of the role of the particular gene product itself, or to illuminate some deeper aspect of the biology of the organism.

One example of this application of transgenics is the production for commercial reasons of plants which are resistant to herbicide. Herbicide resistance is a potentially useful trait in agriculture, since it permits the use of broad-spectrum herbicides (herbicides which are not selective in killing only certain types of plant) to eliminate weeds from a crop without damaging the crop itself. We will look now at two examples of ways in which such resistance has been produced in transgenic plants, shown in Figures 12.3 and 12.4.

The first example is that of glyphosate resistance. Glyphosate is a broad-spectrum herbicide which is toxic to plants but which is rapidly degraded in the soil and has a lower environmental impact than some other herbicides. As it kills all plants, it can generally only be used on fields to remove weeds either before the seed of the crop is sown, or before the seed germinates. The compound, glyphosate, acts by blocking the action of the enzyme 5-enolpyruvoylshikimate-3-phosphate synthetase (abbreviated to EPSPS), which is involved in the biosynthesis of the essential amino acids tryptophan, tyrosine and phenylalanine. This enzyme is not found in animals, and glyphosate is highly specific in inhibiting only EPSPS, which

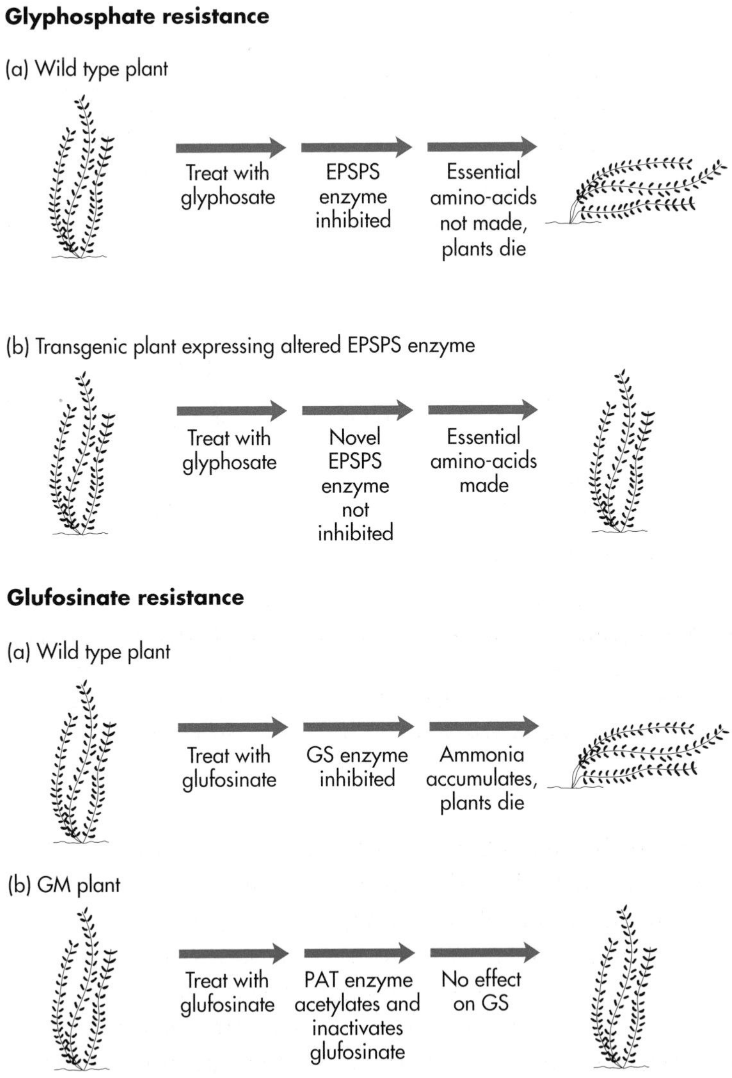

Figure 12.4 Two mechanisms for engineering herbicide resistance into transgenic plants. For glyphosate resistance, the target of the herbicide (in this case an enzyme) is replaced by a variant which is not sensitive. For glufosinate resistance, the novel gene produces a protein which breaks down the herbicide to a non-toxic form.

explains its low toxicity to organisms other than plants. Some bacteria contain a gene for a form of EPSPS which is not inhibited by glyphosate, and when this gene (called *aroA*) is transferred from bacteria into plants (with appropriate signals to ensure its expression; see Section 9.3 and Figure 9.1) the plants become resistant to glyphosate. A more sophisticated approach that has been used to produce glyphosate-resistant plants was to clone the plant's own EPSPS gene, mutate it so that the enzyme it produced was now no longer inhibited by glyphosate, and to return it into the plant. The construction of these plants is described in more detail in Section 12.3.

In the case above, protection from the herbicide is produced by introducing a gene encoding a protein which is resistant to the herbicide's toxic effect. An alternative way of engineering herbicide resistance is to engineer plants to produce a protein that can break down the herbicide, and this approach has indeed been used for producing transgenic herbicide resistant plants. For example, the herbicide glufosinate (which sounds confusingly like glyphosate, but which is not related to it by either structure or mechanism of action) acts by inhibiting the enzyme glutamine synthetase (GS). GS uses ammonia in the synthesis of the amino acid glutamine, and when GS is inhibited, ammonia accumulates and the plants die. The active ingredient in glufosinate (a compound called phosphinothricin) has a structure that resembles glutamine, and it is able to bind to and inhibit GS. To make plants resistant to glufosinate, a gene was identified from a bacterium that encodes a protein which inactivates phosphinothricin by acetylation. The gene, referred to as the *pat* gene as it produces the enzyme phosphinothricin acetyl transferase, was cloned from bacteria, modified for plant expression, and introduced into plants, rendering them glufosinate resistant. Both these modes of resistance are illustrated in Figure 12.4.

Q12.2. Summarize the essential difference between glyphosate and glufosinate resistance. Would you expect either of them to be achievable in plants without the use of genetic modification techniques? Explain your answer.

There are many examples of traits which have been engineered into plants which are to do with the potential uses of the plants in agriculture. Herbicide resistance is one. Engineering resistance to pests is another common one: transgenic plants have been constructed which are resistant to attack by specific viruses, or which produce their own insecticide. Plants are also being developed which are more resistant to various stresses such as high temperature, freezing and high salinity, although none of these are yet grown commercially. More ambitious plans exist for plants with altered properties which enhance their value as food, for example by changing the balance of fatty acids present in the oil that they produce to a more healthy one. Perhaps one of the most celebrated, albeit controversial, examples of a transgenic plant developed to enhance its value as a food is so-called Golden Rice, developed by Ingo Potrykus and Peter Beyer, which has high levels of vitamin A, and which has been touted as part of a solution to the problem of vitamin A deficiency in developing countries, where it causes much preventable blindness. There are of course numerous social, political and ethical issues tied up with the development of GM foods; these are beyond the scope of the current book but you will be able to find plenty of opinions on the topic offered on the internet.

Another example of the use of transgenes to change the phenotype of an organism is in the production of transgenic organisms which can be used as "disease models". These are organisms which have been engineered to display a trait similar to that seen in a human disease. The aim of this approach is to reproduce, in an animal, a disease state which closely resembles the same disease in humans. The transgenic animals can then be studied to try to better understand the biology of the disease, or to test possible drugs which may combat it.

A celebrated and controversial example of this approach is the production of so-called "oncomice". These are mice which have been genetically engineered to express various different human oncogenes (see Figure 12.5). Oncogenes are genes which are present in all human cells where they fulfill various key functions, often to do with cell signaling and events in the cell cycle. Mutation of these oncogenes in certain ways can lead to a much greater likelihood of the cells which express them developing into tumor (cancerous) cells. The oncogenes in the transgenic mice were taken from lines of human cells which were established from tumors where the oncogenes were in one of these mutated forms. Thus, the oncomice develop

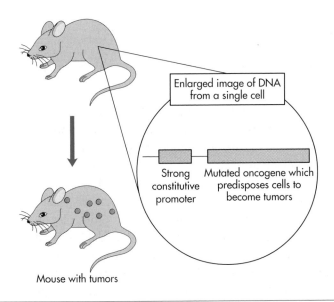

Figure 12.5 The oncomouse: a transgenic mouse predisposed to developing tumors. The expression at high level of an oncogene in all the cells of this mouse increases the frequency with which spontaneous tumors arise, and makes the mouse much more sensitive to the effects of agents that can cause cancer.

tumors with high frequency, either spontaneously or in response to the presence of particular chemicals. Not surprisingly, the development of these mice has led to much debate about the ethical issues involved in this kind of research.

> **Q12.3.** If a mutated oncogene were completely recessive to its wild-type counterpart, would the oncomice have to be heterozygous or homozygous for the introduced gene in order for the trait of high tumor frequency to be seen? Once a line of mice containing a single new mutated oncogene had been constructed, how do you think a homozygous line could be derived from these?

However, there are numerous other examples of the applications of the same principle. Sometimes these have come about from deliberate attempts to devise models of different diseases, or sometimes research has unexpectedly led to an insight into a particular disease which was not anticipated when the research was begun. For example, research on transgenic mice that had been genetically modified to over-produce a neuropeptide called galanin was initiated to look at the effects of this peptide on the mouse reproductive system. However, it was found that such mice are also poor at completing difficult tasks that require a good memory. In humans, this is a classic symptom of Alzheimer's disease, and it has been found that such people have galanin at excess levels in parts of their brains. Thus these mice may serve as a model for understanding, at least in part, the nature of the degenerative changes that occur as Alzheimer's disease progresses.

In all the examples above, transgenic organisms were made by being engineered to make high levels of a particular protein. A related technique is to produce organisms where particular genes are inactivated, and to study the effects of this on the organism. This important methodology will be described in Section 12.5 below.

There are many examples of transgenic animals that have been produced as models of human disease. In addition to diseases such as cancer and Alzheimer's disease, they include rheumatoid arthritis, multiple sclerosis, muscular dystrophy, cystic fibrosis, atherosclerosis and many more.

Using transgenic organisms as bioreactors

In the section above, transgenic organisms were made with the intention of modifying their phenotype in some way. In this section, we will consider a fundamentally different use of transgenic organisms: as "bioreactors" for the production of valuable proteins.

The principle of deriving useful products from plants and animals, rather than using them simply for food, is almost as old as the human race.

Until the development of recombinant DNA techniques, however, such products were limited to those which were naturally produced by organisms. DNA cloning enables the use of organisms to produce particular proteins which are encoded by genes that are introduced into them from other sources. This idea has been developed most with bacteria and with eukaryotic cell cultures, and was discussed in detail in Chapter 9 (see Sections 9.3, 9.4 and 9.5).

In some cases, however, there are distinct advantages to using animals and plants for producing particular proteins, rather than cells grown in culture. These include:

- Some proteins are only active after post-translational modifications (such as cleavage of preproteins, or glycosylation) which do not take place in bacteria
- Living organisms have a distinct advantage over fermentation vessels in that they are already fully adapted to turn their inputs (food, and sunlight in the case of plants) into useful material, and to efficiently get rid of waste
- The technology for widespread growth of animals and plants (farming, in other words) is already in place
- Cost calculations suggest that the use of transgenic organisms to make products which are low in cost but which have a very large potential market could be much cheaper than the use of cell cultures

The technique (sometimes referred to as biopharming) is not yet in commercial use, but numerous projects are currently being evaluated. These include the production in tobacco plants of antibodies against the bacteria that cause tooth decay, and the production in sheep's milk of a protein (AAT, short for alpha-1-antitrypsin) that may be a useful treatment agent for people with cystic fibrosis and some other respiratory disorders (illustrated in Figure 12.6). In these and related cases, the products have to be purified from the organism (or from its milk) before use. An even more radical potential application of transgenic technology is the production of edible plants which contain vaccines, which when eaten would lead to the acquisition of resistance against a particular organism. There are many potential advantages to having such a system for vaccine delivery available. Laboratory-based experiments with animal models and with human volunteers have demonstrated that this approach may indeed work. For example, the gene for a protein from hepatitis B virus has been cloned and expressed in transgenic potato, and when human volunteers ate these potatoes (uncooked), it could be shown that their immune systems were stimulated to produce antibodies against this protein, which in principle should protect against infection by the virus. The details of this are discussed in Section 12.3.

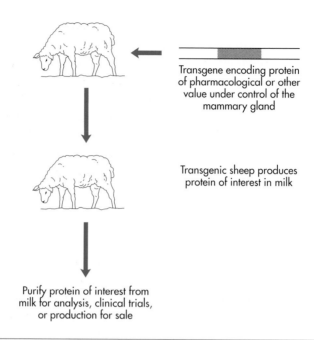

Transgene encoding protein of pharmacological or other value under control of the mammary gland

Transgenic sheep produces protein of interest in milk

Purify protein of interest from milk for analysis, clinical trials, or production for sale

Figure 12.6 The principle of "biopharming". A transgenic organism (in this case a sheep) has been genetically modified to express in its milk a protein which has a pharmaceutical property of research or commercial value.

Q12.4. List as many advantages and disadvantages as you can think of for producing transgenic plants containing edible vaccines.

Turning genes off

The examples above describe situations where the alteration in the phenotype of the transgenic organism comes about because of the novel gene which is being expressed. It is also often important, both for genetic analysis and for applied reasons, to decrease the expression of a gene which is already present in the chromosome of an organism. This kind of gene is often referred to as an endogenous gene. The most effective way of doing this is to remove the gene completely, or at least disrupt it so that it no longer produces a functional protein product. We have already discussed ways of disrupting expression in Chapter 10 with the RNAi approach, but we sometimes want to produce a change which is permanent and heritable – which means deleting or altering the gene. Techniques for doing this are only available for a limited number of organisms, and are described in Section 12.5. An alternative approach to inactivating a gene is to make a transgenic organism where an introduced sequence of DNA is integrated into the gene, thus preventing expression of the functional protein that the gene encodes. Disrupting the expression of a gene enables hypotheses to be tested about the possible function of the endogenous gene (since if the expression of the gene is lowered, this function will no longer be effectively carried out, and the effect on the phenotype of the organism can be observed). From an applied perspective, this approach may be used to turn off the expression of a gene which produces an undesirable trait in the organism.

What kind of DNA sequence could affect the expression of an endogenous gene in a given plant or animal? One possible strategy to down-regulate an endogenous gene would be to produce, in all cells that express the gene, an mRNA copy of the gene in the antisense orientation: in other words, exactly complementary to the mRNA for the gene. This approach has been tried in numerous different organisms, and although it is not always successful, it does represent an important application of transgenic organisms.

Although this technique is widely used, the way in which antisense approaches work is not fully understood. Originally, it was thought that mRNA complementary to a given message would bind to that message and this would prevent ribosomes from binding to the message and translating it into protein. Subsequent work has shown this is an over-simplified view of what actually occurs, and in some cases at least antisense is effective in lowering gene expression because cells contain specific RNA-degrading enzymes that recognize and break down double-stranded (but not single-stranded) RNA (Figure 12.7). Various modifications of the original principle have been used, including the use of catalytic antisense RNAs (called ribozymes) which not only bind to specific mRNAs but also cut them up, and of course the RNAi approach discussed in Section 10.2.

Antisense technology has been widely used, and is particularly celebrated because it was used in developing the first ever GM food to be

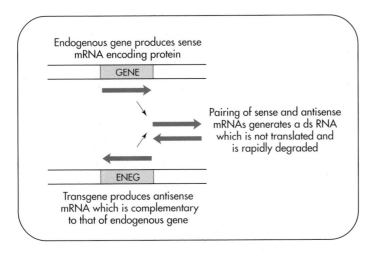

Figure 12.7 Antisense RNA as a method to reduce gene expression.
Pairing of sense and antisense mRNA leads to a double-stranded RNA which is degraded. This reduces or prevents translation of the sense mRNA.

placed on the market: the "Flavr-Savr" tomato produced by Calgene in the USA. Subsequently, tomato paste made from transgenic tomatoes modified using antisense technology was sold in supermarkets in the UK. Although the precise details of how these tomatoes were made differ, the same central principle was used in each case. A key enzyme called polygalacturonase is important in the ripening process of tomatoes and other fruit, causing depolymerization of the pectin present in cell walls and hence leading to softening. Eventually, this leads to fruit becoming very soft (over-ripe) and it was this process that was delayed in both types of transgenic tomatoes. By expressing an antisense version of the gene for polygalacturonase, the amount of mRNA available for translation into active enzyme was markedly reduced and this was effective at delaying softening, although other parts of the ripening process (such as color change from green to red) took place as normal (Figure 12.8).

The Flavr-Savr tomato was a commercial failure, and the tomato paste (although successful) was pulled from supermarket shelves following a strong wave of anti-GM food feeling that swept the United Kingdom in 2000. Whether the failure of Flavr-Savr was due to poor marketing or a poor product is disputed, but the antisense approach clearly does reduce fruit softening and in plants has potential in other areas too such as producing viral resistance.

> **Q12.5.** Can you think of how antisense technology could be used to reduce the expression of an endogenous gene in selected cell types only, rather than in the whole organism?

Analysis of gene expression using reporter genes

Much effort in research is put into understanding the signals that turn genes on and off. The classic example of gene regulation, the study of which has been inflicted upon many generations of undergraduates, is the *lac* operon of *E. coli* . Regulation of gene expression in eukaryotes is a more involved and complex process, and is harder to study, than in bacterial systems where the powerful techniques of bacterial genetics can be applied. Indeed, until the advent of the use of transgenic organisms, information on eukaryotic promoters was sparse. But enormous progress has now been made in this field, and the use of transgenic organisms lies at the heart of this. In particular, the use of so-called promoter probe vectors has been the key to helping unravel the intricacies of eukaryotic transcription. These were described in detail in Section 11.3.

How can transgenic organisms be used with promoter probe technology to study promoters which show tissue specificity? This has been done with numerous organisms, promoters and reporter genes. One example is using the reporter gene GFP to detect muscle-specific expression under the

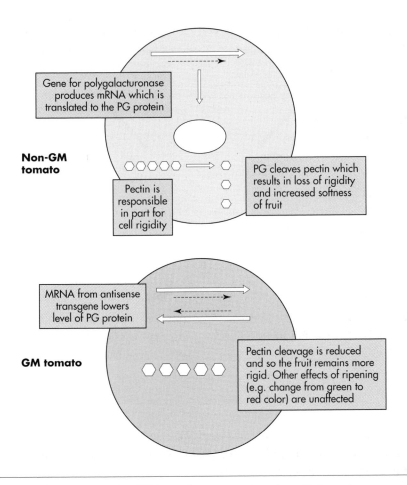

Figure 12.8 Antisense inhibition of tomato ripening. By lowering expression of the polygalacturonase gene, which produces the protein that softens cell walls, tomatoes can be produced that take longer to soften.

control of the promoter for the alpha-actin gene, a gene which is active only in muscle. In these experiments, a DNA fragment containing the promoter for the alpha-actin gene was fused to the coding region of the GFP gene, together with a poly-adenylation signal from the virus SV40. Transgenic zebra fish were produced using microinjection of DNA at the single cell stage. Most of the transgenic fish that resulted expressed GFP only in muscle cells; in other words, the tissue specificity of the promoter was preserved when it was fused to the GFP reporter gene (see Figures 12.9 and 12.10).

12.3 How are Transgenic Organisms Made?

Having discussed many of the reasons why we should want to make transgenic organisms in the first place, we are now going to take a much closer

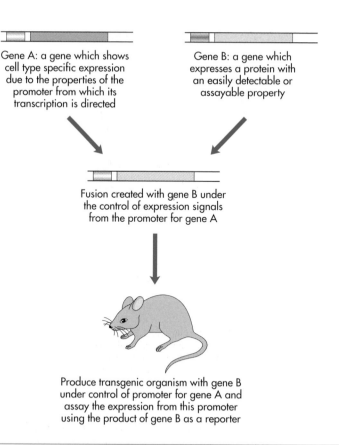

Gene A: a gene which shows cell type specific expression due to the properties of the promoter from which its transcription is directed

Gene B: a gene which expresses a protein with an easily detectable or assayable property

Fusion created with gene B under the control of expression signals from the promoter for gene A

Produce transgenic organism with gene B under control of promoter for gene A and assay the expression from this promoter using the product of gene B as a reporter

Figure 12.9 The concept of reporter genes. Fusion of a promoter to a gene with an easily assayable product enables studies on the regulation of the promoter by studying levels of the reporter gene product.

look at some of the methods by which these organisms are actually produced.

General principles in the production of transgenic organisms

Numerous techniques exist for making transgenic organisms, and these vary a great deal depending on the particular organism concerned. However, all these techniques share some simple general principles, summarized in Figure 12.11 and listed below, which have to be fulfilled in order for the production of such organisms to be possible.

First, the novel DNA which is to be introduced to make the transgenic organism has to be isolated. The techniques discussed in the previous chapters have covered most of the ways in which such isolation is carried out.

Second, the DNA has to be manipulated in such a way that it is likely to do what we want it to do, when it is introduced into the novel organism.

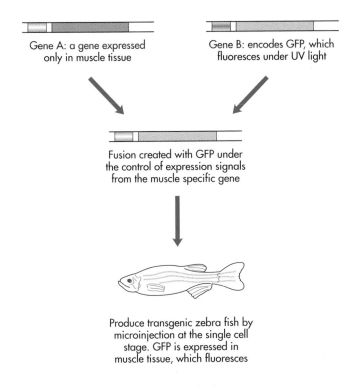

Gene A: a gene expressed
only in muscle tissue

Gene B: encodes GFP, which
fluoresces under UV light

Fusion created with GFP under
the control of expression signals
from the muscle specific gene

Produce transgenic zebra fish by
microinjection at the single cell
stage. GFP is expressed in
muscle tissue, which fluoresces

Figure 12.10 Construction of transgenic zebra fish expressing GFP under a muscle specific promoter. This illustrates how the reporter gene approach can be used in a transgenic organism to study tissue specific gene regulation.

The details of the actual manipulation will depend on the purpose of the experiment, but could include:

- Linking it to a promoter which is specific to the organism or a part of the organism (e.g. a particular tissue or cell type) in question
- Linking it to a reporter gene for studying gene expression
- Altering it in some way to improve its ability to be translated into protein
- Linking it to other DNA sequences which will help target it to a particular location in the transgenic organism.

Third, the DNA must be introduced into one or more cells of the organism in such a way that it will be replicated, and transmitted to both of the daughter cells when the cell divides. This replication and transmission must happen every time the cells grow and divide. As we will see below, the methods used for getting DNA into cells break down into two areas: those

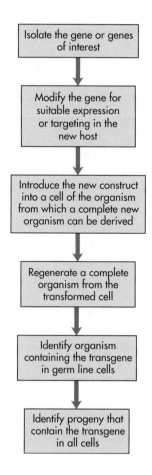

Figure 12.11 General principles of constructing a transgenic organism.

which use a vector of some kind which is based on an entity that as part of its normal life-cycle integrates DNA into the genome, and those which involve putting DNA into the nucleus and relying on recombination systems within the nucleus for getting the DNA integrated into the genome.

Fourth, these cells must be allowed to develop so that a mature organism is formed from them.

Fifth, it must be demonstrated that the newly introduced DNA is successfully transmitted to at least some of the progeny of the organism when it is allowed to reproduce. In other words, the new DNA (the transgene) must be incorporated in the germ cells that produce the gametes that in turn will give rise to the next generation. For this reason, you will often find references in the literature to "stable germ-line transformation", which indicates that not only has DNA been shown to be incorporated in a transgenic organism but also to be passed on in a predictable manner to its offspring.

In the following sections, we will see how these principles are each adhered to in each of several different case studies of producing a transgenic organism.

Supermouse

One of the first examples of a transgenic animal to be produced was also a dramatic example of how the technique of transgenesis can dramatically alter the phenotype of an organism. In this experiment, conducted in the early 1980s by Richard Palmiter, Ralph Brinster and their colleagues, a mouse was produced expressing a rat growth hormone under the control of a promoter that could be turned on by a simple treatment. The effect of expressing the growth hormone was to substantially increase the body size of the transgenic mice.

Cloning and manipulation of the gene

The growth hormone gene was originally isolated from rats using standard techniques. For expression in mice, the gene was cloned into a vector downstream of the promoter for a mouse metallothionein gene. Metallothioneins are cysteine-rich proteins involved in zinc homeostasis and in protecting against the toxic effects of certain heavy metal ions, and the genes that encode them are induced in expression in animal cells by metal ions. They are particularly strongly expressed in liver. By using such a promoter, the growth hormone gene itself could now be regulated by adding metal salts to the diets of the transgenic mice.

Getting the gene into cells

A widely used method for producing transgenic animals is to inject a linear fragment of DNA containing the cloned gene plus the necessary control sequences directly into the nucleus of a fertilized egg cell. For mice, the injection is most effective into eggs which have been very recently fertilized and the two pro-nuclei (one from the sperm cell and one from the egg cell) have not yet fused. The sperm-derived (male) pronucleus is the one most easily seen, and injection is usually into this. Injection is done using a very finely drawn-out glass pipette, using a high precision device called a micro-manipulator. The technique is thus referred to as microinjection. The whole procedure is carried out on a microscope stage and requires a great deal of technical skill. Once in the pronucleus, the DNA fragment will sometimes integrate at a random position in one of the chromatids, and will subsequently be copied and segregated to the daughter cells every time the cell undergoes mitosis. Integration is generally by non-homologous recombination, although we will consider later the special case where homologous or site-specific recombination can be used to target genes specifically to predetermined positions on the chromosome (Box 12.1).

Box 12.1 Recombination: Non-homologous, Homologous and Site-specific

Fundamental to our ability to construct transgenic organisms is the phenomenon of recombination of DNA. Normally when we think of recombination, we think of the process that takes place at meiosis, where chromosomes pair up and physically swap bits of their DNA. The chemistry of this process is complex, but in essence in recombination both strands of two DNA molecules are broken and then rejoined, but rather than being rejoined so that the original molecules are restored, they are swapped so that new molecules result. The outcome of a recombination event depends on the nature of the starting molecules which participated in that event, as shown below.

For example, if a single recombination event takes place between a linear DNA molecule (such as a chromosome) and a circular DNA molecule (such as a plasmid) the effect is to integrate the whole plasmid into the chromosome a). If, however, two linear molecules recombine singly, two linear molecules will result b). Two separate recombination events between two linear molecules will result in two molecules identical to the starting ones except that the region of DNA between the two recombination events will have been swapped c).

The most frequent form of naturally occurring recombination is homologous recombination, which requires the two participating molecules of DNA to possess the same, or nearly the same, DNA sequence. However, in the formation of transgenic organisms, non-homologous recombination can also occur, where there is no significant DNA homology between the novel DNA and the chromosome into which it inserts. Thus, transgenes may insert effectively at random in the chromosome. If we want to direct transgenes to particular places in the chromosome, we must ensure that the transgene or the sequences that flank it have some homology with the target region, and, as such events are relatively rare, we also need a way of selecting for them when they do occur.

Site-specific recombination is a specialized form of homologous recombination. Like homologous recombination, it requires a degree of homology between the two molecules that are doing it, but in this case the length of homology may be very short. The event is catalyzed by a recombinase enzyme which is specific just for that particular homologous sequence. The outcome of the event depends crucially on the orientation of the target sequences with respect to each other.

The other major way in which DNA sequences can join one with another is via the process of transposition. As discussed in Box 6.1, transposons are stretches of DNA which have the remarkable ability (in the presence of the correct enzyme) to "jump" to remote regions of DNA and insert randomly within them. No homology is needed at the new point where the transposon inserts, and the newly inserted transposon may itself transpose again at a subsequent time. Transposons can be adapted to use as powerful tools for DNA manipulation and for generating mutations.

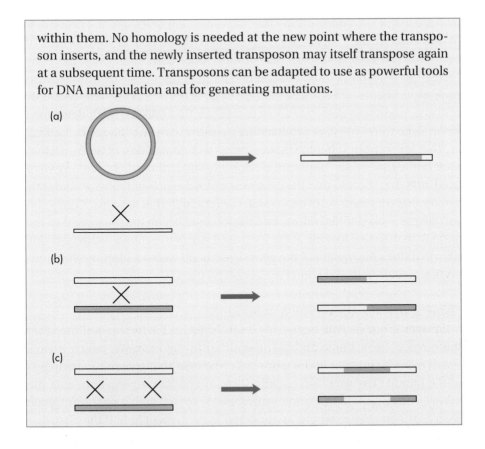

Since the production of the "supermice", an alternative method has been developed for making transgenic mice which utilizes microinjection of embryonic stem cells (ES cells) rather than fertilized egg cells. ES cells are cells isolated from embryos which can be cultivated in tissue culture. They have the very important and useful property that, if injected back into a developing embryo, they can go on to form cells of any cell type. If they contain a transgene, the transgene will also be found in these cell types, and if found in the cells of the germ line, the progeny will now contain the transgene in all their cells. The use of ES cells is discussed further below, when methods for making gene knockouts in mice are described.

Getting an animal from the transgenic cell
Once the fertilized egg has been successfully microinjected, it is transferred to a "pseudo-pregnant" mouse. This is a female mouse which has been mated with a vasectomized male, and which is hence receptive to the new fertilized egg. The egg is transplanted into the oviduct, and if all goes well it will develop into a fetus and eventually a baby mouse. There are many pitfalls in this technique: the new DNA may not integrate into the chromatid, the egg may be damaged by the microinjection process and fail to develop to term, or the baby mice may be abnormal or defective in some way, due probably to

damage to the DNA done during the microinjection or integration processes. In the case of the "supermice", 21 animals developed in total from 170 microinjected eggs, a success rate of little more than 10%. Once the baby mouse has been born, it can be screened for the presence of the transgene (in early experiments by Southern blotting of DNA derived from small amounts of tissue, and these days by PCR, a technique that had not been developed when these mice were first produced). Expression can be analyzed by looking at RNA levels from different tissues using techniques described in Section 11.3; in the case of the "supermice", dot-blotting and northern blotting were used, and this showed that the gene was indeed expressed most abundantly in liver, and that the level of expression varied in different animals. Expression can also be looked for by simply looking for an effect on the phenotype of the animal (in this case, looking at the increased size of the animals when fed zinc salts to induce expression of the genes), although western blotting with antibodies specific to the hormone could also be used.

Showing that the DNA is stably transmitted

A transgenic line has not been truly established until it has been shown that the novel gene is stable and is transmitted to the progeny, which means that it must be present in the germ cells (the cells that give rise to gametes) of the parent. The gene will usually show Mendelian segregation, so if the mouse with a growth hormone transgene is crossed with a non-transgenic mouse, the transgene would be expected to be present in half the offspring (assuming only one copy of the gene was integrated in the original transgenic mouse), and these offspring will themselves be heterozygotes for the gene. If two offspring both of which are heterozygotes for the transgene are crossed, then a 1:2:1 (homozygous for transgene:heterozygous for transgene:no transgene) segregation of the gene will be expected in their progeny. In the case described above, the mice analyzed were the primary offspring from the experiment and no further genetic studies were reported, and interestingly a much later and detailed study on a similar line of transgenic mice showed that the segregation of the transgene did not follow true Mendelian ratios, the likely explanation being that the presence of the transgene reduced the fitness of the mice carrying it *in utero* so that they were under-represented in the litters that were born. Figure 12.12 summarizes the steps in the generation of a transgenic mouse.

ANDi: the first transgenic monkey

The microinjection procedure works well for mice and some other animals, but is not always suitable and is not the only method to choose from for getting genes into cells to produce transgenic animals. An alternative approach is to use a retroviral vector to introduce a novel gene into cells. This method was used in the first (and so far only) reported production of a transgenic primate. The novel gene in this case was for the green fluorescent protein, GFP.

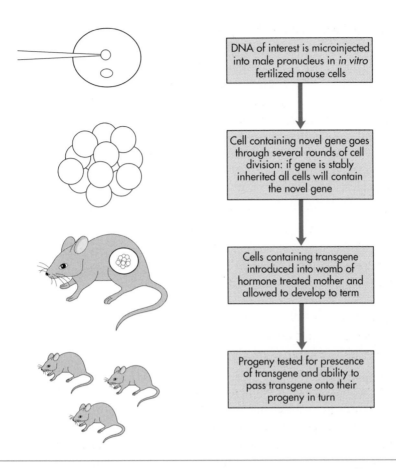

DNA of interest is microinjected into male pronucleus in *in vitro* fertilized mouse cells

Cell containing novel gene goes through several rounds of cell division: if gene is stably inherited all cells will contain the novel gene

Cells containing transgene introduced into womb of hormone treated mother and allowed to develop to term

Progeny tested for prescence of transgene and ability to pass transgene onto their progeny in turn

Figure 12.12 Stages in the production of a transgenic mouse. This method relies upon microinjection of DNA into a fertilized egg which is then allowed to develop. Other methods involve the use of stem cells which are then introduced into early developing embryos.

Cloning and manipulation of the gene

The GFP gene used in these experiments was originally cloned from a fluorescent jellyfish, and identified by its ability to produce fluorescent bacterial cells when expressed within them. To obtain expression in monkey cells, investigators cloned the GFP gene downstream from a promoter for human elongation factor 1, which is strongly expressed in all cells. The whole fusion was then cloned into a vector based on a retrovirus called Moloney murine leukemia retrovirus, or MoMLV. Retroviruses are useful vectors as they transfer their genetic material (RNA) into cells where it is turned into DNA and inserted into the chromosomes, but they are restricted in the range of cells they can infect. Understanding the way they are produced requires an outline understanding of how retroviruses are formed, and this is discussed in Box 12.2. The production of such vectors is shown in Figure 12.13. In order to enable infection of the cells used in this

Box 12.2 The Production of Retroviral Vectors

You have already encountered the RNA viruses known as retroviruses in the chapter on cDNA cloning: they are the source of reverse transcriptase, the enzyme that converts RNA to DNA. The life cycle of a retrovirus is fairly simple: the virus fuses with a cell, and the RNA which it contains enters the cell and is converted to copies of DNA by the reverse transcriptase. These DNA copies go to the nucleus, where some are integrated randomly into host DNA while others are used as templates for the synthesis of more RNA molecules. These RNA molecules contain a minimum of just three genes, called *gag*, *pol* and *env*, coding for the viral capsid protein (*gag*), the reverse transcriptase (*pol*) and a protein found in the viral envelope (*env*). The RNA also contains repeat sequences at either end (called LTRs, or long terminal repeats), which enable the insertion of the DNA copy into the genome. The genome also contains a sequence (ψ) which is required for the packaging of newly synthesized RNA into the viral capsid.

It is possible to produce cell lines containing defective retroviruses by removing this ψ sequence from the viral genome. If such cell lines are subsequently transfected with plasmid DNA containing the LTRs, the viral functions produced by the defective retrovirus which is already present cause the LTRs on the plasmid and any sequence between them to be replicated as if they were viral DNA. If the plasmid also contains a ψ sequence, these RNA molecules will be packaged, along with the reverse transcriptase from the *pol* gene of the defective virus, into the capsids formed from the gag proteins of the defective virus, and thus produce fully infective retroviral particles as shown in Figure 12.13. It is these that are called retroviral vectors. One of the great advantages of this system is that these particles can infect cells, and insert their DNA into them, but cannot form viruses from these infected cells as they do not themselves contain a *gag*, *pol* or *env* gene.

experiment, the investigators coated the retrovirus with a protein coat from another virus (vesicular stomatitis virus, or VSV) which considerably broadens the host range of the retrovirus.

Getting the gene into cells
To infect cells, unfertilized oocytes were injected with tiny volumes of the viral vector, containing the GFP gene. Because of the presence of the GFP gene, the initial success of the microinjection could be judged by observing the microinjected oocytes under a fluorescent microscope: within 4 h, half of them showed some fluorescence.

Getting an animal from the transgenic cell
Oocytes were then fertilized *in vitro* by injecting sperm into them (a technique also used in human *in vitro* fertilization), and the fertilized oocytes

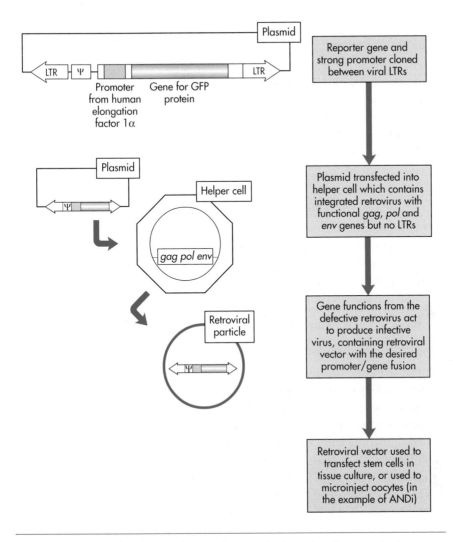

Figure 12.13 Production and use of retroviral vector for producing a transgenic monkey.

were allowed to go through several cell divisions before being transferred to surrogate mothers. The hazards and difficulties implicit in this process can be seen from the fact that of 224 oocytes injected, 40 developed and were transplanted, but only five pregnancies occurred and only one went to full term with the birth of ANDi.

Showing that the DNA is stably transmitted
Presence of the GFP gene was easily demonstrated in ANDi by PCR, and transcription of the gene was demonstrated by RT-PCR. Expression of the protein was not seen in ANDi, although it was seen in other fetuses which were obtained from this experiment but which did not develop to term. But

as to whether the gene is transmitted to his offspring: we will have to wait, patiently, to find out.

A transgenic mosquito

Microinjection of DNA to achieve transformation relies upon recombination of the microinjected DNA into the genome, a process which occurs rather inefficiently and where the DNA is often duplicated or scrambled upon insertion. The use of the retroviral vector in the production of the transgenic monkey ANDi, and other examples, relies upon a cellular process (reverse transcription of the retroviral mRNA and random insertion of the DNA produced into the genome) that is rather more frequent and precise. Another method for carrying DNA into the genome using a cellular process is to put the transgene into a transposable element which randomly jumps into the DNA, and inject it into cells. This method has proven particularly useful for transformation of insects: it is described below in the case of transformation of *Aedes aegypti*, the mosquito that carries yellow fever.

Cloning and manipulation of the gene

The construction of transgenic organisms is often done at first using genes which will give an easily discernible phenotype, rather than one which is experimentally or practically particularly important, so that methods can be quickly optimized. In the case of the first experiments reporting successful germ-line transformation of mosquitoes, investigators began with a line of mosquitoes that had white eyes because they were homozygous for a mutation in the gene for the enzyme kynurenine hydroxylase, which catalyzes a vital step on the way to the formation of the eye pigment. The *cn* gene, which encodes the unmutated enzyme, was cloned from *Drosophila melanogaster*, together with the promoter region for the gene. It was then ligated into a transposable element called Hermes, carried on a plasmid. The Hermes transposable element (transposon for short) was isolated originally from house flies, and, like many transposons, consists of terminal-inverted repeat sequences flanking a transposase gene which itself encodes the protein needed for the element to transpose. In the construct used here, the transposase gene was removed and placed under the control of a heat inducible promoter from *Drosophila melanogaster* (the promoter of the *hsp*70 gene) on a separate plasmid.

Getting the gene into cells

To produce stable mosquito transformants, both plasmids (the one carrying the Hermes transposon with the *cn* gene within it, and the one carrying the Hermes transposase under the control of the hsp70 promoter) were microinjected into insect embryos at an early (preblastoderm) stage of their development, and these were then heat shocked for 1 h. This was to

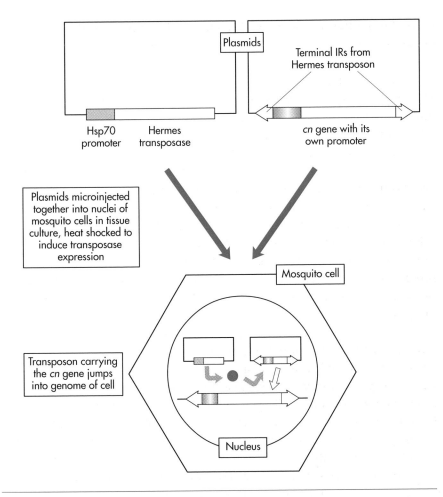

Figure 12.14 Use of the Hermes transposon to create a transgenic mosquito.

induce the expression of the Hermes transposase, which in turn would cause the Hermes element, carrying the *cn* gene, to jump randomly into the DNA of the mosquito. The process is shown in Figure 12.14.

Getting an animal from the transgenic cell
Embryos were then allowed to go on to develop into mature adult insects, which were examined for eye color. Approximately half of the adults that developed had colored eyes, the remainder still had white eyes. It might be thought at first that those with colored eyes would be the transformed animals, and those with white eyes would be those which had not been transformed. However, the situation is more complex: as the microinjection and transposition was done at a stage where the insects were already multicellular, it is very unlikely that every single cell would have become

transformed. The insects would be likely to be mosaics, in which some cells would be transformed and some not. Thus it would be a matter of chance whether the cells that go to make up the eye would contain the transgene (and hence would be colored), and it would be entirely feasible for insects to have white eyes and yet to contain the transgene in their germ line cells.

> **Q12.6.** Given that eye color cannot be used as an indicator of successful transformation in the first generation of progeny produced from the microinjected embryos, how can the presence of the transgene in the germ line be detected?

Showing the that DNA is stably transmitted

To show that a transgene has been incorporated into the germ line, insects (both those with white eyes and those with colored eyes) arising from this experiment were crossed with a line which was homozygous for white eyes. In two cases the resultant progeny, when crossed with their siblings, produced colored eye:white eye progeny in a ratio of 3:1, consistent with single integrations of the Hermes transposon carrying the *cn* gene having taken place in the initial experiment. In subsequent tests the colored eye phenotype segregated as a single Mendelian gene. Thus this method can, with appropriate tests, be used to construct lines of mosquitoes which stably express a novel transgene.

> **Q12.7.** One interesting feature of this experiment is that a transposon from a house fly (the Hermes element) was used, together with its own transposase expressed from a *D. melanogaster* hsp70 promoter. What does this tell you about the ability of these various genetic elements to function in different organisms?

Potatoes containing edible vaccines

We will take a look next at some of the ways in which plants can be transformed. Plants have a number of advantages and a number of drawbacks when compared with animals in this regard. On the plus side, they are generally easier to work with in tissue culture. In particular, good methods are available for developing complete plants from small amounts of plant tissue, even from single cells. However, they often tend to have large and complex genomes compared with animals, and are less amenable to techniques such as microinjection.

Early experiments in plant transformation developed a method that is still widely used for dicotyledonous plant species (dicots; such as tomato, potato and oil-seed rape) and also for some monocotyledonous plants (monocots; such as rice, wheat and barley). This relies on the properties of a bacterium called *Agrobacterium tumefaciens*. This bacterium is a plant pathogen that can infect virtually all dicots, and (under appropriate laboratory conditions) some monocots as well. When susceptible plants are infected with the bacterium they produce a lump of tissue at the infection site called a crown gall. The plant cells within this crown gall are induced to alter their normal developmental pathway by the synthesis of plant hormones (auxins and cytokines) which, remarkably, are encoded not by the plant but by the bacterium. They also have their metabolism altered to produce modified amino acids, called opines, the synthesis of which is also encoded by the bacterium, which are extruded into the soil where other *A. tumefaciens* cells can use them as food. The most extraordinary and most useful feature of the *A. tumefaciens* life cycle is that the novel proteins which it encodes are produced by the plant itself, and this means that the DNA which encodes them has to be transferred from the bacterium to the plant.

In infection with *A. tumefaciens*, this is precisely what happens. *A. tumefaciens* cells contain plasmids called Ti (for tumor-inducing) plasmids, and on infection of a plant by *A. tumefaciens*, a section of DNA from that plasmid, referred to as the T-DNA, is cut from the plasmid, transferred into the plant cell, and inserted randomly into the plant's DNA. The genes on this T-DNA then cause the production of opines, and also produce hormones that causes the plant cell to grow and divide rapidly, producing the crown gall. As the T-DNA is integrated into the genome, all the cells in the gall will themselves contain the T-DNA.

The genes required for the excision of the T-DNA from the Ti plasmid, for the entry of the T-DNA into the cell, and for the insertion of the T-DNA into the plant's genome (collectively referred to as the *vir* genes) are also not found within the T-DNA itself, but are encoded elsewhere on the Ti plasmid. Moreover, early experiments showed that removal of the genes for synthesizing opines and plant hormones from the T-DNA did not affect its ability to insert into plant DNA. All that was needed was the presence of two border sequences (which define the two ends of the T-DNA) which are 25 base pair imperfect inverted repeats.

T-DNA thus is clearly an ideal candidate for using as a vector. Any DNA can be inserted between the T borders, and this DNA would then be transferred into the plant cell and integrated stably within its genome, as shown in Figure 12.15. We'll see how this is used in practice in the next section.

Cloning and manipulation of the gene
The example we will look at here is the construction of transgenic potato plants which express an edible vaccine; that is, a protein that when eaten causes an immune response which protects against subsequent infection by the organism from which the protein was derived. The protein in question is the major surface antigen (referred to as HbsAg) of the hepatitis B virus.

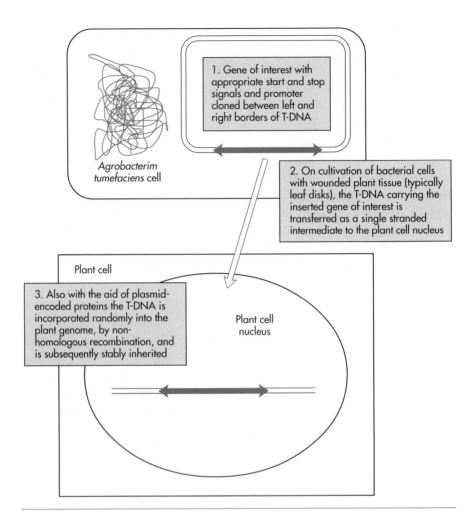

Figure 12.15 Principles of *Agrobacterium*-mediated plant transformation.

This gene was originally cloned from the genome of the human hepatitis B virus. To engineer the gene for use in transformed plants, the gene had to be put under the control of a promoter that is highly active in plant tissue, so that high levels of expression would result. The promoter used was a promoter derived from a plant virus (cauliflower mosaic virus) which normally drives the transcription of an mRNA with a sedimentation coefficient of 35S, and which is hence referred to as the cauliflower mosaic virus 35S promoter, or CaMV 35S for short. This promoter is active in most plant tissues, and the gene for HbsAg was cloned downstream from it. At the end of the HbsAg was placed a terminator sequence, called *nos*, which was derived from one of the opine genes of wild-type T-DNA. This prevents read-through from the strong CaMV 35S promoter into other genes, which could potentially cause a problem in plant transformation. The entire DNA – promoter, HbsAg gene, and *nos* terminator – was cloned between a pair of T-DNA borders. Also present between these borders was another gene, driven this time by the *nos* promoter and followed by another *nos* terminator, encoding a protein called neomycin phosphotransferase II. The significance of this will be seen shortly.

The *vir* genes required for excision of the T-DNA, and its entry into the plant cell and integration into the plant genome, are generally expressed on a separate plasmid to the one carrying the T-DNA. This is because these genes take up a lot of DNA, and as we have seen in earlier chapters, it is better to keep cloning vectors as simple as possible. This system is often referred to as a binary vector system, as it requires two plasmids to be present in the *A. tumefaciens* strain. In practice, the manipulations are done on the cloning vector (which contains the T-DNA borders) using *E. coli* as a host. Once the correct vector has been constructed, it is transferred into an *A. tumefaciens* strain carrying the plasmid with the vir genes.

Getting the gene into cells

To get a T-DNA containing the genes of interest into plant cells, tissue from the plant to be transformed is simply incubated in a culture of *A. tumefaciens* for 1 or 2 days. During this time, T-DNA is transferred to some of the potato cells and inserts into the genome. Since *A. tumefaciens* can only infect wounded tissue, sections of leaf are cut using a sterile scalpel, or sometimes leaf disks are punched out using a hole punch and then sterilized in ethanol, and these are then dipped into a culture of *A. tumefaciens* carrying the necessary vectors. Transformation can also be done on tissue from tubers or stems.

Getting a plant from the transgenic cell

A method is needed to distinguish cells which have taken up and integrated the T-DNA from those which have not. The genes which cause the rapid growth of plant cells that lead to crown gall formation have been removed in T-DNA which is used for plant transformation, so another phenotypic marker must be used. Typically, the selection process is done by putting the leaf disks onto a growth medium which favors the production of undifferentiated plant cell mass, referred to as callus. In addition, the medium contains the antibiotic kanamycin, which is toxic to eukaryotic cells as well as bacterial cells as it inhibits protein synthesis. Transformed cells should express not only the virus antigen but also the neomycin phosphotransferase II protein mentioned above, which inactivates the kanamycin. Thus in theory, only transformed cells can give rise to healthy green callus tissue on this initial medium.

To get back to a complete potato plant is now a skillful exercise in tissue culture. By manipulating the plant hormone content of the medium, callus can be induced to redifferentiate to give rise to shoots and roots, and ultimately to regenerate and form a new plant.

Showing that the DNA is stably transmitted

A significant issue in the transformation of eukaryotes, including plants, is that even after successful transformation the level of expression of the transgene can vary enormously from one transformant to the next. There are several reasons for this. In the case of *Agrobacterium*-mediated transformation of plants, this is thought to be in part because integration of T-DNA is a random process, and the position in the genome where a given T-DNA becomes inserted can have a large effect on how well that DNA is expressed, notwithstanding the fact that all are being driven by the same promoter. Thus an early step in developing a transgenic line is to screen a large number (typically 20) of transformants for the expression of the desired gene, using either an RNA detection technique or a method for detecting levels of protein, such as a western blot.

Transformation can also lead to the presence of multiple integrations in a single transformed line. This may be desirable in the short term (since it may favor higher expression), but when these lines are used to raise progeny, the inserted transgenes will segregate to produce non-identical individuals where levels of transgene expression will now be different. Generally, it is better to work with lines where the inserted DNA is present as a single copy, and this is easy to detect. Seeds from a cross where the transgenic line is one of the parents should segregate the transgene in a 1:1 ratio, and this can rapidly be checked by germinating the seeds on medium containing kanamycin. Seeds which do not contain the kanamycin resistance gene will germinate on this medium, but they grow poorly and remain white, whereas the kanamycin-resistant seedlings are green. Thus, if the gene is segregating stably and as a single locus, roughly half the seedlings will be green when germinated in the presence of kanamycin.

The production of glyphosate-resistant maize

In Section 12.1 we looked at the example of a transgenic maize that had been made resistant to the herbicide glyphosate. This was done by introducing into it a gene for an enzyme which is resistant to the effects of the herbicide. This example is explored in more detail below, and this will illustrate several further key points about the construction of transgenic plants.

Cloning and manipulation of the gene

The gene used to make herbicide resistant maize encodes the enzyme 5-enolpyruvoylshikimate-3-phosphate synthetase (abbreviated EPSPS). This enzyme is essential in plants and is inhibited by glyphosate. In the case that we will look at, the gene was cloned from maize itself, and modified by site-directed mutagenesis so that it would produce a protein that was resistant to glyphosate but was still able to carry out its essential function in amino acid biosynthesis effectively.

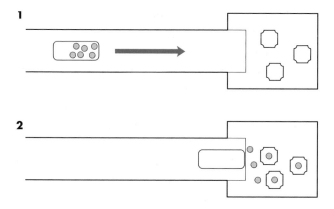

Figure 12.16 Plant cell transformation by particle bombardment. DNA (either a linear fragment or plasmid) containing the gene of interest and usually a selectable marker is coated onto microscopic particles of gold or tungsten, which in turn are placed onto a plastic projectile. This is fired at a gauze which allows the microparticles through, and has plant tissue on the other side. Transformed cells can be visualized if a suitable reporter gene has been used, or can be selected if a selectable marker has been incorporated with the DNA.

The promoter used to express the gene was that of an actin gene from rice which was shown to be highly active in all tissues when used in maize, and the *nos* terminator was placed downstream of the EPSPS coding region to prevent read-through into any adjacent genes.

Getting the gene into cells
Although *A. tumefaciens* can be used to transform monocotyledonous species such as maize and rice, another method is widely used for these sorts of species. This involves coating very small inert particles, typically of tungsten or gold, with the appropriate DNA and firing them into the tissue which is to be transformed. This method is generally called particle bombardment, but is also referred to as biolistics, and the devices for doing it are sometimes called gene guns. It is shown diagrammatically in Figure 12.16. The particles are generally deposited onto a projectile made of plastic which is accelerated using compressed gas and hits a gauze which stops it but allows the DNA-coated gold particles through. On the other side of the gauze is the tissue which is to be transformed. Remarkably, in many cases the DNA from the particles is then integrated into the genome of the cell into which the particles have been fired.

Getting a plant from the transgenic cell
In the case of glyphosate-resistant maize, the "target" for particle bombardment was an embryogenic cell line; that is, a line of maize cells that could be induced to differentiate into embryos and ultimately into

complete plants by manipulation of the make up of the media in which they were grown. Selection of transformed cells was achieved using glyphosate itself, since only those cells that expressed the new glyphosate-resistant EPSPS protein would be able to grow: a useful trick which removed the need to use an antibiotic resistance marker gene in the transformation process. Regenerating an intact maize plant from transformed cells is an impressive, but fairly routine, exercise in cell culture, with changes of medium needed as the regenerating tissue passes through different stages of development.

Showing that the DNA is stably transmitted
As with the cases described above, the initial transformants made using this process were screened to find those where expression was high and these were then grown to maturity and back-crossed (i.e. crossed with the wild-type parents from which the transgenic plants had first been derived). A single inserted transgene would behave like a single dominant Mendelian locus in this cross, and plants where this was true were identified. Repeated back-crosses of progeny with the same wild-type parental strain showed that even after several generations the novel gene still segregated in an entirely predictable way.

Q12.8. Why would you expect the new EPSPS gene to be dominant in a back-cross?

12.4 Drawbacks and Problems

Unfortunately, it is sometimes possible to give the impression that the generation of transgenic organisms is a straightforward and trouble-free technology. However, anyone who has spent any time working in a laboratory where such organisms are being produced or studied would probably disagree, and they might give the following three points as reasons.

The first is that they all require a very high level of technical skill and, frequently, sophisticated apparatus. Manipulations such as microinjection or plant tissue culture take months or years to learn and to perfect. There are many points where these techniques, which after all involve manipulating living tissue, can go awry, and anyone who is doing them routinely will have a whole set of tricks that they use to get things to work as well as possible.

The second point is that there are many pitfalls in producing transgenic organisms even if the initial manipulations work well. With animal transgenics, processes such as implantation and development are ones which, even under completely natural conditions, have a significant failure rate. The frequency of success is further reduced with embryos that have been manipulated in the Petri dish: microinjection, for example, is a stressful experience for a cell and many cells will either die shortly after microinjection, or will fail to develop fully as embryos. Thus these experiments have a very high failure rate and large numbers of microinjections may be needed at the start of the experiment to produce a few successfully transformed lines at the end of it. With plant transgenics, too, there are several potential problems. In particular, the process of taking plants through a tissue culture stage and then reinducing them to form normal plants again generates a lot of variation – called somaclonal variation – in the final plants, which has nothing to do with the presence of the transgene.

Even when lines have been produced, transgenes are sometimes unstable, being lost from the genome, or may lose expression even though the genes are still present. Genes may show very complex patterns of inheritance which makes it impossible to determine how many transgenic loci are present. The way in which DNA can become integrated can vary a great deal, for example with several copies of a transgene integrated at one place or with copies at different places within the genome. Other alterations in the target DNA can also occur as a result of the integration process, in ways that are still far from understood. It has also been found, at least in mice, that the expression of a large number (as much as 4%) of genes in the animals which have been cloned by nuclear transfer (see Section 12.3) are altered compared with controls; so although the animals may have been expected to be identical apart from any deliberately introduced changes, this is not necessarily the case. This can complicate the interpretation of experiments carried out using these animals.

Third, there are limitations to the transgenic approach, because some of the questions that one would like to ask experimentally about gene function are not answerable by simply integrating genes into the genome at random. Indeed, one of the commonest manipulations that you might want to carry out is not to introduce a new gene, but to remove, or at least turn off, a pre-existing one. Classical genetics has indeed always proceeded this way: by screening or selecting for mutations and then studying the phenotype and pattern of inheritance of the mutation in the organism concerned. With the rise of genomics, one of the commonest questions that will be asked about a particular gene is, what is its role in the organism? And the most straightforward way to answer this is to find out what happens to the organism if this gene is no longer present.

This requires methods for either turning pre-determined particular genes off, or disrupting the genes so that they are not expressed. In Section 12.3 above, the use of antisense technology was described as an example of the first method. In the next section, we will see how the science of producing transgenic organisms has been developed to make the latter approach possible. This describes the production of knockout organisms, which are ones where one or more genes have been inactivated, either by disruption with unrelated DNA, or by deletion of all or part of the coding sequence for the gene.

12.5 Knockout Mice and Other Organisms: The Growth of Precision in Transgene Targeting

The use of ES cells to produce knockout organisms

Microinjection of DNA into cells normally leads to non-homologous recombination of the DNA into random or semi-random positions in the genome. However, homologous recombination can occur if the injected DNA contains regions homologous to those in the genome which is being targeted, although it does so at a lower frequency. If there are two regions on the incoming injected DNA which are homologous to two regions in the target DNA, and recombination occurs between both of them, the net effect is to swap the DNA on the genome which is flanked by these two regions for the equivalent piece on the microinjected DNA. This is one basis for making gene knockouts.

Recombination of microinjected DNA with the genome is rare enough, and homologous recombination is even rarer. How then can such events be detected? Standard methods rely on two selections: one to select for any recombination event, and a subsequent one to select against any non-homologous event. In addition, we need a method to get from cells (selected in a tissue culture dish) back to a complete animal. This requires the use of embryonic stem cells (ES cells) which are cells isolated from early embryos that can be grown in tissue culture but which when put back into early developing embryos can go on to give rise to all the different tissues of the body, including (crucially) the germ-line cells.

Let us suppose that we wish to disrupt a gene which we will call X. The first step is to clone this gene from a library of the organism or by PCR. Then into the gene we clone a selectable marker in such a way as to disrupt the coding sequence so that the gene will no longer produce a functional protein. Typically, the *neo* gene which codes for resistance to the antibiotics neomycin and G418 is used for this. Next we clone, downstream from gene X and the *neo* insert, another gene: the *tk* (thymidine kinase) gene from herpes simplex virus. The reason for this will shortly become evident.

Linear DNA containing the disrupted gene and the *tk* gene is purified and mixed with ES cells, and a short electric pulse is applied. The pulse transiently opens pores in the cell membranes, allowing the DNA to enter the cells, a technique known as electroporation. Cells are then plated on growth medium containing neomycin. Only the small minority of cells that have taken up the DNA and integrated it into their genomes will be able to survive and give rise to colonies. As shown in Figure 12.17, these cells will contain a mixture of those where the DNA has integrated by non-homologous recombination (the majority) and a minority where a double homologous recombination between gene X on one of the chromosomes and the disrupted gene X on the incoming DNA has led to the insertion of the disrupted gene X onto the chromosome. In the former case, the *tk* gene will also be present, but in the latter it will usually be lost, as it lies outside the regions required for homologous recombination to occur.

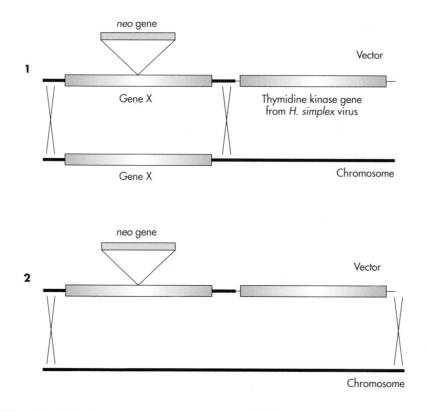

Figure 12.17 Production of and selection for cells carrying homologous recombination events leading to gene knockouts. The crosses show the sites of cross-over events, which will lead to integration of the vector DNA between the cross-overs into the chromosome. Event (1) (homologous recombination) will lead to the insertional inactivation of the chromosomal copy of gene X, leading to a neo^R phenotype for the cells which contain this event. The cells will be resistant to gancyclovir as the *tk* gene will not be integrated. Event (2) (non-homologous recombination) will lead to cells which are neo^R but also gancyclovir-sensitive, and these will not grow on medium containing both compounds.

Fortunately, cells containing the *tk* gene (which are not wanted in this experiment) can be selected against by adding the chemical gancyclovir. The thymidine kinase if present and expressed converts this into a nucleotide analogue which is incorporated into replicating DNA, an event which is lethal for the cell in which it occurs. Thus the combined presence of neomycin and gancyclovir in the growth medium selects for all cells where an insertion of novel DNA has occurred, and against those where the insertion is non-homologous, leaving those where the insertion is homologous and has thus disrupted the target gene.

There is much more work to do before a transgenic animal with the specified gene knockout results. First, the transformed ES cells must be introduced back into early embryos (typically at the 32 cell stage) which themselves are then transplanted into a pseudo-pregnant host mother. When and if these embryos develop to term, the resulting progeny will be chimeras, containing some cells with two copies of the normal gene and some with one copy of the normal gene and one with the disrupted gene. This is because they arose from a mix of these two cell types. In some of the progeny, with luck, there will be some where the germ-line cells contain a copy of the disrupted gene. These will thus pass a copy of this disrupted gene on to the next generation.

The animals produced by this method are heterozygous, with one normal and one disrupted copy of gene X. If mated to a strain which is wild-type for gene X, the resulting progeny will be half wild-type and half again heterozygous. In most cases, the effect of disrupting the gene will only be seen if the animal is homozygous for the disrupted gene, since the product of the single copy of the wild-type gene will compensate for the loss of the disrupted gene. To obtain homozygous lines in which both copies of gene X are disrupted, two heterozygous lines must be crossed; homozygotes should arise in 25% of the progeny by normal Mendelian inheritance. If the gene is essential no viable homozygotes will be produced. More commonly, the homozygotes will show a variety of changes in their phenotype, and it is these that will give us a clue as to the function of the wild-type gene.

This procedure is now routine in mice. Indeed, large collections of knockout mice are commercially available, both heterozygous and homozygous for a large number of disrupted genes, and several on-line databases exist to try to keep track of the large number of mutants that have been studied.

To illustrate the power of this technology, let's consider two examples of knockout mice where the phenotype of the mutation has given useful information about the functions of the genes concerned.

The first example is of a protein which is essential for the regulation of muscle development. Myogenin is a regulatory factor which is required for the development of muscle cells from precursor cells (myoblasts). Mice with knockouts in a myogenin gene were produced by the methods described above. Mice with one of their two myogenin genes disrupted are viable and fully fertile, showing that sufficient myogenin is produced by the expression of one gene for development to be normal. However, when a heterozygous male is mated with a heterozygous female, only wild-type and heterozygous progeny survive. The homozygotes, lacking a functional myogenin gene, show very little development of muscle tissue, and contain undifferentiated myoblasts. They develop *in utero*, but die very shortly after birth as the lack of intercostal muscles mean that they are unable to breathe independently.

The second example is of a protein required for tolerance to toxic metals. As discussed in Section 12.3, many organisms produce proteins called metallothioneins, which are cysteine-rich proteins with the ability to bind ions of toxic metals such as mercury or cadmium: essentially, they act as "molecular sponges" which sequester these metals from other proteins which could be damaged by them. The proteins have no other known function. Mice contain two genes for metallothionein proteins (MTI and MTII) which are quite close (about 6 kb apart), so the knockout construct was made by inserting oligonucleotides containing stop codons into exons for both genes and cloning a *neo* marker into the DNA between the two genes. After selection for resistance to neomycin and gancylcovir, as described above, PCR with oligonucleotides for both the genes was used to detect clones where both genes now contained the stop codons. Mice were generated from these cells, and, after identification of a line where the double knockout was transmitted to progeny, a homozygous line was established. This line was normal for growth and fertility, but, as predicted, was much more sensitive to the toxic metal ion Cd^{2+} than the wild-type. The procedure is illustrated in Figure 12.18.

Q12.9. How could you produce mice which are "double knockouts"; that is, carrying disruptions at two different genetic loci?

Knockout organisms produced following cloning by nuclear transfer

The success of the method described above for targeting specific loci has the potential drawback that it depends on the availability of a suitable line of ES cells plus the technology to get these cells back into a developing embryo and to get the embryo to grow to term. It is also the case that, as the initial progeny produced are mosaics (with some cells containing the novel gene disruption and some not), a second generation must be bred to get animals which carry the transgenes in all their cells. This is acceptable in the case of mice, which have a very short generation time, but for livestock animals which may take several years to reproduce it imposes a long delay on these types of experiments. Yet there is much interest in making knockout livestock, for various reasons. An alternative route to obtaining transgenic livestock with specific genes disrupted is that used in the production of Dolly the sheep, the first cloned mammal, namely cloning by nuclear transfer.

Cloning by nuclear transfer does not, in itself, involve any deliberate genetic manipulation. The principle is very simple (although the practice is very difficult): nuclei from a cell line of an organism are isolated and injected into a cell that has had its own nucleus removed, and the cell is allowed to divide and then implanted into a surrogate mother. If the cell develops successfully to a mature animal, that animal should be effectively a clone of the animal from which the nucleus first originated. (However, remember the point made earlier that gene expression in animals generated from nuclear transfer may for some genes vary quite considerably from control animals.)

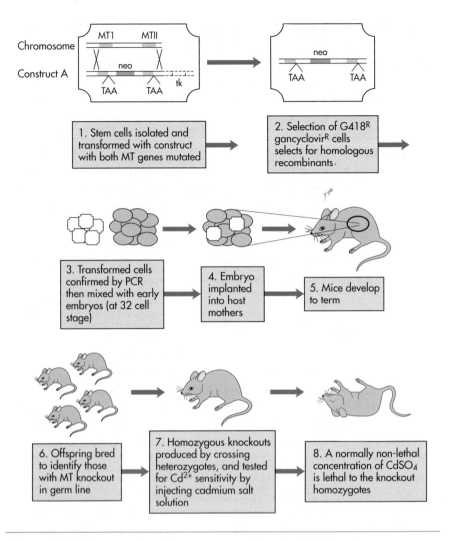

Figure 12.18 Transgenic mice constructed to be deficient in metallothionein genes are more sensitive to cadmium poisoning.

Clearly, if the cell line which is used as the nuclear donor is first genetically altered to disrupt a gene, then this gene disruption will be present in the cells of the mature organism that is produced. This has been the basis for several reports of knockout sheep and pigs, and although the technology is still very new, it is of such potential importance that we'll consider it briefly here.

One application of gene knockout technology in livestock is the possibility of producing pigs whose organs could be used for transplants into humans (this process of transplanting organs from one species into another is called xenotransplantation). There are many problems associated with such possibilities, including the fact that when such transplants are carried out, a process called hyper-acute rejection occurs. The immune system of the recipient very quickly spots the new organ as being foreign, and mounts a spectacular attack against the organ which can kill it in a matter of minutes. The main reason for this is that cells from species such as pigs carry a particular sugar on the glycosylated proteins on the cell surface, which is absent from human and Old World monkey cells. Humans carry antibodies against this sugar, so when a tissue expressing it is exposed to the immune system it is rapidly recognized and destroyed.

For this reason, much effort has focused on producing pigs which lack the enzyme which puts this sugar onto cell surface proteins, referred to as α-1,3-galactosyl transferase, encoded by a gene called GGTA1. A recent report described how such a line of pigs was produced, using nuclear transfer. In the first part of the experiment, part of the cloned GGTA1 gene (exon 9) was disrupted with a *neo* gene, and this construct was then transferred into a pig cell line by electroporation. Rather than using counter-selection with gancyclovir, the researchers in this experiment first selected for cell lines where integration of the sequences had occurred (on G418) and then used RT-PCR to screen these lines for a precise integration event.

Once a suitable line had been found, it was used to clone piglets by nuclear transfer, transferring the nucleus (containing both a wild-type and a disrupted GGTA1 gene) into oocytes which had had their nuclei removed, allowing some cell division to occur, and then transferring the early embryos into surrogate mother pigs. A total of seven piglets were born to three surrogates, of which three died fairly shortly after birth. The remaining four all carry the disrupted GGTA1 gene.

Selective knockouts engineered using cell and site-specific recombination

Powerful though knockout technology is, there are limitations in its use. In particular, many genes which scientists wish to study in adult animals are required for embryonic growth, and consequently it is impossible to create animals which are homozygous for knockouts in such genes. A useful way round this would be to be able to knock out genes selectively, either at a particular stage in the animal's growth or even in specific tissues. One route for doing this is to use antisense technology, as described above, where the antisense mRNA is expressed under the control of a suitable inducible or cell-specific promoter and reduces the level of expression of the protein encoded by the gene to which it is targeted. However, the degree of inhibition of gene expression obtainable with this method is variable. A more robust method is now used in mice which can actually lead to deletion of a given gene but only in the cell type of choice.

This method relies on the form of recombination called site-specific recombination, referred to in Box 12.1. In site-specific recombination, a specific recombinase enzyme catalyzes strand exchange between two specific homologous sequences. There are many examples found in bacterial cells, and one has been adapted for use in transgenic mice: it is referred to as the Cre-*loxP* system. In this system, a recombinase (Cre) can catalyze recombination between two specific DNA sequences (called *loxP* sequences). If the two sites are facing in the same direction as each other, the effect of this recombination is to delete the region of DNA between the two sites, leaving one copy of the site in the genome. This is shown in Figure 12.19. The *loxP* sequence is a fairly small sequence (39 bases long) which, as long as its position is chosen with care, can have minimal effect on gene expression. By controlling the expression of Cre, the recombination between the *loxP* sequences can be controlled, and so the deletion that such a recombination causes can be limited to specific cell types.

How does this work in practice? The first example of an endogenous gene being deleted in this way was the *polβ* gene, the gene for a DNA polymerase which was believed to have an important role in the development of T cells. Studies showed that embryos which lacked a functional *polβ* gene were unable to develop, making it impossible to investigate the role of this gene in adult mice. Consequently, a strain of mice was engineered where the gene was only lost in T cells. First, standard homologous recombination methods were used to construct a line of mice which had an intact and functional *polβ* gene integrated at the normal *polβ* locus, with the

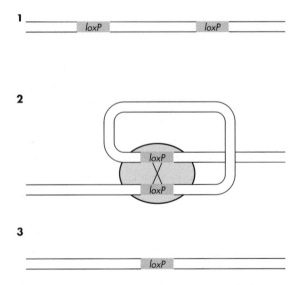

Figure 12.19 How the Cre recombinase can cause deletion of the DNA between two lox sites. Two *loxP* sites in direct repeat (1) can be aligned by the Cre recombinase (shown as a gray oval), looping the DNA between them (2). Cre then catalyzes strand exchange between the two. The intervening DNA (shown with a dashed line) will be lost as it has no origin of replication, leaving a single *loxP* site in the chromosome (3).

selective HSV-*tk-neo* cassette upstream of the gene. Flanking both of these inserts (the new copy of the functional *polβ* gene and the selective cassette) were two copies of the *loxP* site, in the same orientation as each other. (This is often referred to as "floxing" the gene). Mice carrying this construct were fully viable. They were then mated with mice carrying a single deletion of the *polβ* gene (which were hence viable, as the other copy of the gene was functional) and a copy of the *cre* gene driven by a promoter which was only active in T cells. The prediction was that some of the progeny of this mating would now not contain a *polβ* gene at all in the T cells, since they would inherit a chromosome carrying the *polβ* deletion from one parent, and the *polβ* on the other chromosome would be deleted by site-specific recombination when the Cre protein was expressed in the T cells. This was indeed found to be the case.

The use of T-DNA to produce knockouts in plants

It is easy to think of examples of potential benefits if genes could be similarly knocked out in plants. First, there is the obvious point that exactly as with transgenic mice, studies on the phenotypes of gene knockouts is the most obvious way to understand the function in the whole organism of the gene concerned. But in addition there are many examples of where the ability to remove genes selectively from crop plants could alter the properties of those plants in potentially useful ways. Enzymes are responsible for much food spoilage (when foods become over-ripe), for the synthesis of harmful products which prevent plants being used as foods, or which at least make them less healthy than they might be, and other proteins encode receptors which are recognized by plant pathogens such as viruses and bacteria. Removal of any of the genes encoding these proteins could potentially improve the quality or shelf life of the food.

Annoyingly, plant cells appear to be much less good at carrying out homologous recombination, at least under laboratory conditions, than animal cells. So far there are no reliable and broadly applicable ways of targeting a chosen gene in the ways described above for mammalian cells, although a new system which appears to work in rice has promise. Instead, a different approach to making gene knockouts is used, and this is to date largely limited to one of plant biology's favorite model organisms, the thale cress *Arabidopsis thaliana*. As we saw above, *A. tumefaciens* acts by transferring a piece of DNA (T-DNA) into plant cells and integrating it at random positions of the host cell DNA, a property which has been neatly exploited for plant transformation. However, since T-DNA inserts randomly in the DNA, it can be used as a tool to disrupt genes in much the same way as transposons are used (Box 6.1). In a project to generate a library of knockout mutants in *A. thaliana*, large numbers (tens of thousands) of transformed Arabidopsis plants were generated, each from a unique T-DNA insertion event, and the positions of the T-DNA inserts are now being mapped by inverse PCR sequencing of the DNA into which they have inserted. T-DNA insertions in over 21,700 different genes were identified in this way (*A. thaliana* is estimated to have a total of about 29,500 genes). To find a knockout in his or her gene of interest, a researcher would search the database of these T-DNA knockouts with the sequence of the gene in which they are interested, to see whether a T-DNA has already (by chance) been mapped in that gene. Seeds from the plant in which the T-DNA was mapped can then be sent to the researcher, who can grow them and, by crossing, producing homozygous gene knockouts in the next generation.

There are two major drawbacks to this approach. First, not every gene was "hit" by T-DNA in this project. More seriously, these knockouts are available only in the model organism *A. thaliana*, so although invaluable as a research tool, they have little direct practical use, although of course once genes have been identified using this method, homologues from related and commercially important plant species can be identified.

12.6 Is the Technology Available to Produce Transgenic People?

An obvious question that arises from the development of transgenic technologies such as those described above is, are we in a position yet where we could produce transgenic humans? This is essentially a sub-set of the debate about whether we should allow the production of cloned humans; that is, humans derived by nuclear transfer. To go one step further and produce transgenic humans, one would need a number of different steps to be possible.

The steps would essentially be as follows:

1. A method would be needed for isolating a suitable transgene, and placing it under the control of promoters likely to enable its expression in the right place at the right time.
2. This gene would then have to be introduced into the nucleus of a human cell grown in tissue culture, and cells that had received the gene would need to be selected for in some way. If the experiment were a simple "addition" experiment, this would be sufficient; if the aim was to generate a knockout mutation, additional negative selection against clones where integration had been non-homologous would be needed.
3. Then, a nucleus from the transgenic human cell would have to be transferred into an enucleated egg cell, and the cell then allowed to grow and divide for several rounds.
4. Finally, the developing embryo would need to be transplanted into a womb to grow to term.

You should by now realize that all these individual steps are all technically already either routine in humans, or at least feasible based on their success in other organisms. IVF treatment in particular has led to refinement of the ability to put early embryos (typically at the 16-cell stage) into wombs and to get successful live births from them. Step 1 is routine DNA manipulation. Steps 2 and 3, while not yet attempted in humans, are routine in other organisms and have at least been shown to work in primates in the case of ANDi, discussed above.

The possibility that humans may eventually achieve direct control over their germ line is seen by some as compelling and exciting but by many others as chilling and unacceptable. The fact that we lack the space here to discuss this topic, and other controversial topics raised by the methods described in this chapter and elsewhere in this book, does not imply that we regard these issues as unimportant, and we encourage you to discuss them with your fellow students and teachers.

Questions and Answers

Q12.1. In Figure 12.1, all the progeny of a cross between a transgenic dark-coated parent and a non-transgenic white-coated parent have dark coats. Assuming that dark coat color is controlled by a single gene, and that individuals from the white-coated parental non-transgenic line only give rise to white progeny when crossed with each other, what does this tell us about the transgenic dark-coated parent? If the progeny of this cross were again crossed with a non-transgenic mouse, would the progeny of this second cross also all be dark-coated?

A12.1. *The simplest explanation is that the dark-coated parent must have been homozygous for the dark-coated gene, since if it had had only one copy of this gene, half the progeny would have been predicted to be white. All the progeny will therefore contain a single copy of this gene, and if crossed again with a white-coated parent, they would be predicted to give rise to 50% dark-coated and 50% white progeny. More complex explanations involving more than one copy of the transgene in the transgenic line are possible, in which case the simple 1:1 ratio would not be seen in the next generation.*

Q12.2. Summarize the essential difference between glyphosate and glufosinate resistance. Would you expect either of them to be achievable in plants without the use of genetic modification techniques? Explain your answer.

A12.2. *Glyphosate resistance involves expressing a variant of the protein which is normally inactivated by the herbicide glyphosate, whereas glufosinate resistance involves the introduction of a new enzyme which can metabolize the herbicide. Glyphosate resistance might therefore be expected to be attainable through traditional methods of plant breeding, since mutant forms of the plant enzyme might be found that were not susceptible to the herbicide. It is less likely that glufosinate resistance could be achieved by* *traditional methods, at least in the short term, as it involves the introduction into the plant of a gene which encodes a completely novel enzyme.*

Q12.3. If a mutated oncogene were completely recessive to its wild-type counterpart, would the oncomice have to be heterozygous or homozygous for the introduced gene in order for the trait of high tumor frequency to be seen? Once a line of mice containing a single new mutated oncogene had been constructed, how do you think a homozygous line could be derived from these?

A12.3. *They would have to be homozygous in order for the oncogenic pheno-type to be observed. To establish a homozygous line, a transgenic mouse with one copy of the oncogene (identified by PCR and Southern blotting) would be bred with a non-transgenic line to generate offspring, some of which (approximately 50%) would also carry the transgene. Two mice of opposite gender, both carrying the transgene, would then be crossed. 25% of their progeny should now be homozygous for the transgene. These could be identified by further breeding experiments as well as by molecular methods.*

Q12.4. List as many advantages and disadvantages as you can think of for producing transgenic plants containing edible vaccines.

A12.4. *Some possible advantages are (1) the vaccines can be easily delivered without requiring local experts; (2) the vaccines do not need to be shipped and stored: plants expressing the vaccines could be grown locally; (3) there may be less resistance to eating a food than to having an injection (some religions specifically forbid injection); (4) production could be very cheap. You may be able to think of other advantages. Some disadvantages are (1) people may not accept the idea of eating GM plants; (2) the efficacy of the vaccines has not yet been established and may be lower than those delivered by injection; (3) large studies would be needed to confirm that there were no deleterious side-effects. Again, you may be able to think of additional answers.*

Q12.5. Can you think of how antisense technology could be used to reduce the expression of an endogenous gene in selected cell types only, rather than in the whole organism?

A12.5. *The first way to try this would be to identify a promoter which is only active in the cell types of interest. Once this has been done, the gene encoding the antisense mRNA could be placed downstream from such a promoter and a transgenic organism made containing this construct. Although the construct will be present in every cell, the antisense will only be made in those cells where the promoter is active, and so the target gene will only be turned off in these cells. (Note that in plants, mRNA can spread from cell to cell in some tissues and so this approach might not always be successful.)*

Q12.6. Given that eye color cannot be used as an indicator of successful transformation in the first generation of progeny produced from the microinjected embryos, how can the presence of the transgene in the germ line be detected?

A12.6. *If a transgene is present in the germ line, it will be transmitted to some of the progeny of the transgenic animal. The number of progeny that receive the transgene will depend on how many copies of the transgene have inserted at different loci in the initial integration event; if it is only one copy, then only half the progeny (on average) will receive the transgene if mated with a non-transgenic parent. Thus by following inheritance in the next generation, the present of the transgene in the germ line cells can be proven. A more direct approach would be to do PCR on tissue isolated from germ cells of the organism to confirm the presence of the transgene. This would have to be done without damaging the capacity of the organism for reproduction, and in some organisms it may not be technically feasible.*

Q12.7. One interesting feature of this experiment is that a transposon from a house fly (the Hermes element) was used, together with its own trans-posase expressed from a *D. melanogaster* hsp70 promoter. What does this tell you about the ability of these various genetic elements to function in different organisms?

A12.7. *This shows that transposons can function in organisms other than the one in which they originate, as long as the appropriate enzymes are expressed. Transposons are very useful as genetic tools, and this ability to move them from one organism to another has been exploited on numerous occasions.*

Q12.8. Why would you expect the new EPSPS gene to be dominant in a back-cross?

A12.8. *As long as the plants contain some of the resistant EPSPS protein, they will not be killed by glyphosate. Thus, they only need one copy of the gene to survive herbicide treatment, even if the other copy of the gene is wild-type (i.e. encodes the normal herbicide-sensitive protein).*

Q12.9. How could you produce mice which are "double knockouts"; that is, carrying disruptions at two different genetic loci?

A12.9. *The simplest way is by crossing two initial lines of mice which are homozygous for different gene disruptions. This will yield progeny which are all heterozygous for both knockouts, and further interbreeding of these individuals can be used to generate progeny which are homozygous for both knockouts. An alternative method would be to isolate an ES line from one of the single knockout lines and use this to start to generate the other gene knockout, but in practice, this route would be much harder and is not used.*

Further Reading

There are vast numbers of papers describing methods for the generation of transgenic organisms, and exploring the possible uses of such organisms. Those below are a small selection; all describe cases that have been specifically discussed in this chapter.

Dramatic growth of mice that develop from eggs microinjected with metallothionein-growth hormone fusion genes (1982) Palmiter RD, Brinster RL, Hammer RE, Trumbauer ME, Rosenfeld MG, Birnberg NC, *et al.*, Nature, Volume 300 Pages 611–615.
The production of "supermice" – a classic paper that helped lead to the establishment of the use of transgenic mice as major research tools

High-frequency generation of transgenic zebrafish which reliably express GFP in whole muscles or the whole body by using promoters of zebrafish origin (2001) Higashijima S, Okamoto H, Ueno N, Hotta Y and Eguchi G, Developmental Biology, Volume 192 Pages 289–299.
The use of reporter proteins to monitor gene expression is described here for a system in zebrafish

Transgenic monkeys produced by retroviral gene transfer into mature oocytes (2001) Chan AW, Chong KY, Martinovich C, Simerly C and Schatten G, Science, Volume 291 Pages 309–312.
The experiments that led to the birth of ANDi, the transgenic monkey

Oral immunization with hepatitis B surface antigen expressed in transgenic plants (2001) Kong Q, Richter L, Yang YF, Arntzen CJ, Mason HS and Thanavala Y, Proceedings of the National Academy of Sciences of the United States of America, Volume 98 Pages 11539–11544.
One day, will all our vaccines be provided in mashed potato? Read this paper and decide for yourself!

Production of alpha 1,3-galactosyltransferase-deficient pigs (2003) Phelps CJ, *et al.*, Science, Volume 299 Pages 411–414.
A large multi-author paper describing the successful generation of pigs lacking GGTA1 gene, an important milestone on the route to using pigs as potential organ donors for humans

13 Forensic and Medical Applications

Learning outcomes:

By the end of this chapter you will have an understanding of:

- DNA profiling and of how it can be used in forensics
- a range of techniques used for molecular diagnosis of inherited disease
- the current role of gene cloning in the production of pharmaceutical products and the potential of pharmacogenomics in the development of new pharmaceuticals
- how genetic tests can help in the diagnosis of infectious disease, and what advantages they have compared with conventional techniques
- how genetic tests can be used in the diagnosis and management of cancer

13.1 Introduction
In this book we have introduced the main technologies by which genes can be manipulated; we have also discussed many applications of this technology to illustrate the different sections. In this final chapter we will look at key areas which have not been covered in detail elsewhere in which this technology is used in important ways, which affect our daily lives.

13.2 Forensics
One of the most high profile applications of DNA technology is in forensic science; you cannot read a newspaper article about a murder hunt, or watch a crime series on the television without hearing mention of DNA profiling. The original technique of DNA fingerprinting was developed in 1985 by Alec Jeffries and involved Southern hybridization. This has now been superseded by a PCR-based technique which is used by many police forces across the world. The key to both of these techniques is that they offer a way of looking at genetic loci which vary between individuals, and use these characteristics to identify individuals. The genetic loci used are regions of repetitive DNA (Section 2.3).

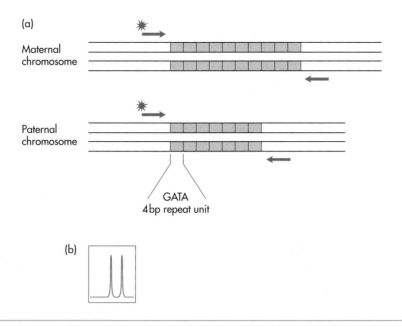

Figure 13.1 Amplification of a heterozygous microsatellite locus from homologous chromosomes. a) The blue boxes indicate the repeat sequence, in this case GATA. In this example the maternal chromosome has 10 repeats while the paternal chromosome has seven. b) PCR amplification of this microsatellite locus yields two products of different lengths. One of the PCR primers is labeled with a fluorescent dye so that when the products are separated by capillary electrophoresis two peaks are seen.

Q13.1. Name the two main classes of repetitive DNA found in the human genome and briefly describe each one.

DNA profiling makes use of regions of the genome called microsatellite or short tandem repeat (STR) DNA. In these regions a short DNA sequence, of 5 bp or less, is repeated many times resulting in a repeat region of up to 150 bp. These STR loci are highly polymorphic with a wide variability in the number of repeats between individuals. Because of the problems encountered by DNA polymerase in copying highly repetitive DNA sequences these regions are copied incorrectly more often than non-repeat DNA. This slippage results in variation in the number of repeats within each microsatellite between individuals. DNA profiling uses these highly polymorphic STR loci in order to generate a pattern which can be used to identify individuals.

PCR has been key to the forensic analysis of variation between humans from the very beginning. It is amplification by PCR that makes it possible to analyze DNA found in small samples left inadvertently at a crime scene. Forensic samples are often blood, semen, hair or skin cells and without

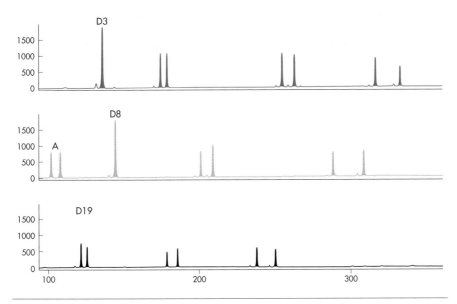

Figure 13.2 DNA profile showing the 11 microsatellite loci used in the UK for forensic DNA profiling. The peaks corresponding to each STR locus are identified by size. Fluorescent labels of different color are used to distinguish between products from loci with overlapping size ranges, e.g. D3S1358 (D3), D8S1179 (D8), and D19S433 (D19). The profile is from a male, with two alleles for the amelogenin gender identification locus (A). This individual is heterozygous at all loci except D3S1358 and D8S1179 where there is only one peak.

PCR it would usually be impossible to obtain sufficient DNA to perform tests; however in the case of DNA profiling PCR plays an even more integral role.

13.3 DNA Profiling

DNA profiling is basically a technique designed to measure the number of repeats present at specific highly polymorphic STR loci. The procedure involves PCR amplification of the target STR using primers designed to hybridize to conserved flanking regions (Figure 13.1). One of the primers is labeled with a fluorescent dye or fluorophore (Box 5.3), so that the PCR products can be detected after being separated by capillary electrophoresis (Box 7.2).

Because humans have two copies of each chromosome, one from each parent, an individual could be homozygous for one of these loci, if both copies have the same number of repeats, or more likely heterozygous, if each chromosome carries a different allele with a different number of repeats. By examining a number of these polymorphic STR loci a unique pattern can be identified for an individual (Figure 13.2).

Standard DNA profiling procedures used by the Forensic Science Service in the UK involve analysis of 10 STR loci located on autosomal chromosomes together with the amelogenin locus (discussed below) which is used to determine gender. The FBI in the United States uses 13 STR loci and the amelogenin locus. One of the criteria used to select the loci to be examined is that these are highly polymorphic loci showing a wide range of different alleles. The frequency with which each allele of a particular STR locus occurs varies, so the probabilities are different for different profiles. It is possible to calculate the statistical likelihood that two unrelated people would by chance have the same DNA profile; this is usually in the region of 1 in 10^{12} for the 13 commonly used STR loci.

In deciding what repeat length is most suited to forensic analysis there is a trade-off between the overall length of the PCR product and the ability to differentiate alleles. Short repeats will give rise to short PCR products that are easier to amplify, especially from degraded samples. On the other hand alleles of STRs with very short repeats may be difficult to distinguish from one another as they may only differ by a few nucleotides. STR loci comprising tetranucleotide repeats (Table 13.1) are optimal for DNA profiling. These are relatively common in the human genome and generate short PCR products (Table 13.1). The alleles differ by multiples of 4 bp and these can be accurately detected by capillary electrophoresis. In choosing STR loci for forensic profiling it is also important that they are distributed throughout the genome (Table 13.1) to ensure that each is inherited independently.

In addition to the autosomal loci an additional locus, part of the amelogenin gene, is amplified. This gene is found on both the X and the Y chromosomes, but the Y chromosome allele has a deletion in the first intron and hence yields a smaller PCR product. This marker is used to determine the sex of the individual from whom the DNA sample was taken.

13.4 Multiplex PCR

A single DNA profile requires 11 PCR reactions in the UK, or 14 reactions in the USA, one for each of the STR loci to be tested. In order to make the process of generating DNA profiles more efficient and cost effective it is necessary to combine the PCR reactions; this process is called multiplexing. Because of the specificity of the PCR reaction it is possible to include more than one primer pair in a single reaction and hence to amplify more than one product at a time. To do this successfully it is necessary to design primers with similar optimal reaction conditions and which will give rise to products whose lengths are sufficiently different to be clearly distinguishable. It is also possible to multiplex using STR loci which yield alleles with overlapping size ranges by using different fluorescent labels to distinguish the products (Figure 13.2). The multiplex PCR system used by the Forensic Science Service in the UK amplifies all 11 loci in one reaction and uses three different fluorophore labels; there is a similar kit available for the 14 loci used in the USA.

Table 13.1 Properties of loci used in forensic DNA profiling

Locus	Repeat Sequence	Chromosomal Location	Product Size
TPOX†	[AATG]$_n$	2p	218–242
D2S1338*	[TGCC]$_n$[TTCC]$_n$	2q	289–341
D3S1358	[TCTA][TCTG]$_{1–2}$[TCTA]$_n$	3p	114–142
FGA	[TTTC]$_3$TTTT TTCT[CTTT]$_n$CTCC[TTCC]$_2$	4q	215–353
D5S818†	AGAT	5q	135–171
CSF1PO†	[AGAT]$_n$	5q	281–317
D7S820†	[GATA]$_n$	7q	258–294
D8S1179	[TCTR]$_n$	8q	128–172
TH01	[AATG]$_n$	11p	165–204
VWA	TCTA[TCTG]$_{3–4}$[TCTA]$_n$	12p	157–209
D13S317†	GATA	13q	206–234
D16S539	[AGAT]$_n$	16q	234–274
D18S51	[AGAA]$_n$	18q	273–345
D19S433*	[AAGG][AAAG][AAGG][TAGG][AAGG]$_n$	19q	106–140
D21S11	[TCTA]$_n$[TCTG]$_n$ [TCTA]$_3$TA[TCTA]$_3$ TCA[TCTA]$_2$TCCATA[TCTA]$_n$	21q	187–243
AMELX		Xp	107
AMELY		Yp	113

The table shows the properties of 15 autosomal STR loci and the amelogenin locus used in DNA profiling. There are eight autosomal loci common to the set of STRs used by the UK Forensic Science Service and the FBI in the USA. An additional two (*) are only used in the UK and an additional five (†) are only used in the US. Most consist of simple tetranucleotide repeats; however, some have more complex repeat patterns. The chromosomal location indicates the chromosome number; the letters p and q indicate the short and long arms of the chromosomes respectively. The product sizes quoted are calculated from kit allelic ladders and may not represent the full range of sizes observed in populations.

Q13.2. What considerations have been taken into account in choosing the STR loci for forensic analysis?

13.5 Samples for Forensic Analysis

The vast majority of the DNA profiles kept in the major DNA databases NDNAD and CODIS (Box 13.1) are derived from samples taken from individuals who have been arrested for, or convicted of, serious crimes. These samples, usually either blood or buccal cells from the inside of the cheek,

---

Box 13.1 DNA Profile Databases

The primary DNA profile database in the UK is the National DNA Database (NDNAD) which is run by the Forensic Science Service on behalf of the Association of Chief Police Officers. In the United States the equivalent database is the Combined DNA Index System (CODIS) which is maintained by the FBI. Each of these databases contains over 2.5 million DNA profiles, most of which have been obtained from individuals arrested for, or convicted of, serious offences. A smaller proportion of the profiles in each database are forensic profiles obtained from crime scenes. These databases can be searched for matches to samples obtained from crime scenes and have become a very important tool for the police in obtaining evidence which may lead to the identification and eventual conviction of offenders.

---

are taken and stored in controlled conditions and yield reliable high quality DNA profiles. Samples collected from crime scenes may be of blood, semen, hair or skin cells, and while DNA is a remarkably robust molecule, the conditions before collection are seldom ideal and many samples are degraded. It is here that one of the strengths of the STR-based DNA profile system comes into play, as it is possible to amplify the short STR loci by PCR from samples of relatively poor quality. In fact DNA profiling has been used to revisit unsolved cases and analysis of forensic samples from over 20 years ago has successfully been used to obtain convictions.

13.6 Obtaining More Information from DNA Profiles

A DNA profile from a sample left at a crime scene is only really useful either if there is a match to it in one of the DNA databases or if you have a profile from a suspect against which to compare it. The inclusion of the amylogenin locus in the standard DNA profile makes it possible to determine the gender of the donor. However, it would be useful in the investigation of crime to be able to deduce other physical characteristics such as race, hair and eye color, height and age. Two services currently offered by the Forensic Science Service test for red hair and infer ethnicity.

The red-hair test is based on detecting mutations in the *MC1R* gene which encodes the melanocortin 1 receptor. *MC1R* acts as a molecular switch that affects hair color. The test looks at 12 variants of this gene, eight of which are associated with red hair. The test based around detecting these can identify 84% of red heads.

Inference of ethnicity is based on studies showing that the frequency of the STR alleles, used for genetic profiling, differs between populations. Humans are thought to have evolved from early hominids living in Africa who then moved out of Africa and dispersed across the world. Over time, differences have arisen between the geographically isolated gene pools and these differences can be used to discriminate between five different ethnic groups: white-skinned European, Afro-Caribbean, Indian subcontinent, South-East Asian and Middle Eastern.

13.7 Other Applications of DNA Profiling

DNA profiling has another important application in establishing family relationships. This may be in cases where paternity is in question but it is also used by immigration authorities to establish family relationships in cases where documentary evidence is not available. Because we inherit one of each pair of homologous chromosomes from each parent, we also inherit one STR allele from each parent. By comparing the DNA profile of a child and its parents and applying the principles of Mendelian genetics it is possible to determine paternity.

Q13.3. (a) The table below shows the genotypes of John and Mary and their two children Ann and Peter. Each STR locus is given a number representing the number of repeat units it consists of. John suspects that one or other of the children may not be his. By examining the data can you decide if his suspicions are correct?

	John	Mary	Ann	Peter
TPOX	14, 17	15, 16	14, 15	15, 18
D3S1358	7, 12	9, 11	11, 12	7, 9
FGA	15, 15	15, 15	15, 15	15, 15
D5S818	15, 15	13, 15	15, 15	13, 15
CSF1PO	11, 14	10, 12	11, 12	9, 10
D7S820	8, 8	7, 7	7, 8	7, 9
D8S1179	9, 9	9, 12	9, 9	10, 12
TH01	5, 6	4, 7	4, 5	6, 7
VWA	16, 19	15, 19	15, 19	18, 19
D13S317	8, 11	5, 6	6, 11	5, 10
D16S539	9, 11	9, 9	9, 11	9, 12
D18S51	13, 16	12, 16	16, 16	12, 17
D21S11	7, 9	4, 5	4, 7	5, 9
AMEL	X, Y	X, X	X, X	X, Y

(b) The most likely explanation for Ann's profile is that John is her father, but can you think of any other possibility?

Although forensic analysis of human identity has been the main driving force in the development of DNA profiling technology, it can of course be used with other animals. Many areas of research as diverse as animal behavior, ecology and conservation biology make use of DNA profiling to study population diversity and relatedness between individuals. Samples can be taken from hair or even feces of animals and the DNA extracted and profiled, allowing estimates of the numbers of individuals within populations and also of the genetic diversity in populations. This information is of interest in its own right and can be very important in devising conservation strategies. DNA profiling can also be used to monitor the sale of meat from endangered species. Profiles derived from whale meat on sale in a market place can establish how many individual whales are represented, and hence be used to check that quotas are being adhered to. Because some STR alleles are more common in particular populations it may also be possible to determine the geographical location where the whale was caught.

13.8 Medical Applications

The medical applications of the technologies described in this book are wide-ranging and diverse. Molecular techniques are already routine in the diagnosis of genetic disease and a whole range of pharmaceuticals are produced using gene cloning technologies. The potential of the human genome project in allowing us to understand the functioning of the human body in new ways is only beginning to be realized and many significant medical advances are bound to result.

Many important pharmaceutical products are proteins and gene cloning is already used extensively in their production. In Chapter 9 we discussed how bacteria, yeast and tissue culture cells can be used in the production of proteins; many of the successful applications of this technology have been in the production of pharmaceuticals (Table 9.1) where the use of cultured cells, whether bacteria or eukaryotic cells, offers a number of advantages. In many cases the only other possibility is to isolate the proteins from humans or animals. Proteins from humans carry the risk of transmitting disease, as in the case of HIV passed to hemophiliacs in factor VIII blood clotting factor. Proteins from other animals may not behave in exactly the same way as the human protein as in the case of insulin, one of the first pharmaceuticals to be produced by genetic engineering techniques. Gene cloning has proved to be especially valuable in the production of safe and effective vaccines. Using this technology it is possible to clone specific proteins from a pathogen, for instance viral coat proteins from hepatitis B, produce these proteins in cultured cells and use them as a vaccine to elicit a protective immune response against the whole organism. These so-called subunit vaccines avoid many of the problems encountered with killed or attenuated vaccine preparations; the absence of whole organisms avoids any risk of causing an infection and also reduces side effects.

In Chapter 12 we introduced the idea of biopharming; of using transgenic plants or animals to produce pharmaceutical proteins. This technology is in its infancy, but it holds the promise of fields of plants or flocks of sheep cheaply and efficiently producing valuable pharmaceutical products. In most cases it is envisaged that the pharmaceutical product would be harvested and purified, although in the case of edible vaccines (Section 12.2) it is hoped that it will be possible to deliver the vaccine orally by eating the plant. The advantages of this approach are not only in the acceptability of this method of delivery to people who have an aversion to hypodermic needles, but also in being able to grow, harvest and store vaccines using low-tech farming techniques rather than expensive high-tech pharmaceutical technology. Another example of using transgenic animals to produce therapeutic products is the idea of "humanizing" the pig as a donor of organs for use in xenotransplantation which is discussed in detail in Section 12.5.

Gene therapy is perhaps the ultimate medical application of gene cloning. The approach here is to genetically modify the cells of the patient for therapeutic purposes. This is different from the idea of making a transgenic human discussed at the end of Chapter 12 because only some of the cells in the patient are targeted, and not those of the germ line. Gene therapy is intuitively applicable to inherited disorders, which stem from mutations in our genes and hence invite a genetic approach to therapy, but it has also been proposed as a way of treating both cancer and infectious disease. Gene therapy is an exciting area of research but it is a very wide-ranging field and it is outside the scope of this book.

Our greater understanding of the human genome makes it possible to begin to understand the genetic components of more complex diseases and this should help to identify new drug targets and lead to more intelligent drug design. Understanding how an individual's genetic make-up affects the way in which they respond to drugs is a powerful new approach to the development and use of pharmaceuticals; this rapidly developing field has been dubbed pharmacogenomics. The aim of pharmacogenomics is to identify specific markers associated with the differing responses of individuals to drugs.

Because responses to drugs are complex, probably involving more than one gene, the approach taken here is to look for SNP markers spread throughout the genome. SNPs can be detected using a technique developed from the Sanger DNA sequencing protocol. This technique uses dideoxy dye terminators like those used in the automated sequencing protocol described in Section 7.8. Addition of one these dideoxynucleosides not only terminates DNA synthesis but also labels the product with a specific fluorophore. Synthesis is primed using an oligonucleotide which will hybridize to the sequence just before the position of the SNP to be typed (Figure 13.3); the primer is then extended by a single dye terminator

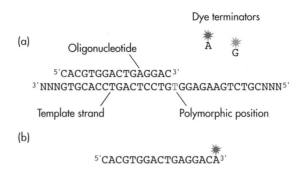

Figure 13.3 Single nucleotide primer extension. a) Test DNA is hybridized with an oligonucleotide primer, which is complementary to part of the sequence to be assayed, but where the 3' end is complementary to the nucleotide before the polymorphic position. This is used in primer extension, in combination with dye terminators (fluorescently labeled dideoxynucleoside triphosphates) to add a single additional nucleotide. In the example shown here, there can be either a T or a C in the polymorphic position. b) The products are analyzed in the same way as the products of a sequencing reaction and the color of the fluorescent dye will indicate the character of the polymorphism. In this case an A incorporated by primer extension indicates a T in the template DNA.

nucleotide at the 3' end. The product is analyzed using a fluorescent sequencer and the color of the fluorescence indicates which nucleotide is present in the SNP. This minisequencing procedure can be used in a multiplex format to detect a range of SNPs in a single test. Multiplexing is achieved by adding tails of varying lengths to the 5' end of the sequencing primers so that the products representing each SNP are of different lengths. This has been used successfully to test for 15 common SNP mutations in the β globin gene, which give rise to β thalassemia, in a single multiplex minisequencing reaction.

For pharmacogenomics to become a reality it will be necessary to develop techniques which permit the scanning of a patient's genome for many SNP markers in a rapid and cost effective way. Minisequencing can be combined with DNA array technology (Section 11.6) to provide a high throughput technique for SNP detection. In this procedure hundreds of oligonucleotides primers, each specific for the sequence before a particular SNP, are immobilized on a DNA chip. Test DNA is extracted and hybridized with the primers, minisequencing reactions are carried out and each spot on the array will fluoresce with a color which indicates which nucleotide has been added to the 3' end, thus indicating which allele is present for each SNP.

The HapMap project has completed the mapping of one million SNP markers in the human genome. Combined with minisequencing on a chip this should provide an efficient way of scanning an individual's genome to find correlations between blocks of markers and phenotypic characteristics such as disease susceptibility or response to a drug or vaccine. Understanding, at the level of the genes, why individuals respond differently to drugs is likely to have an impact both on drug development and on the way in which drugs are prescribed. Drug development is a costly process and many drugs fail in the later stages of the approval process. It is hoped that pharmaceutical companies will be able to develop drugs more easily using genome targets, and to target them to the groups of people who are most likely to be able to benefit. Some drugs have failed in drug trials because of adverse side effects experienced by some individuals; however, these compounds may still be effective treatments in other groups of patients. Clinical trials routinely use genetic tests to screen for variants of the cytochrome P450 family of liver enzymes which are responsible for the breakdown of many classes of drugs. This allows for more accurate calculation of doses and more appropriate targeting of drugs. In the future it should be possible for doctors to analyze a patient's genetic profile and to use this information to inform choices about which drug to prescribe and also about dosage. This will reduce the risk of side effects and ensure that the most appropriate treatment is prescribed first time. In addition to the obvious benefits to the patient this should improve the cost effectiveness of the treatment.

It is clear that with the advent of pharmacogenomics, diagnosis is going to make use of a wide range of DNA-based tests. This technology is currently most highly developed and routinely used in the area of diagnosis of inherited disease but has many important applications in the diagnosis of infectious disease and cancer.

As with gene therapy, it is intuitive that gene-based tests are the ideal way in which to diagnose an inherited disorder. After all we know that the root cause of the disorder is a genetic mutation, so detecting the presence or absence of the mutation is an ideal approach to diagnosis. Gene-based tests have a number of additional advantages. They can detect the presence of disease genes in people who do not display symptoms and they can also differentiate between different disease states with similar symptoms. Being able to detect a disease mutation in people before they begin to display symptoms is particularly important in cases where it is possible to reduce the severity of the disease if intervention occurs early. A classic example of the benefits of early intervention is in the case of phenylketonuria, although in this case the test that is used routinely is not a genetic one. This inherited metabolic disorder can result in toxic levels of phenylalanine in the body resulting in mental retardation; but the disease can be successfully managed, if detected early, by adherence to a diet with carefully controlled levels of phenylalanine. A very successful strategy of testing newborn babies for phenylketonuria was introduced in the UK in the 1970s and patients suffering the adverse effects of this disorder are now rare.

With the advent of cheap and reliable genetic tests for a range of common inherited disorders it has been suggested that newborns should be routinely tested for the presence of a range of common disease causing mutations including cystic fibrosis, β thalassemia and sickle cell anemia. This would be most advantageous in a disease like cystic fibrosis where early intervention can improve the management of the disease and improve quality of life. For many other common inherited diseases, however, there are no effective treatments.

The other group of people who have a disease gene but do not display symptoms are carriers of recessive disorders. Because they are heterozygous for the genetic mutation they often show no symptoms of the disease but are at risk of having affected offspring if their partner is also a carrier. In this case tests can be performed on blood samples to determine carrier status allowing couples to make informed decisions about the risk of having a child affected by a genetic disease.

DNA-based tests, however, can also be useful in the diagnosis of other types of disease including cancer and infectious disease. In the case of cancer, DNA-based tests can aid in diagnosis where disease states have similar clinical manifestations but different genetic causes. This is, of course, most important where different treatments are appropriate in each case. A case in point would be the hereditary form of breast cancer which has very similar manifestations to sporadic breast cancer but responds differently to treatment. DNA-based molecular tests have also had a major impact in the diagnosis of infectious disease, particularly of viral disease where it is often much easier to amplify viral DNA than to culture the virus.

13.9 Techniques for Diagnosis of Inherited Disorders

There are at least 10,000 genetic disorders which are known to result from a mutation in a single gene, and although most of these are individually rare together they affect about 1% of the population. Many of the more common of these conditions have been well understood for many years. This is at least in part because, as they are caused by a mutation in a single gene, they show simple patterns of inheritance and can be traced through families. Some of the less common disorders have only been fully understood with the help of modern methods of genetic analysis.

In addition to the single gene disorders, there is a whole range of complex disorders which result either from more than one mutation, or require environmental or lifestyle factors in addition to the genetic ones. Many common problems are thought to fall into this category, including diabetes, heart disease, autism and Alzheimer's disease. These diseases cluster in families but do not show a clear pattern of inheritance making it more difficult to track down the genes involved. In the example of diabetes, scientists have identified at least five genes that may be involved in the juvenile form of the disease and a further three involved in the adult form.

Experiments with mice have revealed 15 genes on 10 different chromosomes where mutations increase the risk of developing diabetes. And in addition to these genetic factors lifestyle, especially diet, is known to be important in the development of the disease. It is likely that the genetic basis of many of these complex, multifactorial diseases will be elucidated over the next few years. Once the genes and mutations involved are known, it will be possible to test for the genetic disposition to these diseases using the same range of tests that are currently used for single gene disorders.

Because of the wide range of different types of mutation that give rise to genetic disease (Box 13.2), the constraints imposed by the nature of the samples to be tested (Box 13.3) and the requirements for robust easily automated test formats, there is no single type of test that is suitable for diagnosis of all inherited disorders. Most routine tests are based on PCR amplification of the region of the genome where the gene is found and often hybridization with a DNA probe. The samples used for genetic diagnosis are similar to those used for forensic analysis and include blood samples and buccal swabs, which yield sufficient cells to extract DNA for PCR-based tests. One major application of genetic testing is in prenatal and pre-implantation genetic diagnosis. In this case the sampling procedures (Box 13.3) only yield a very small number of cells and without PCR there would be severe limitations on the number and type of tests that could be performed.

Allele specific PCR amplification

This test makes use of the fact that PCR amplification is particularly sensitive to a mismatch between the template and primer at the 3′ end of the primer. Allele specific primers ensure that amplification only occurs when there is a perfect match between template and primer (Figure 13.4).

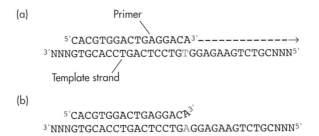

Figure 13.4 Primer specific PCR amplification. A PCR reaction is performed using a primer specific for the allele being tested for, with the variable nucleotide at the 3′ end of the primer. a) When the primer and the template are exactly complementary the PCR reaction can proceed as normal. b) However, if there is a mismatch at the 3′ end of the primer DNA polymerase is unable to extend the primer and there will be no amplification.

Box 13.2 Genetic Abnormalities Giving Rise to Disease

There are a wide range of different types of genetic abnormality which can give rise to diseases. These fall broadly into two categories: those that involve large changes, often resulting in a visible alteration to the chromosomes and smaller changes usually referred to as molecular abnormalities. However, the development of technology is blurring this distinction as it becomes possible to "see" changes on a much smaller scale.

Large-scale abnormalities include numerical chromosomal abnormalities which involve the gain or loss of a whole chromosome (see Section 2.10) and also translocations where parts of non-homologous chromosomes are joined together. So-called microdeletions involve the loss of segments of the chromosome of several million base pairs. A classical example is the deletion of 4 Mbp from the long arm of chromosome 15. This gives rise to Prader–Willi and Angelman syndromes.

The simplest type of molecular abnormalities are substitutions of one nucleotide for another. This type of mutation, called a SNP (Section 2.3) is responsible for much of the phenotypic variation between individuals; however, a small proportion of SNPs have harmful effects. For example, sickle cell disease is caused by a simple mutation of a single A to a T in the gene encoding β globin. This mutation results in the replacement of a glutamic acid residue with a valine, the effect of which is to cause the hemoglobin molecules to aggregate into long fibers which distort the red blood cells giving them the characteristic "sickle" shape.

Small insertions and deletions are also commonly observed. For instance the most common mutation in people with cystic fibrosis, ΔF508 is the deletion of three nucleotides which results in the loss of a single phenylalanine residue in the CFTR protein. Larger deletions are also seen for example in muscular dystrophy, here deletions are very large and can span several exons. In broad terms muscular dystrophies fall into two types. The milder Becker form is generally characterized by mutations which preserve the reading frame (Section 8.1). These result in a shortened form of the dystrophin protein. Duchenne muscular dystrophy, where symptoms are severe, generally results from deletions which cause a frame shift and premature termination of the protein (Question 13.6).

The other main type of mutation responsible for genetic disease is the expansion of repeat sequences. These sequences are basically microsatellite or STR sequences where the number of repeat sequences affects the phenotype. Expansion of trinucleotide repeat sequences was first identified in Huntington's disease. The number of repeats of the sequence CAG (which encodes glutamate) at the C terminus of the Huntington's gene varies in the normal population from six to 35. However, once the number of repeats exceeds this threshold symptoms of Huntington's disease will develop. In diagnosis of Huntington's disease it is therefore essential to be able to determine the number of repeats as this correlates with age of onset and severity of the condition.

With single gene disorders the mutation is confined to a specific region of the human genome, a gene or its regulatory sequences. However, there may be many different mutations possible within this region, some of which are common and others much less so. For example in the case of cystic fibrosis the ΔF508 mutation is by far the most common mutation being found on 75% of CF chromosomes in the UK. The next most common mutation G551D is found in only 3% of CF chromosomes. There are some 20 mutations found routinely in the UK and hundreds more which are only rarely found. The incidence of the ΔF508 mutation also varies, occurring in only 20% of CF chromosomes in Turkey but in 90% in Denmark. Another common inherited disorder, β thalassemia, results from point mutations and small deletions in the β globin gene, causing anemia of varying degrees of severity. It is prevalent in people who originate in the Mediterranean, the Middle East, India and South-East Asia. Again there is a variety of mutations and their prevalence varies between ethnic groups.

Usually, primers specific for both normal and mutant alleles are used in separate PCR reactions so that both alleles can be detected. This makes it possible to identify heterozygous carriers as well as affected individuals. This technique can be used to detect a range of different types of mutations which give rise to genetic diseases including deletions and insertions as well as single nucleotide polymorphisms as they will all create mismatches between primer and template.

The sensitivity and efficiency of this technique can be improved by labeling the allele specific primer with a fluorophore (Box 5.3). As with DNA profiling the amplification products can then be analyzed by capillary electrophoresis, this improves detection of weakly amplified products. By labeling each allele with a different fluorophore it is possible to detect both alleles in a single reaction.

Fluorescent PCR for detection of expanded trinucleotide repeats
This is in fact exactly the same technique as that used in DNA profiling (Section 13.3). The expanded trinucleotide repeats responsible for diseases like Huntington's disease (Box 13.2) are in effect STRs and testing for them involves PCR amplification of the repeat region in a PCR reaction with one fluorescently labeled primer. The sensitivity gained by using fluorescent PCR and capillary electrophoresis is particularly important in the context of Huntington's disease as it is important to be able to determine the

Box 13.3 Methods of Obtaining Samples for Prenatal Genetic Diagnosis

In order to assess the genetic make up of a fetus it is necessary to obtain some fetal cells; samples are obtained by amniocentesis or chorionic villus sampling (CVS). Amniocentesis involves inserting a hollow needle through the abdominal wall and taking a sample of the amniotic fluid that surrounds the fetus in the womb; this fluid contains a small number of fetal cells. Amniocentesis cannot be performed until 15 weeks of gestation and as the yield of fetal cells is very low there is usually a delay of several days, while the cells are grown in the laboratory, before tests can be performed. CVS involves taking a small part of the placental tissue; which has the same genetic make up as the embryo. This can be performed at 10 weeks gestation and usually yields sufficient material for direct testing. Both of these sampling procedures involve a small increase in the risk of miscarriage and so are usually only offered to people who have a known increased risk of having a fetus affected by a genetic disease. This could be women over the age of 35 where there is a risk of Down's syndrome, couples who already have a child or other close relative affected by a genetic disease or cases where other routine prenatal screening procedures have detected possible abnormalities.

The possibility of a non-invasive approach to prenatal diagnosis has come about as a result of recent advances in cell sorting technology. A very small number of fetal cells cross the placenta and are found in maternal blood and these include immature and hence nucleated red blood cells. Cell sorting techniques use antibodies to specific cell surface antigens to tag particular cell types prior to sorting and can successfully enrich a sample many fold for these fetal cells. Unfortunately, as this procedure is only an enrichment procedure, and cannot eliminate maternal cells altogether, at the moment the samples are only suitable for testing for chromosomal abnormalities by procedures like FISH (Section 13.9).

One of the major drawbacks with prenatal diagnosis is that in most cases the only possible intervention is termination of the pregnancy. However, as *in vitro* fertilization (IVF) becomes a more reliable and routine procedure, it is possible for couples who know that they are at risk of having a child with a serious genetic disease, to opt for IVF and pre-implantation genetic diagnosis. In this case the fertilized egg is allowed to develop in culture until there are six to 10 cells, at which point one cell, called a blastomere, is removed from the embryo. The blastomere is then processed for either FISH or PCR depending on the genetic condition to be tested for. Once the results of the genetic tests are known, healthy embryos can be selected and used in IVF.

number of repeats as this correlates with age of onset and severity of the condition.

Multiplex PCR in genetic diagnosis

As with DNA profiling PCR reactions can be multiplexed for genetic diagnosis (see Section 13.4); different regions of the genome can be amplified in a single PCR reaction allowing tests for a whole range of disorders to be performed on a single sample. Multiplex PCR is also useful where it is necessary to screen for several different mutations in the same gene in order to diagnose a disease (Box 13.2).

Multiplex ligation-dependent probe amplification (MLPA)

One of the problems encountered with multiplex PCR is in ensuring uniform amplification of many different PCR products. MLPA addresses these problems by using the ligation of sequence specific probes to generate a more uniform target for PCR which can be amplified by a single pair of primers (Figure 13.5). The probes hybridize to sequence adjacent to each other in the target region. If they are perfectly complementary to the test sequence they can be joined by DNA ligase and the product formed can be amplified by PCR. If there is a mismatch at the position of the join DNA ligase will not join the fragments. The successful ligation products all have the same PCR primer sites at their ends and hence can be amplified by a single pair of primers. Probes specific for different sequences can be labeled with stuffer sequences of different lengths and the products can be separated by size using capillary electrophoresis. This technique can be used to amplify each of the 79 exons of the dystrophin gene (Box 13.2) in a single test and can theoretically detect all deletions of one or more exons. This test is also able to detect differences in the amount of DNA and can be used to detect deletions and duplications both of larger regions and even whole chromosomes.

Allele specific oligonucleotide (ASO)

An ASO is simply a short synthetic oligonucleotide, which represents the sequence of either the wild type gene or the mutation which is to be tested for. They are used in the same way as DNA probes (Section 5.5) to detect complementary sequences by hybridization. By controlling the hybridization conditions it is possible to ensure that the oligonucleotide will only form a stable duplex with a sequence that is exactly complementary and hence to differentiate between alleles that differ by as little as a single nucleotide. ASOs are usually between 15 and 20 nucleotides long and, unlike allele specific PCR primers, designed so that the single nucleotide difference is in the center of the molecule. ASOs can be used in a dot blot format (Figure 13.6) or when it is important to determine the size of a region of the genome in Southern hybridization (Box 5.2).

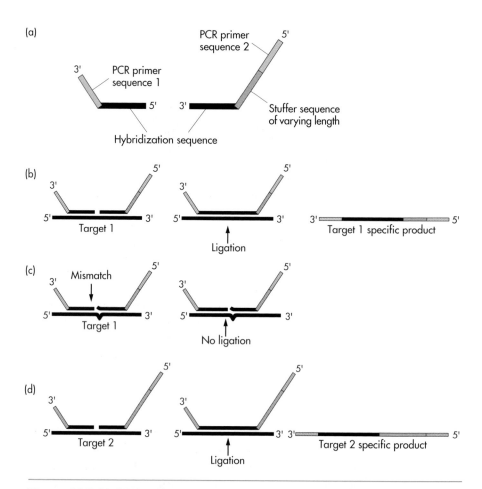

Figure 13.5 Multiplex ligation-dependent probe amplification (MLPA).
a) Probes consist of a hybridization sequence specific for the target region
(black), a variable length stuffer sequence (gray) and a PCR primer sequence
(blue). b) The hybridization regions are designed so that two probes hybridize
immediately adjacent to each other with the join at the site of the possible
mutation. If the probes are perfectly complementary to the target sequence DNA
ligase will be able to join them together to form a single molecule; this molecule
has both PCR primer sequences and hence can be amplified by PCR. c) If there
is a mismatch, ligation will not occur, there will be no template for the PCR
reaction and no product will be formed. d) Different target sequences can be
investigated using probes with the same PCR primer sequences but different
hybridization sequences tagged with stuffer sequences of different lengths.

Q13.4. What factors in a hybridization experiment can be altered to ensure
that the oligonucleotide will only form a stable duplex with a sequence that
is exactly complementary?

A very straightforward example of the use of ASOs in detecting disease
mutations is in the case of sickle cell disease. As there is only one mutation

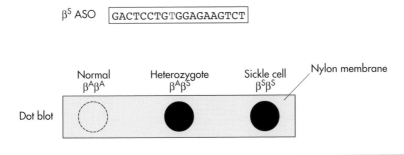

Figure 13.6 Use of allele specific oligonucleotides (ASOs) in a dot-blot assay. In a dot blot an aqueous solution of DNA from the patient is applied directly to a piece of nylon membrane and allowed to dry. The DNA is denatured either by heating prior to application or exposure to alkali once it is on the filter, fixed to the membrane and probed with labeled allele specific oligonucleotides (ASOs). Stringent hybridization conditions are used so that the ASO will only hybridize when it is a perfect match.

In the example shown the β^S ASO represents the sequence around the sickle cell mutation in the human β globin gene. If this is used to probe human genomic DNA it will hybridize to any sample of DNA with the sickle cell mutation, giving a positive result for the sickle cell homozygote and the heterozygote (filled circles) and a negative result for normal individuals (dashed unfilled circle).

which gives rise to this condition (Box13.2) it is easy to design two ASOs, one specific for the sickle cell mutation and one specific for the wild type. These can be used in combination not only to detect the presence of the sickle cell mutation but also to differentiate the heterozygous carrier from the homozygotes (Figure 13.6).

In order to make these tests cost effective and efficient it is necessary to develop formats by which the test procedure can be automated and where many samples can be tested at the same time. Several of the developments of ASO-based tests, including molecular beacons and the fluorogenic 5′ exonuclease assay, which are described in the following sections, involve formats where the test can be performed in solution rather than on a solid membrane. This makes it possible to perform the tests in a 96-well plate which allows for automated sample handling. Another common theme here is to use ASO probes labeled with fluorescent markers (Box 5.2) which can be read in an automatic plate reader.

Molecular beacons

Molecular beacons are dual-labeled molecules which only emit fluorescence when they are bound to their target DNA sequence. In the absence of target sequence they adopt a hairpin loop structure due to the complementarity of the sequence at the 5′ and 3′ ends (Figure 13.7a). In this conformation they do not fluoresce. However, in the presence of a complementary

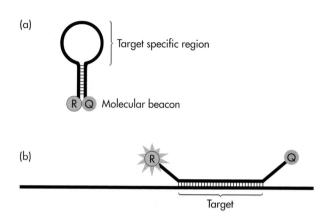

(a)

Target specific region

R Q Molecular beacon

(b)

R

Q

Target

Figure 13.7 Using a molecular beacon to detect a disease gene. a) The molecular beacon is a single-stranded DNA molecule with a fluorophore reporter covalently linked to one end (R in the blue circle) and a quencher to the other (Q in the gray circle). The central target specific region is an ASO complementary to the target. Flanking this region are short sequences that are complementary to each other. The complementary sequences cause the molecule to adopt a hairpin loop structure which brings the reporter and quencher into close proximity and prevents the reporter from fluorescing. b) Once mixed with test DNA the target specific region hybridizes with the target region on the test DNA; this puts enough distance between the reporter and quencher to allow the reporter to fluoresce. If there is no complementary target DNA sequence the molecular beacon will not anneal and there will be no fluorescence.

target sequence the target specific region will hybridize causing a conformational change (Figure 13.7b). The fluorophore will now emit fluorescence. In designing a molecular beacon the sequences are chosen so that the perfectly complementary probe-template hybrid is more stable than the hairpin loop structure. This means that the molecular beacons will preferentially form the probe template hybrid in the presence of target sequence. Like ASOs, molecular beacons are able to distinguish between alleles which differ by as little as one nucleotide. If there is a single mismatch between probe and target, the hybrid is less stable and the hairpin loop conformation is favored, so no fluorescence is emitted. To distinguish between homozygous and heterozygous individuals two molecular beacons are used, each one specific for one or other allele and each labeled with a fluorophore of a different color. If the test results in fluorescence of only one color then the sample was from a homozygous individual, if both colors are produced then the individual is heterozygous.

Q13.5. What parameters need to be considered when designing the sequence of the molecular beacon to ensure that the probe target hybrid is more stable than the hairpin structure?

As with most diagnostic tests molecular beacons are usually used in conjunction with a PCR reaction. Amplification is required to produce enough DNA to hybridize with sufficient of the beacon molecules to yield a detectable level of fluorescence. In fact the molecular beacon can be added to the assay mixture before PCR amplification and fluorescence measured in real time in a development of real time PCR (Section 3.18). The number of cycles required to reach a given level of fluorescence can be used as a measure of the amount of target sequence in the original sample. This provides a method not only for detecting the target sequence but also for quantification. Using this quantitative approach it is possible to distinguish between homozygous and heterozygous individuals using a single probe as the homozygotes will have twice as much of one allele as the heterozygote. Molecular beacons used in conjunction with real time PCR can also be useful in the diagnosis of infectious disease (see Section 13.11).

Molecular beacon assays offer an additional safeguard in diagnostic testing because they require both PCR amplification and annealing of the molecular beacon to the product in order to generate a signal. In other words they require three oligonucleotides to anneal, the two PCR primers and the molecular beacon, reducing the chances of a false positive resulting from mispriming by PCR primers.

Because fluorophores are available which emit a range of different colors of fluorescence (Box 5.2), it is possible to use molecular beacons in conjunction with multiplex PCR in order to test for several different mutations at the same time. Each different mutation is tagged with a different fluorophore allowing the results of the different tests to be distinguished from each other. Most common genetic disorders are caused by more than one type of mutation, either in a gene or the upstream sequences regulating its expression. In this case it is necessary to test for a range of mutations in order to accurately diagnose the condition. For a disease like cystic fibrosis (Box 13.2) it is possible to test for seven of the most common mutations in a single multiplex PCR reaction using molecular beacons labeled with seven different fluorophores.

Fluorogenic 5′ exonuclease assay (TaqMan®)

This assay also makes use of dual-labeled fluorescent probes (Figure 13.8) which are commercially available as TaqMan® probes. The target DNA is amplified by PCR and the TaqMan® probe is included in the reaction. When the target DNA becomes single-stranded during the denaturation step the probe hybridizes to its complementary sequence but cannot fluoresce because of the proximity of the quencher. However, as PCR proceeds the probe is degraded by the 5′ exonuclease activity of *Taq* polymerase, releasing it into solution where it emits a fluorescent signal. In each successive cycle of the PCR more probe molecules will anneal, and then be degraded releasing the fluorophore and increasing the intensity of the

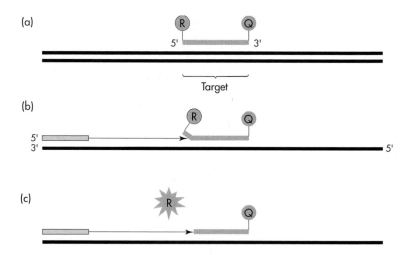

Figure 13.8 Fluorogenic 5′ exonuclease TaqMan® assay. a) This assay uses a special DNA probe (shown in blue) labeled at the 5′ end with a fluorophore (R in the blue circle) and at the 3′ end with a quencher (Q in the gray circle). The probe is also blocked from extension at the 3′ end to prevent it from acting as a PCR primer. When the template strands (black lines) separate in the denaturing step of the PCR reaction the probe hybridizes with its complementary sequence. b) As PCR amplification proceeds one of the PCR primers (shown in gray) will initiate copying of the template strand to which the probe is hybridized. In this case the probe will be degraded by the 5′ exonuclease activity of *Taq* polymerase. c) This releases the fluorophore reporter into solution where it is able to fluoresce.

signal. As with the molecular beacon assay this allows the assay to be performed in solution making automation more practical. It also offers many of the same advantages including the added specificity gained by using thee oligonucleotides and the ability to multiplex and use probes labeled with different fluorophores.

Mutation scanning

In some situations it is not sufficient to test for specific mutations but becomes necessary to scan for the presence of any mutation in a particular gene. This may be because the condition that is being tested for results from a very wide range of mutations, none of which is very common, or it could be where a mutation is suspected which has not previously been characterized. A range of techniques has been developed for high throughput mutation scanning which can detect small differences between DNA molecules without sequencing. One such method relies on the observation that when DNA is denatured the single-stranded molecules adopt specific conformations, called single-stranded conformational polymorphisms (SSCPs), which are very sensitive to the precise sequence of the molecule.

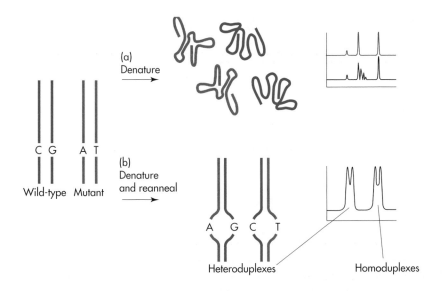

Figure 13.9 Techniques which detect any mutation in a region of DNA.
a) Single-stranded DNA molecules adopt a specific conformation in the same way
that RNA molecules do because of base pairing between sections of
complementary DNA. These conformations are specific to particular sequences
and a single base pair substitution can cause a change in the conformation
adopted. The conformation affects the mobility of the molecules and can be
detected by electrophoresis under non-denaturing conditions. b) Heteroduplexes
formed between wild type and mutant DNA molecules are less stable because
there are fewer hydrogen bonds between the two strands in the heteroduplex.
In denaturing high-performance liquid chromatography (dHPLC) the molecules are
analyzed under partially denaturing conditions and a temperature gradient is
applied; the less stable heteroduplex molecules are eluted at a lower temperature
than the homoduplexes.

These molecules can be separated in non-denaturing conditions either on
a slab gel or by capillary electrophoresis (Figure 13.9a) where a mutation
will show up as a different conformational polymorphism with altered
electrophoretic mobility. SSCPs can detect the presence of mutations, and
can be used to characterize mutations by comparison with known mutations but sequencing is necessary to characterize a novel mutation
detected in this way. Other techniques detect the instability of heteroduplexes formed between wild-type and mutant DNA molecules. The test
molecule is mixed with a wild-type control sequence, heated to separate
the strands and allowed to re-anneal. Some heteroduplex molecules will be
formed comprising one wild-type and one mutant strand. The resulting
molecules are analyzed under partially denaturing conditions where the
heteroduplexes are unstable because they have fewer hydrogen bonds
between the strands. Techniques such as denaturing high-performance

liquid chromatography (dHPLC) and temperature gradient capillary electrophoresis use a temperature gradient to elute the heteroduplex molecules before the homoduplexes.

These high-throughput techniques are able to detect differences between wild-type DNA sequences and those carrying mutations. Once calibrated with known mutations they are also able to identify specific mutations. However, even when these techniques are used it may be necessary to sequence the whole of the gene, and its upstream regulatory sequences, to precisely characterize novel mutations. One further problem thrown up here is that simply because a difference is found between the wild-type and the test sequence, whether by sequencing or one of the other techniques, this does not prove that the mutation gives rise to a faulty protein which is responsible for an inherited disease. If the mutation introduces a stop codon (Section 8.1) the protein product will be truncated at this point and it will probably be pathogenic, otherwise it may simply be part of the natural variation between individuals. Deciding whether a particular sequence change is pathogenic is a major challenge in diagnostic laboratories.

Q13.6. What types of mutation are likely to introduce stop codons into a gene?

Fluorescent in situ hybridization (FISH) for detection of chromosomal abnormalities

This technique uses a fluorescent DNA probe to detect specific sequences in chromosome preparations; usually denatured metaphase chromosomes where the DNA is maximally contracted and the chromosomes are easy to see. Probes can be either specific for a particular part of the chromosome (Figure 13.10a) or, in the case of chromosome paints, for particular chromosomes (Figure 13.10b). Chromosome paints are made by fluorescent labeling of a mixture of DNA clones derived from many positions on the chromosome of interest, often from chromosome specific libraries (Figure 13.10b).

Chromosome painting has had a major impact on the diagnosis of numerical chromosomal abnormalities (Box 13.2) such as Down's syndrome. It is much faster than traditional karyotyping, taking between 24 and 48 h compared with some 10 days for traditional methods. It is also more precise because it can unequivocally distinguish between chromosomes with similar appearance. Combined tests are available for the chromosomes most commonly found in numerical abnormalities (namely 13, 18, 21, X and Y) which use differently colored probes to differentiate the chromosomes. Chromosome painting is also useful in detecting translocations (Box 13.2); in this case probes for two different chromosomes, labeled with different fluorophores are used. If a translocation event has occurred

(a) Locus specific probe

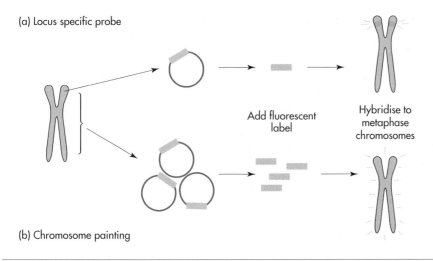

Add fluorescent
label

Hybridise to
metaphase
chromosomes

(b) Chromosome painting

Figure 13.10 Fluorescent *in situ* hybridization (FISH). These techniques are used to label specific sequences in chromosome preparations. a) Locus specific probes are made by cloning the region to be investigated and labeling with a fluorophore. They can be used to probe for complementary sequences in preparations of whole cells and tissues. b) Chromosome paints contain a mixture of labeled probes representing positions throughout the chromosome, usually made from chromosome specific libraries. They can also be used to probe whole cell and tissue preparations.

parts of chromosomes colored with different probes will be found joined together. Unfortunately, there are not enough fluorescent dyes to color each of the 24 different human chromosomes a different color. However, a sort of multiplex chromosome paint has been developed where some probes are labeled with more than one fluorophore. Automated digital image analysis is used to analyze these and to assign pseudocolors to each individual chromosome resulting in a multicolored molecular or spectral karyotype.

Locus specific probes (Figure 13.10a) can be used for more detailed analysis of chromosomes. These are, in the main, designed to detect microdeletions (Box 13.2). Probes are usually designed to hybridize to sequences within the deletion boundary. Normal cells will show a characteristic double spot, as the sister chromosomes are labeled, which will be absent from abnormal cells.

Q13.7. Why is multiplex PCR an important tool in genetic diagnosis?

13.10 Whole Genome Amplification
The availability of good quality genomic DNA in reasonable quantities is fundamental to genetic testing, this is particularly critical with techniques

such as pre-implantation genetic diagnosis (Box 13.3) where the sample may consist of one or two cells. There are also many research situations where the availability of adequate quantities of genomic DNA may be limiting, these include the analysis of forensic and archaeological specimens and preserved clinical samples such as dried heel prick blood samples from newborns. The aim of whole genome amplification (WGA) is to develop a technique for high yield, high-fidelity amplification of genomic DNA to allow multiple assays to be performed on DNA from a limited sample. A number of techniques have been described for WGA based on PCR using random or degenerate primers. These techniques tend to generate DNA fragments of relatively short length and to show preferential amplification of some sequences and under-representation of others.

These problems have led to the development of an alternative approach called multiple displacement amplification (MDA). This technique results in exponential amplification of template DNA without temperature cycling. It makes use of a DNA polymerase from the bacteriophage φ29. There are three key features of this enzyme: First, φ29 polymerase can synthesize DNA over long distances producing fragments of up to 70 kb. Second, it has a low error rate as it has proofreading activity. Finally, φ29 DNA polymerase displaces any double-stranded DNA that it encounters during synthesis. In MDA, DNA synthesis is primed at many positions on the genomic DNA template using multiple random primers. As synthesis proceeds strand displacement of newly synthesized DNA provides new single-stranded DNA to act as template for further priming events. Eventually, a hyperbranched structure is formed (Figure 13.11) leading to exponential amplification of the genomic template. Unlike PCR there is no temperature cycling which can damage template DNA. MDA yields fragments with an average size of 10 kb and produces up to 100,000-fold amplification of genomic DNA without significant amplification bias, providing ample DNA on which to perform numerous diagnostic tests.

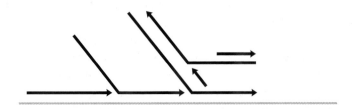

Figure 13.11 Hyperbranched structure formed during multiple displacement amplification (MDA). DNA synthesis (black) is primed at many positions on the genomic DNA template (blue). φ29 polymerase displaces double-stranded DNA as it encounters it during DNA synthesis, the displaced DNA becomes template for further DNA synthesis and a hyperbranched structure develops.

13.11 Diagnosis of Infectious Disease

Traditional techniques for identification of micro-organisms in clinical samples, to aid in diagnosis of infectious disease, usually involve culturing organisms from the original specimen and then performing a battery of biochemical tests. The culturing step is usually necessary to purify organisms which may be present in mixtures and to provide sufficient organisms for testing. Micro-organisms have a very limited range of physical characteristics and so most of the tests used in identification rely on examining the biochemistry of the organisms. There are several drawbacks to these traditional approaches, including the time taken to culture organisms, the possibility that some organisms may grow better than others and even the possibility, particularly in the case of viruses, that they are difficult to grow in the laboratory and in some cases cannot be grown in the laboratory at all. Although these traditional techniques form the 'gold standard' of testing in hospital microbiology laboratories a range of molecular techniques are being developed, often to address particular problems.

Molecular diagnostics of bacterial pathogens

We have already seen in Chapter 8 how 16S ribosomal RNA sequences can be used to study evolutionary relationships between bacteria (Section 8.17). Although, in general, the sequence of the 16S molecule is highly conserved (Box 8.4) it has regions that are very highly conserved and other more variable regions. The sequences in these variable regions are species specific and can be used to identify bacteria. In this approach PCR primers, complementary to flanking conserved sequences, are used to amplify the variable regions. The product is then sequenced and the sequence compared against a database of 16S sequences to identify the bacteria it was derived from. This approach to identification of bacteria has particular advantages with organisms which cannot be easily cultured in the laboratory, as the DNA is amplified by PCR rather than the organisms being amplified by growing in culture.

Amplification of 16S genes by PCR can be very effectively combined with molecular beacon technology (Section 13.9) to identify bacteria in mixtures. A PCR reaction with a single set of primers complementary to conserved sequences will amplify species specific sequences from a range of different bacteria in a mixture. These can then be probed with molecular beacons complementary to the individual species specific sequences and labeled with different fluorophores. This makes it possible to identify more than one type of pathogenic bacterium in a mixture. By measuring the fluorescence levels in real time (Section 3.18) it is also possible to make quantitative measurements and to detect the presence of a rare pathogen in the presence of a more abundant one.

One topical application of this technology is in devising tests to detect and identify bioterrorism agents. In the case of suspected bioterrorist attack there is a need for a robust and rapid assay for a selection of possible bioterrorist agents. A real-time PCR assay has been devised which can simultaneously detect four bacteria which are considered to have the potential to be used as bioterrorism agents, using a single set of PCR primers and four species-specific molecular beacons. In the case of a bioterrorism incident it would be vital to be able to identify the bacteria concerned very rapidly. Work is under way on minimizing the time required for each of the separate stages in identification of pathogens, from sampling and sample preparation, to PCR, detection and analysis and has led to the proposal that a 3 min detection and identification system based on PCR could be developed for bioterrorism agents.

This same technology has obvious applications in routine clinical laboratories where patient care could be improved by reducing the time taken to identify pathogens to hours rather than days. Tests designed to identify pathogens could be combined with tests to look for mutations associated with resistance to particular antibiotics, providing simultaneous pathogen identification and resistance characterization. This should result in a reduction in the use of antibiotics and in length of stay in hospital for many patients.

Molecular diagnostics in virology

The impact of molecular diagnostics in the field of virology has been more immediate than in bacteriology as molecular tests are able to overcome many of the problems inherent in identification of viruses. One of the major problems with clinical diagnosis of viral infections is that it is difficult, and in some cases impossible, to grow many viruses *in vitro*. As a result many tests rely on detecting the presence of virus specific antibodies which may not be detectable in the early stages of infection. Qualitative and quantitative tests for viral DNA or RNA provide rapid and highly sensitive tests without the need to culture the viruses. These nucleic acid tests (NATs) are also sensitive enough to distinguish between different subtypes of viruses.

Detection and identification of viruses is vitally important not only in diagnosis but also in the management of viral disease. In patients with chronic infections, such as HIV and hepatitis B and C, viral load and emergence of antiviral resistant strains can be monitored by NAT. This allows precise monitoring of the effectiveness of treatment and alterations to antiviral therapy in the case of the emergence of resistance. NATs are also now used in a number of countries for routine screening of blood and organ donations for the presence of HIV and hepatitis C.

Molecular diagnostic tests for viruses are usually based on PCR amplification of nucleic acids from clinical samples and the use of fluorescence labeled probes, both molecular beacons and TaqMan® probes (Section 13.9) are used. Unlike with molecular tests for the identification of bacteria where a single target, the 16S ribosomal RNA gene, can be used for the identification of all bacteria, there are no sequences which are conserved amongst all viruses. Indeed some viruses show extreme genetic variability, making it difficult to devise multiplex tests capable of detecting a range of viruses. In the case of HIV, for example, a real-time TaqMan® test requires nine primer and probe sets to detect all of the common subtypes.

It is often necessary to be able to test for several different viruses that can be responsible for a particular disease. For example, there are 12 viruses responsible for common respiratory infections which may be very difficult to differentiate on the basis of symptoms. DNA micro-arrays (Section 11.6) have been used to develop a single test for these viruses. Oligonucleotides representing signature sequences (sequences found frequently in the group of viruses being tested for and not in other viruses) are spotted on a micro-array. Nucleic acid is amplified by PCR, using random primers, from clinical specimens, labeled and used to probe the array in order to identify which virus is present.

The recent discovery of a new human coronavirus responsible for severe acute respiratory syndrome (SARS) is a particularly good example of the impact of molecular techniques in virology. Molecular techniques, including reverse transcriptase PCR and sequencing played a vital part in determining the cause of this newly recognized disease, and shortly thereafter a molecular diagnostic test was available for the virus which was instrumental in bringing the epidemic to an end.

> **Q13.8.** What advantages do DNA-based diagnostic tests for bacterial and viral infections offer over conventional tests?

13.12 Diagnosis and Management of Cancer

As a general rule a clinical diagnosis of cancer is usually confirmed by cytological analysis of biopsy material. However, cancer arises as a result of mutations, which allow groups of cells to proliferate in an uncontrolled way, and analyzing these mutations can provide much additional information about the nature of the cancer. In many cases molecular diagnostic techniques can distinguish between subtypes of cancers which are difficult to differentiate by other means. This can be important, as there may be implications for prognosis and for management of the cancer, especially if the different subtypes respond differently to treatment. Molecular diagnostic tests are also used in monitoring the response of cancers to treatment in a similar way to monitoring the progress of viral diseases (Section 13.11). The molecular diagnosis of cancer involves the same range of techniques that are described for inherited disease. FISH is a particularly key technique in diagnosis of cancer because it is often necessary to perform tests on biopsy material, and also because large scale chromosomal rearrangements are characteristic of a number of cancers.

Breast cancer is a good example with which to illustrate many of the areas where molecular diagnostic tests contribute to the diagnosis and management of cancer. Once the initial diagnosis has been confirmed by cytology it is important to establish whether the cancer is over-expressing the human epidermal growth receptor *HER-2*. This oncogene is over-expressed in 20–30% of invasive breast cancers and this subtype of the cancer is susceptible to treatment with Herceptin®, a monoclonal antibody specific for the *HER-2* oncogene protein. Over-expression of the *HER-2* gene results from gene amplification and can be detected by FISH (Section 13.9) analysis of biopsy material.

While most cases of breast cancer are sporadic, arising as a result of mutations accumulated in the specific tissue over a period of time, there is also an inherited form of the disease. Women with breast cancer who have a family history of the disease tend to have a mutation in one or other of the tumour suppressor genes *BRCA*1 and *BRCA*2. It is important to be able to distinguish this particular group of women who have a greater than average risk of developing further tumors and may opt for more radical surgery. Unfortunately, there is an immense range of mutations which can give rise to the oncogenic forms of *BRCA*1 and *BRCA*2. A commercial test is available which detects five specific large deletions in *BRCA*1, but these mutations only account for 15–20% of inherited breast cancer. Mutation scanning techniques such as dHPLC or SSCP analysis (Section 13.9) are often used as pre-screening techniques followed by sequencing of samples that show altered patterns. Deciding whether a particular sequence change is pathogenic is a major problem unless it has been previously reported, or can be clearly seen to have a serious effect on the protein, for instance if it introduces a premature stop codon (Question 13.6). Mutations in *BRCA*1 have also been found in up to 30% of cases of sporadic breast cancer, where the mutation has arisen spontaneously. There is increasing evidence that cancers associated with mutations in *BRCA*1, whether inherited or sporadic respond differently to commonly used chemotherapy treatments, so again it is important to be able to establish a molecular diagnosis in order to inform treatment strategies.

Breast cancer is a complex disease involving genetic factors, both inherited and acquired, as well as environmental factors. In fact, at the molecular level it is not one single disease at all but a collection of diseases with differing molecular characteristics. Being able to distinguish between these subtypes is having a major impact on management of certain subtypes of the disease although the majority of breast cancers are still difficult to differentiate. Expression profiling using micro-array technology (Section 11.6) is beginning to have an impact here. In one experiment patterns of expression of 25,000 genes in 98 primary breast cancers were analyzed.

From this experiment 70 genes were identified whose expression patterns, taken together, can be used to identify tumors likely to respond to chemotherapy after surgery and radiotherapy. This type of analysis should help to target treatments appropriately, and avoid patients being subjected to unpleasant treatments which are likely to be ineffective.

> **Q13.9** Explain how molecular techniques are used in the diagnosis and management of cancer, in each case give examples of the techniques used.

Questions and Answers

Q13.1. Name the two main classes of repetitive DNA found in the human genome and briefly describe each one.

A13.1. *About 46% of the human genome is made up of repeat sequences. These fall into two main classes, transposon derived repeats and tandemly repeated or satellite DNA. The various types of repeat sequences, which remain in the human genome as a result of transposon activity in the past, are described in detail in Section 2.3. Satellite DNA is composed of short DNA sequences which are repeated many times, it is classified into three types, satellite, mini- and microsatellite depending on the length and number of repeats.*

Q13.2. What considerations have been taken into account in choosing the STR loci for forensic analysis?

A13.2. *The STR loci need to be highly polymorphic showing wide allelic variability, and the length of the PCR products needs to be in the range 90–500 bp to ensure that they can be reliably amplified even from degraded samples. Tetranucleotide repeats are preferred as these yield alleles which differ in length by 4 bp and can be easily distinguished by capillary electrophoresis. In order for a set of STR loci to produce a useful profile for forensic purposes they need to be dispersed through the genome. They also need to be amenable to multiplex PCR so it is important to consider whether primers with compatible reaction conditions can be designed and whether the range of product sizes is such that products of different STR loci can be distinguished from each other.*

Q13.3. (a) The table below shows the genotypes of John and Mary and their two children Ann and Peter. Each STR locus is given a number representing the number of repeat units it consists of. John suspects that one or other of the children may not be his. By examining the data can you decide if his suspicions are correct?

(b) The most likely explanation for Ann's profile is that John is her father, but can you think of any other possibility?

	John	Mary	Ann	Peter
TPOX	14, 17	15, 16	14, 15	15, 18
D3S1358	7, 12	9, 11	11, 12	7, 9
FGA	15, 15	15, 15	15, 15	15, 15
D5S818	15, 15	13, 15	15, 15	13, 15
CSF1PO	11, 14	10, 12	11, 12	9, 10
D7S820	8, 8	7, 7	7, 8	7, 9
D8S1179	9, 9	9, 12	9, 9	10, 12
TH01	5, 6	4, 7	4, 5	6, 7
VWA	16, 19	15, 19	15, 19	18, 19
D13S317	8, 11	5, 6	6, 11	5, 10
D16S539	9, 11	9, 9	9, 11	9, 12
D18S51	13, 16	12, 16	16, 16	12, 17
D21S11	7, 9	4, 5	4, 7	5, 9
AMEL	X, Y	X, X	X, X	X, Y

A13.3. (*a*) *The DNA profile confirms that John's suspicions are correct; half the bands in Ann's profile are also present in John's which is what you would expect if he was her father. However, you can clearly see that at some loci Peter has alleles which he cannot have inherited from either Mary or John. For instance he is 15, 18 at the TPOX locus, Mary is 15, 16 so he has inherited the 15 allele from his mother, the 18 allele must have come from his father but John does not have this allele. In fact Peter has paternal alleles at 7 loci (TPOX, CSF1PO, D7S820, D8S1179, VWA, D13S317 and D18S51), which cannot have come from John, making it highly unlikely that John is his father.*

(*b*) *If John had an identical twin brother, he would have the same DNA profile and he could in fact be Ann's father.*

Q13.4. What factors in a hybridization experiment can be altered to ensure that the oligonucleotide will only form a stable duplex with a sequence that is exactly complementary?

A13.4. *Remember from Chapter 5 that DNA duplexes are less stable at higher temperatures, in low salt concentrations and higher concentrations of SDS. All of these factors can be used to create high stringency conditions where hybridization of exact matches is favored.*

Q13.5. What parameters need to be considered when designing the sequence of the molecular beacon to ensure that the probe target hybrid is more stable than the hairpin structure?

A13.5. *The overall length and the GC content are the two key factors here (Box 3.6). The longer the region of complementarity and the greater the proportion of GC residues, the more stable the hybrid will be.*

Q13.6. What types of mutation are likely to introduce stop codons into a gene?

A13.6. *Any single base pair substitution which changes an amino acid encoding codon into a stop codon, for example TTA or TTG (which encode leucine) need only a single substitution of a T for an A in the middle position to give the stop codons TAA and TAG, respectively. Also, any mutation which results in the loss or gain of a number of nucleotides that is not divisible by three, will almost certainly give rise to a stop codon. Deletion of three nucleotides (or multiples of three) results in the loss of a single amino acid; unless this particular amino acid has a critical role in the protein the protein may still be able to function normally. However, deletion of other numbers of nucleotides results in a shift in the reading frame. This means that from this point onwards a different sequence of codons is being read which will encode a nonsense protein. However, because stop codons are relatively common in non coding reading frames (see Figure 8.2) the most likely effect is in fact that a stop codon will be encountered shortly after the point of the mutation and a shortened or truncated protein will be produced.*

Q13.7. Why is multiplex PCR an important tool in genetic diagnosis?

A13.7. *The ability to amplify DNA by PCR is the key to genetic testing, particularly where the sample has been taken from a fetus or early embryo and may consist of very few cells. Multiplex PCR makes it possible to test for a range of different mutations in the same PCR reaction, permitting testing either for several different diseases or for a range of mutations which cause the same disease.*

Q13.8. What advantages do DNA-based diagnostic tests for bacterial and viral infections offer over conventional tests?

A13.8. *In both cases DNA-based tests overcome some of the problems associated with the requirement of conventional tests to culture the organisms. In the case of bacteria problems may arise because some bacteria are hard to culture, because bacteria are present in mixtures or simply because of the time and expertise required to grow bacteria in the laboratory. Many viruses cannot easily be cultured in the laboratory and diagnostic tests often rely on detecting antibodies in patient serum and these may not be detectable in the early stages of infection. In DNA-based tests PCR amplification of DNA replaces laboratory culture. Techniques such as the 5′ exonuclease (TaqMan®) and molecular beacon assays allow for quantification in addition to identification. This can be very useful to detect the presence of small numbers of one pathogen in the presence of a more abundant one and also* *monitoring the progress of a disease and how it is responding to treatment. Molecular tests may also be able to provide additional information about resistance to antimicrobial agents, which is useful in determining an appropriate course of treatment and in monitoring its effectiveness.*

Q13.9. Explain how molecular techniques are used in the diagnosis and management of cancer, in each case give examples of the techniques used.

A13.9. *Initial diagnosis would probably require a test that can identify genetic changes in a biopsy sample. FISH would be an ideal technique in this situation because it can detect translocations and other changes such as gene amplification. Any of the techniques which are used in the diagnosis of single gene disorders using allele specific PCR primers or ASOs could also be used to identify specific mutations known to be associated with the particular type of cancer. The same techniques are invaluable in the management of cancer as it is important to be able to differentiate subtypes to enable treatments to be targeted. Expression profiling using micro-array technology shows promise of being able to identify subtypes where many genes are involved. Finally, molecular techniques are invaluable in monitoring the success of treatment and in looking for the development of resistance to chemotherapeutic agents to help in the management of the disease.*

Further Reading

The PowerPlex™ 16 System (2000) Sprecher C, Krenke B, Amiott B, Rabbach D and Grooms K, Profiles in DNA, Pages 3–6.
This article in Promega's in-house magazine describes a commercial system for detecting commonly used forensic markers. The article can be accessed using the following url: http://www.promega.com/profiles/401/ProfilesinDNA_401_03.pdf

Moving towards individualized medicine with pharmacogenomics (2004) Evans WE and Relling MV, Nature, Volume 429 Pages 464–468.
This short "insight review" article from Nature summarizes pharmacogenomics in an easy to read format

Molecular beacons for multiplex detection of four bacterial bioterrorism agents (2004) Varma-Basil M, El-Hajj H, Marras SAE, Hazbón MH, Mann JM, Connell ND, *et al.*, Clinical Chemistry, Volume 50 Pages 1060–1062.
This is a mainly technical paper but it also discusses important general principles

Molecular diagnostics in virology (2004) Vernet G, Journal of Clinical Virology, Volume 31 Pages 239–247.
A comprehensive review of the area

Gene expression patterns of breast carcinomas distinguish tumor subclasses with clinical implications (2001) Sorlie T, Perou CM, Tibshirani R, Turid A, Geisler S, Johnsen H, *et al.*, Proceedings of the National Academy of Sciences of the United States of America, Volume 98 Pages 10869–10874.
This is a primary research paper which shows how expression profiling is used to try to explain differences in recovery from breast cancer

Glossary

Activator A protein which increases the level of expression (i.e. the amount of mRNA produced) of a gene or operon by binding to the upstream region.

Adaptor A short double-stranded synthetic oligonucleotide with one blunt end and a nucleotide extension that can base pair with a sticky end. See linker.

Allele Any of two or more alternative forms of a gene that occupy the same locus on a chromosome.

Alpha helix A protein secondary structure in which the peptide backbone adopts a helical conformation.

Amino terminus/N terminus One end of a linear polypeptide that has an amino acid with a free NH_2 or amino group. This is the end where protein synthesis begins. See carboxyl terminus.

Aneuploid More or less than the correct number of chromosomes.

Anneal The process by which two complementary strands of a DNA molecule hybridize to form a duplex with hydrogen bonds forming between complementary bases.

Antibody A protein produced by the immune system of animals in response to a specific antigen, to which it then binds with high affinity.

Antigen A substance that stimulates the production of antibodies by the immune system.

Antisense A DNA or RNA molecule which is complementary to the coding strand.

Auxotroph A mutant which cannot synthesize a particular compound that it requires for growth, for example an amino acid or a sugar, although the parent from which the mutant was derived can synthesize this compound.

Bacterial artificial chromosome (BAC) Artificially constructed bacterial plasmids which can accommodate very large DNA inserts of at least 300 kb.

Bacteriophage A virus which infects bacteria. It is often referred to as a phage.

Beta sheet A protein secondary structure in which the polypeptide chain is fully extended and two or more adjacent strands are hydrogen bonded together.

Biopharming Genetic engineering of a livestock animal or crop plant to produce a commercial product, often a pharmaceutical.

Carboxyl terminus/C terminus One end of a linear polypeptide which has an amino acid with a free COOH or carboxyl group. This is where protein synthesis ends. See amino terminus.

cDNA DNA synthesized using, and hence complementary to, an mRNA template by the action of reverse transcriptase.

Centromere The region of the eukaryotic chromosome where spindle fibers attach during mitosis, leading to alignment of chromosomes on the cell equator followed by their movement to opposite poles of a dividing cell.

Chromatin The complex of DNA, histones and other proteins which make up eukaryotic chromosomes.

Chromosome jumping Chromosome walking (see below) at speed: a method that moves from a known to an unknown piece of DNA but without the need to clone every piece of DNA between the two.

Chromosome walking A method for moving from a piece of DNA of known sequence (typically a polymorphism) to a nearby piece of DNA of unknown sequence, by cloning a series of overlapping pieces of DNA between the two.

Codon   Three nucleotides in a nucleic acid sequence which encode an amino acid or indicate the end of a coding region.

Codon usage/preference   The tendency displayed by all living things to preferentially use one of a range of alternative codons to encode a particular amino acid.

Competent   Cells are said to be competent if they are able to take up foreign DNA.

Complementary   Two nucleotides or polynucleotide chains (DNA or RNA) that can base-pair.

Complementation   Classically, this refers to the ability of recessive mutations in two different genes to produce a wild-type phenotype in a diploid organism. In the context of gene cloning it usually refers to a technique used to identify a gene by its ability to enable the growth of an organism in which the same gene is defective.

Concatemer   A polynucleotide molecule made up of repeated units of the same sequence joined end to end.

Consensus sequence   A derived sequence showing the most common residue (nucleotide or amino acid) at each position in a set of aligned sequences.

Conserved domain   A structural region seen in a group of proteins, usually with a distinct evolutionary origin and function.

Contig   One of a series of overlapping DNA fragments produced during the course of a sequencing project, made up by assembling many smaller DNA sequences.

cos site   A sequence in a bacteriophage DNA molecule that is cut during phage maturation to produce cohesive, single-stranded extensions located at the ends of the linearized genome.

Cosmid   An artificially constructed plasmid cloning vector which contains a bacteriophage *cos* site enabling the vector to be packaged into bacteriophage λ for infection into *E. coli*, enabling relatively large DNA fragments (up to about 50 kb) to be cloned at high efficiency.

Degenerate   The genetic code is said to be degenerate because there is more than one codon for most amino acids. The term is also used to refer to a mixture of oligonucleotides representing all of the possible sequences that could encode a particular polypeptide.

Denature   To cause a loss of structure. For polynucleotides, this means to cause the two strands of a double-stranded DNA or RNA molecule to separate by disrupting the hydrogen bonds between complementary bases. This is usually done by heating or using alkali. In the case of proteins it means a loss of secondary and tertiary structure.

Diploid   A cell which has two sets of homologous chromosomes. See haploid.

DNA library   A collection of DNA clones containing the whole genome of an organism, part of the genome or all of the genes expressed in a particular tissue at a particular time.

DNA ligase   An enzyme used in gene cloning to covalently join two pieces of DNA.

DNA polymerase   An enzyme which synthesizes a complementary DNA strand from a DNA or RNA template.

dNTP   Abbreviation for deoxynucleoside triphosphate, the building block of DNA.

Duplex   In gene cloning this refers to a double-stranded DNA or RNA molecule. See also heteroduplex.

Dye terminator   A dideoxynucleotide with a fluorophore attached. When used in a DNA sequencing reaction the incorporation of a dye terminator will simultaneously terminate the reaction and label the product.

Electropherogram   The output from an automated sequencing machine. It shows a series of colored peaks where each peak represents the presence of a particular nucleotide in the sequence.

Electrophoresis   Separation of charged molecules (typically nucleic acid or protein) by an electric field. The molecules move through a porous gel which separates them according to relative size or charge.

Electroporation The process of inducing cells to take up DNA by subjecting them to a brief pulse of electric current.

Embryonic stem cells (ESCs) Undifferentiated cells isolated from an embryo which have the potential to differentiate into any tissue or cell type in the adult organism.

Enhancer A region of eukaryotic DNA that contains binding sites for transcription factors. These regions may be several kilobases upstream or downstream of the core promoter.

Eukaryote An organism whose cells contain a membrane bound nucleus and other organelles. Includes higher animals, plants, fungi, and protists. See prokaryote.

Exon The region of a eukaryotic gene that is found in the mature mRNA, this normally includes the protein coding region. There may be many exons in a given gene, separated by introns. See intron.

Expressed sequence tag (EST) Partial sequence of a cDNA molecule. The sequence is long enough to be unique to a particular gene, and as it is derived from cDNA it represents part of a gene that was expressed under the experimental conditions.

Fluorescent *in situ* hybridization (FISH) Use of fluorescently labeled nucleotide probes to label whole chromosomes or chromosome regions in preparations of interphase or metaphase chromosomes.

Fluorophore The part of a fluorescent dye molecule which can be excited by light to emit light at a different frequency.

Fusion protein A protein expressed from a recombinant DNA molecule containing some amino acids from one protein and some from another.

Genetic map A diagram of all or part of an organism's genome showing the relative locations of specific genetic markers.

Genome The total complement of genetic material for a given organism.

Genomics The study of the sequence of the whole genome of an organism.

Genotype The genetic make-up of an organism. This term is often used to refer to the combination of alleles at a particular locus or loci.

Glycosylation An important post-translational modification of proteins in eukaryotic cells where carbohydrate groups are added to specific amino acids in protein molecules.

Gram negative A bacterium which does not stain purple with Gram's iodine. These organisms usually have a relatively thin layer of peptidoglycan and an outer membrane.

Gram positive A bacterium which stains purple with Gram's iodine. These organisms have a thick robust outer layer of peptidoglycan but no outer membrane.

Haploid A cell with a single set of chromosomes. See diploid.

Hapten A small molecule which can be detected by an antibody when it is attached to a large molecule.

Heteroduplex A duplex formed by annealing two nucleic acid molecules that are not perfectly complementary in sequence.

Heterologous The word means "different from". In genetics, it is used to describe chromosomes in a cell which do not contain the same genes (i.e. are not homologous). It is also used to refer to a gene or protein which is cloned into a species of organism different from the one from which it originated.

Heterozygous Having two different alleles at the same genetic location. See homozygous.

Homoduplex A duplex formed by annealing two exactly complementary DNA or RNA molecules together. See heteroduplex.

Homologous A set of genes, proteins, DNA or amino acid sequences which have a common evolutionary origin.

Homologous recombination Exchange of genetic material between two similar DNA segments by breakage and reunion, with each other, of the two DNA strands.

Homology Similarity resulting from a common evolutionary origin. Homologous DNA or protein sequences have a high degree of similarity, as they have evolved from a common ancestral sequence.

Homozygous Having the same allele at a genetic location. See heterozygous.

Human Genome Project An international collaborative research project to map and sequence the human genome.

Hybridization The process by which a single-stranded polynucleotide molecule (DNA or RNA) base pairs with a molecule with its complementary sequence to form a double-stranded molecule.

Hydrophobic Water hating. Hydrophobic (non-polar) molecules are relatively insoluble in water.

In vitro From a Latin phrase meaning "in glass" this is scientific shorthand for experiments performed in an artificial environment such as a test tube. See *in vivo*.

In vivo From a Latin phrase meaning "in life", this is scientific shorthand for experiments performed in living cells or organisms. See *in vitro*.

Inducer A small molecule that acts either on an activator to make it bind DNA and activate gene expression, or on a repressor to make it leave DNA so that repression is relieved.

Intron Region in a eukaryotic gene which does not code for amino acids in the final protein, and is spliced out of the mRNA after transcription and before transport of the mRNA to the cytosol. See exon.

Isoschizomer Restriction enzymes which recognize the same DNA sequence. They may cut at the same or at different positions within it.

Karyotype The entire chromosomal complement of a cell as seen at metaphase. This is often shown as a photomicrograph of the chromosomes arranged as homologous pairs of decreasing size.

Kilobase One thousand bases. The abbreviation is kb.

Knock-out Inactivation of a particular gene by deletion or insertion of foreign DNA. Also, an organism in which one or both copies of a specific gene have been inactivated.

Lambda A particular bacteriophage that infects *E. coli*. It has been very widely used both to answer many basic questions about cell biology, and as the basis of numerous cloning vectors.

Linkage The phenomenon where two different genes are inherited together more than would be expected by random chance, due to their being physically close together on a piece of DNA.

Linker A short double-stranded synthetic oligonucleotide containing one or more internal restriction enzyme recognition sites.

Locus A specific location on a chromosome.

Log phase Usually used to refer to single-celled organisms growing in culture, this term describes the phase of growth where the organisms are growing at the maximum rate of division possible in a particular set of growth conditions. Also called exponential phase. At this point the growth rate is proportional to the number of organisms present.

Map-based cloning A cloning method that relies on detailed genetic mapping of mutant alleles of a gene as the first stage towards cloning the gene.

Marker rescue Restoration of function by the replacement of a defective gene with a functional copy, via homologous recombination.

Megabase One million bases. The abbreviation is Mb.

Melting temperature (T_m) The temperature at which half the bonds in a double-stranded DNA molecule are broken.

Mendelian The pattern of inheritance of genes first described by Gregor Mendel in which genes segregate independently, one copy being inherited from each parent. Note this only occurs when genes are far apart on a chromosome, or on separate chromosomes.

Micro-array A large number of oligonucleotides (typically corresponding to individual known open reading frames in a given genome) spotted at high density and at specific locations on a solid support, usually a glass slide or silicone wafer.

Micro-injection The process by which DNA can be introduced into the nucleus of a cell through a fine glass micro-capillary tube; used in the construction of some transgenic organisms.

Motif (sequence motif) A part of a DNA or protein sequence which is conserved in a group of sequences. The occurrence of a motif usually correlates with a particular function.

Multiple cloning site In a cloning vector, a short synthetic region containing the recognition sequences for several restriction enzymes. Also called a polylinker.

Multiplex (PCR) A single PCR reaction in which several different regions of the same template are amplified using different pairs of primers.

Mutagen Anything that causes a heritable alteration in DNA sequence.

Mutagenesis Treatment of either an organism or isolated DNA with a mutagen.

N terminus See amino terminus.

Non-homologous recombination Exchange of genetic material between two DNA segments by recombination without there being any significant sequence similarity between the regions.

Oligonucleotide A short, synthetic, single-stranded DNA or RNA molecule; often shortened to oligo.

Oncogene A gene which when mutated or expressed inappropriately predisposes cells to become transformed into tumor cells.

Oncomice Transgenic mice genetically engineered to express an oncogene, which hence have a very high rate of tumor formation.

Open reading frame A region of DNA sequence that does not contain any in frame stop codons; it can potentially encode a protein but does not necessarily do so.

Operon Two or more adjacent genes transcribed from the same promoter as a single mRNA molecule.

Orthologues A set of homologous genes, proteins, DNA or amino acid sequences which have evolved from a common ancestor by speciation; they commonly have the same function. For example, β globin from mouse and man are orthologous. See paralogues.

Paralogues A set of homologous genes, proteins, DNA or amino acid sequences which have evolved from a common ancestor by gene duplication. They commonly have evolved to perform different but related functions. β globin and α globin are paralogous. See orthologues.

Partial digest Restriction enzyme digestion of DNA under sub-optimal conditions resulting in DNA fragments with some restriction enzyme recognition sites which have not been cut.

PCR Polymerase chain reaction: a method for the amplification of a specific DNA sequence using oligonucleotides that flank that sequence and a thermostable DNA polymerase.

Periplasm The compartment in a Gram negative bacterium between the outer and the inner (or cytoplasmic) membrane.

Phage A virus which infects bacteria. See bacteriophage.

Pharmacogenomics The study of how an individual's genetic make-up influences the way that they respond to pharmaceuticals.

Phenotype The observable characteristics of an organism. The phenotype results from genetic or environmental factors, or the interactions between the two.

Phylogenetic tree A branching diagram showing the evolutionary relationships between organisms or molecules.

Plaque The clear area seen in a bacterial lawn where bacteria have been infected and lysed by bacteriophage.

Plasmid A small circular non-essential extra chromosomal DNA molecule that replicates autonomously. Plasmids are very widely used as cloning vectors.

Polylinker In a cloning vector, a short synthetic region containing the recognition sequences for several restriction enzymes. Also called a multiple cloning site.

Polymorphism The occurrence of two or more variants of a gene or DNA sequence in a population at a frequency that cannot be accounted for by recurrent mutation. A variant that occurs in less than 1% of the population is considered a mutation rather than a polymorphism.

Polyploid Having more than two complete sets of chromosomes.

Positional cloning Another name for map-based cloning.

Pre-implantation Refers to an embryo fertilized *in vitro* before it is transferred into the uterus.

Prenatal After conception and before birth.

Primer A short single-stranded DNA molecule, often a synthetic oligonucleotide, which provides a 3' hydroxyl for the initiation of DNA synthesis by DNA polymerase.

Probe A short single-stranded DNA molecule, often a synthetic oligonucleotide, which is labeled and used to detect a complementary sequence to which it hybridizes.

Profile In bioinformatics this is a description of the members of a protein family in the form of a matrix in which the frequency of each amino acid at each position is recorded.

Prokaryote Organism without a membrane-bound nucleus or other membrane bound organelles. Bacteria and archaea are prokaryotes. See eukaryote.

Promoter DNA sequence recognized by RNA polymerase for subsequent initiation of mRNA synthesis. Now often used to include the regulatory sequences that affect RNA polymerase binding.

Promoter probe A plasmid containing a reporter gene without a promoter. Cloning of a DNA fragment encoding a promoter upstream of the reporter will result in expression of the reporter allowing the identification of promoter sequences and study of their properties.

Proteome All of the proteins expressed by a cell, tissue or organism at a particular time under specific conditions.

Proteomics The large-scale analysis of all the proteins in a cell or tissue at a specific time and under specific conditions.

Reading frame A series of triplet codons starting from a particular nucleotide. There are three possible forward and three reverse reading frames.

Recombinant Used in gene cloning to describe a new combination of DNA fragments.

Recombination The process of breaking and rejoining of DNA strands usually resulting in the exchange of genetic material.

Regular expression Used in bioinformatics to describe a string of characters that describes a series of related DNA or amino acid sequences.

Reporter gene A gene encoding a protein with an easily detectable property or effect on the organism's phenotype. Used to detect and measure gene expression and in promoter probe vectors.

Reporter protein The protein product of a reporter gene.

Repressor A protein which decreases the level of expression of a gene or operon by binding to the upstream region.

Restriction endonuclease A naturally occurring bacterial enzyme which recognizes and cuts DNA molecules at specific sites. Type II restriction endonucleases are used extensively in gene cloning. Also called a restriction enzyme.

Restriction fragment A piece of DNA created by cutting DNA with a restriction endonuclease.

Restriction map A diagram showing the relative positions of sites for restriction enzymes on a DNA molecule.

RFLP Restriction fragment length polymorphism. A difference in the length of a specific restriction fragment found in different members of the same species. Usually caused by a small mutation which creates or destroys a restriction endonuclease recognition site, or by a variation in the number of repeats of a tandemly repeated sequence.

RNA interference/RNAi Reduction of expression of a particular gene by expression of double-stranded RNAs complementary to part of the gene.

Segregation The separation of chromatids (identical replicated copies of chromosomes) in mitosis or meiosis.

Shine–Dalgarno sequence Another name for the ribosome binding site on bacterial mRNA, named after the people who first described it.

Signal sequence/signal peptide A sequence of 15–30 amino acids at the N terminus of a secreted protein which is required for transport through a membrane and is cleaved off after secretion.

Start codon The codon (usually AUG) in an mRNA molecule that codes for the first amino acid of the protein.

Stop codon A codon in mRNA that does not encode an amino acid but signals the end of protein translation. Also called a termination codon. The codons UGA, UAG and UAA are the three stop codons in nearly all organisms.

Suicide plasmid A plasmid which can only replicate in some cell types. If it is introduced into an incompatible host it must integrate into the chromosome or be lost as the cell divides.

Supercoil Double-stranded circular DNA in which the molecule is twisted on itself as a result of over- or under-winding.

Telomere A region of repetitive sequences at the ends of the eukaryotic chromosome, important in replication and stability.

Template The name given to the DNA molecule which is copied by DNA or RNA polymerase.

Terminator A DNA sequence that causes the dissociation of RNA polymerase from DNA and hence terminates transcription of DNA into mRNA.

Trait An inherited characteristic.

Transcription The process by which DNA is copied into complementary mRNA prior to protein synthesis.

Transcriptomics The study of actively transcribed genes in a particular cell or tissue at a particular time by studying the mRNA content, usually detecting the expression of specific genes using a micro-array.

Transformation Alteration in the genetic make-up of a cell by introduction of exogenous genetic material.

Transgene A gene from one organism introduced into another.

Transgenic An organism containing genes from another organism.

Translation The process by which mRNA is decoded by ribosomes and proteins are synthesized.

Transposon A mobile genetic element capable of excision from one location and integration at another location.

Vector   A self-replicating DNA molecule into which DNA can be cloned. The vector can be introduced into a host organism and the cloned DNA will be replicated.

Wild type   The normal or most common genotype or phenotype of a cell or organism. See also polymorphism.

Xenotransplantation   A transplant of an organ or tissue from one species to another, usually from an animal into a human.

Yeast artificial chromosome (YAC)   Artificially constructed cloning vector containing all of the sequences required for replication of a chromosome in yeast. YACs can stably maintain very large DNA inserts.

Yeast two-hybrid   A cloning method that reveals potential protein–protein interactions by the ability of such interactions to activate expression.

Zooblot   A Southern blot in which a probe from one organism is hybridized with genomic DNA from a range of other species.

Index

Entries which are simply page numbrs refer to the main text. Entries with g after the page number are in the glossary.

A

α globin, 103, 118
α helix, 445g
α-1-antitrypsin, 373
β galactosidase, 55, 56, 96, 336, 337
β globin, 420, 424, 425, 429
β sheet, 445g
β thalassemia, 420, 422, 425
16S rRNA, 242, 437
454 sequencing, 197–202
Ac elements, 157
Activator, 445g
Active DNA transposons, 15
Adaptor, 106, 445g
Adherence genes, 149
Aedes aegypti
   genetic manipulation of, 388–390
Affinity purification, 350–351
Agrobacterium tumefaciens
   use in construction of transgenic plants,
     391–393
Alkaline phosphatase
   calf intestinal alkaline phosphatase (CIAP),
     51–52
Allele, 445g
Allele specific oligonucleotide, 427
Allele specific PCR amplification, 423
Alpha-globin *see* α-globin
Alpha helix *see* α helix
Alpha-1-antitrypsin *see* α-1-antitrypsin
Alternative splicing, 12, 216
Alzheimer's Disease, 372
Amelogenin, 414, 415
Amino terminus, 445g
Amniocentesis, 426
ANDi, 384–388
Aneuploidy, 30, 445g
Anneal, 445g
Antibiotic resistance, 26
   in selection for transformants, 37, 43–44
   insertional inactivation with, 53–54
Antibody, 120, 342, 445g
Antigen, 445g
Antisense, 445g
Arabidopsis thaliana, 20
   knockouts in, 405–406
   RNAi in, 288
Archaea, 242
Artemis, 213

ASO *see* allele specific oligonucleotide
Autosome, 9
Auxotroph, 119, 269,283–284, 293, 299, 445g
Avirulence (avr) gene, 157

B

BAC *see* bacterial artificial chromosome
BAC clones, 192
BAC library, 192
Bacterial artificial chromosome, 100–101, 102, 445g
Bacteriophage, 445g
Bacteriophage λ
   P$_L$ promoter, 256
   cloning with, 97
   expression vector, 96
   genome, 26
   in vitro packaging, 94, 95–96
   in whole genome sequencing, 189
   insertion vector, 95, 105
   lysogenic pathway, 26
   lytic pathway, 26
   plating, 94
   replacement vector, 95
   size constraints on packaging, 97, 100
   transfection, 95–96
   vector for library construction, 26, 27, 93
Bacteriophage M13, 179–181
Baculovirus, 273
Beta-galactosidase, *see* β galactosidase
Beta-globin *see* β globin
Beta sheet *see* β sheet
Beta-thalassemia *see* β thalassemia
Biopharming, 373–374, 419, 445g
Bioterrorism agents
   detecting, 438
BLAST, 226, 228
Blasticidin-S, 273
Blocks, 225
Blosum 62, 221, 222
BRCA1, 440
BRCA2, 441
Breast cancer, 422, 440

C

Caenorhabditis elegans, 20, 156, 288–289
   genome wide survey of gene function,
     288–289, 367
Calcium phosphate mediated transfection, 272,
   339

Capillary electrophoresis, 185–186
Carboxyl terminus, 445g
Carrier status, 422
CD4, 309
cDNA, 11, 103, 214,215, 323–326 445g
    cloning, 105–107
    library *see* cDNA library
    rapid amplification of cDNA ends (RACE), 323–325
    sequencing, 182–183
    vectors for cloning, 105
cDNA library, 101–109, 167, 299
    constructing, 103
    differential screening, 132
    screening, 118–123, 353
    starting material, 102–103
    vectors for cloning, 105
    yeast one-hybrid screen, 351–353
Celera, 196–197
Centromere, 28, 445g
cf-9 gene, 157–159, 172
CFTR gene, 226
Chimpanzee, 21, 23
Chinese hamser ovary cells *see* CHO cells
CHO cells, 270
Chorionic villus sampling (CVS), 426
Chromatin, 340–341, 445g
Chromosome, 7
    aberration, 30
    acrocentric, 28, 29
    autosome, 9
    centromere, 28
    chromosomal location, 30
    chromosome jumping, 161, 163, 445g
    chromosome walking, 163–166, 445g
    eukaryotic, physical characteristics, 28
    G band, 28, 29
    interband, 28
    metacentric, 28, 29
    number, 19
    painting, 434–435
    sex, 9
    size, 28
    sub-metacentric, 28, 29
Cladosporium fulvum, 157, 172
Cloning
    bacterial adherence genes, 149–152
    bacterial genes with difficult phenotypes, 152–156
    cDNA, 105–107
    CF gene, 168–169
    by complementation, 119–120
    gene tagging, 142–159
    genomic DNA, 36–45
    genomic DNA libraries, 87–89

    in cosmids, 99–100
    map based, 159–167
    PCR products, 74–76
    plant resistance genes, 156–159
    by signature tagged mutagenesis, 152–156
ClustalW, 233, 234
CODIS *see* Combined DNA Index System
Codon, 446g
Codon preference *see* codon usage
Codon usage, 211, 212, 260, 446g
Co-immunoprecipitation, 301–302
Colinearity, 21, 24
Combined DNA Index System, 416
Comparative biology, 22
Comparative genomics, 22
Competent cells, 43–44, 446g
Complementary, 446g
Complementary DNA *see* cDNA
Complementation, 446g
Concatemer, 97, 446g
Consensus sequence, 233, 446g
    intron-exon boundary, 215
Conservative substitution, 220
Conserved domain, 230, 232, 446g
Contig *see* contiguous sequence
Contiguous sequence, 189–190, 446
Correlation score, 214
Cos site, 93, 94,99–100, 446g
Cosmid, 98–100, 446g
Cosmid library, 150–151
CpG island, 31–32, 215
Cre-*loxP* system, 404, 405
C-value paradox, 8
Cycle sequencing, 185–188
Cystic fibrosis, 142, 226, 422, 424–243
    cloning of CF gene, 168–169, 172
    diagnostic test, 431
    mutations giving rise to, 425
cytocrome P450, 421

D
Danio rerio, 20
Database
    combined DNA Index System, 415, 416
    DNA sequence, 225
    National DNA Database, 415, 416
    primary sequence, 225
    protein sequence, 225
    secondary sequence, 225
DDBJ, 225
DEAE-dextran mediated transfection, 339
Deletion analysis
    example of use, 347–350
Denaturing high performance liquid chromatography, 433, 440

Denaturing polyacrylamide gel electrophoresis, 178,185
dHPLC *see* denaturing high performance liquid chromatography
Dicer, 287
Diploid, 446g
DNA
   electroporation, 272, 339
   introducing into eukaryotic cells, 272, 339
   labeling *see* labeling
   liposome-mediated transfer (lipofection), 272, 339
   sequencing, 173–203
   transfection, 272, 339
   transformation, 43–44
DNA array, 108, 133, 354–357, 420, 440, 449g
DNA fingerprint, 414
DNA library, 446g
   cDNA library, 101, 132
   expression library, 105
   genomic library, 87–89
   sequencing, 189
   subtractive, 133
DNA ligase, 446g
   T4 DNA ligase, 41, 51, 427
DNA macroarray *see* DNA array
DNA microarray *see* DNA array
DNA polymerase, 446g
   DNA polymerase I, 65
   from bacteriophage F29, 436
   Klenow, 65, 181
   reverse transcriptase, 103
   Taq polymerase, 65
DNA preparation
   chromosomal, 88
   miniprep, 46
   plasmid, 45–47
DNA profile, 413
DNA sequencing, 173–203, 197–203
   chemical method (Maxam & Gilbert), 174
   cycle sequencing, 185–188
   dideoxynucleoside triphosphate concentration, 182
   DNA polymerase, 181–182
   enzymatic method (Sanger), 174–182
   genomes, 188–202
   hierarchical shotgun sequencing, 192–196
   high-throughput, 184–188,197–202
   physical gaps, 191–192
   primer walking, 182–183
   pyrosequencing, 197–202
   sequencing gaps, 190–191
   shotgun sequencing, 183–184,189–197
   strategies, 182–184,188–189
   template, 179–182

   whole-genome shotgun sequencing, 189–192, 196–197, 201–202
DNAse I
   footprinting, 344–346
   hypersensitivity assay, 340, 341
   nick translation labeling, 130
dNTP, 446g
Dot blots, 326–328
Down's syndrome, 30, 426, 434, 435
*Dpn*I
   use in site directed mutagenesis, 307
Draft genome sequence, 196
Drosophila melanogaster, 20, 156, 288, 367
Ds elements, 157–159
Duplex, 446g
Dye terminator, 446g

E
E value, 228
Electromobility shift assay (EMSA), 340–343
   antibody, 342–343
   comparison with footprinting, 345
   competition assays, 349–350
   deletion analysis, 347–350
   site-directed mutagenesis, 347–350
   supershift, 342–343
Electropherogram, 186, 446g
Electrophoresis, 446g
   2-dimensional, 121
   agarose, 59–64
   capillary, 185–186, 413, 425, 427, 433
   denaturing polyacrylamide gel, 178,185
   isoelectric focusing, 121
   measuring DNA sizes, 59–63
   polyacrylamide, 59
   preparative, 63
   running an agarose gel, 59–61
   SDS PAGE, 121, 259
   size fractionation, 92
   two-dimensional gel electrophoresis, 358–359
Electroporation, 44, 272, 339, 447g
   see also transformation
EMBL, 225
EMSA, *see* electromobility shift assay (EMSA)
Embryonic stem cells (ESC), 398–400, 447g
End labelling DNA, 320–321, 345
Enhancer, 316–318, 338, 340, 447g
EPSPS, 394
Erythropoietin, 251
Escherichia coli, 20, 239
   as host for cloned DNA, 36
   as host for protein production, 252
   genome, 24–25
   model organism, 20
   transformation, 44

EST *see* expressed sequence tag
Euchromatin, 30
Eukaryote, 242, 447g
Exon, 11, 210, 447g
Expressed sequence tag (EST), 108, 447g
Expression library
    screening, 353
Expression profile, 440

F
Factor VIII, 250, 251
Fasta, 226–228
Fingerprint
    protein family signature, 237
    clone contigs, 193
Finished genome sequence, 196
FISH *see* fluorescent *in situ* hybridisation
Flavr-Savr tomato, 376
Fluorescent *in situ* hybridisation, 426, 434–435,
    439, 447g
Fluorophore, 447g
Footprinting, DNA–protein interactions, 343–346
    comparison with electromobility shift assay,
    345
Forensic science, 414–418
Fugu rubripes, 23
Fusion protein, 264, 447g

G
G-418, 273
gal4, 298
Gancyclovir, 399
GC content, 27, 211
Gel electrophoresis *see* electrophoresis
Gel retardation assay *see* electromobility shift
    assay (EMSA)
Gel shift assay *see* electromobility shift assay
    (EMSA)
GenBank, 225
Gene
    characteristics, 208–210
    eukaryotic, features, 210
    eukaryotic, identifying, 214–217
    number, 16, 17
    prokaryotic, features, 210
    prokaryotic, identifying, 208–214
Gene expression, 326–350
    determination of level, 326–338
    global studies, 353–361
    identifcation of regulatory proteins, 350–353
    identification of regulatory regions, 338–351
Gene expression, identifying regulatory regions
    in DNA
    using the DNase I hypersensitivity assay,
    340–341

Gene knockouts
    construction of, 284–287, 398–406
Gene tagging, 142, 144–149
Gene therapy, 419
Genetic code, 209
    degeneracy, 211
Genetic map, 159, 447g
Genetically modified organism *see* GMO,
Genome, 7, 19, 447g
    bacterial, 24
    human, 9–19
    human chimpanzee comparison, 23
    non-human eukaryotes, 19–24
    sequencing, 188–202
    size, 8, 23
    viral, 26–27
Genomic library, 87–89
    constructing, 87–89
    number of clones, 89
    storage, 92
    use in transposon tagging, 145
Genomics, 447g
Genotype, 447g
GFP, 336
    translational fusion with, 282
    used to detect transgenesis, 386–387
    used to study gene expression, 377
    used to study protien localisation, 291
GGTA1, 403
Giemsa stain, 28
Glivec, 250
Glucagon, 251
Glufosinate resistance, 369–370
Glycosylation, 447g
Glyphosate resistance, 368–370, 394–396
GMO, 365
Golden Rice, 370
gp120, 309
Gram negative, 447g
Gram positive, 447g
Green fluorescent protein *see* GFP
Gribscov plot, 213

H
Haploid, 447g
Hapten, 447g
Hepatitis, 438
Hepatitis B vaccine, 250, 251, 391, 392, 418
HER-2, 440
Herbicide resistance, 368–370,
Heterochromatin, 30
    constitutive, 31
    facultative, 31
Heteroduplex, 447g
Heterologous, 447g

Heterozygous, 447g
Hierarchical shotgun sequencing, 192–196
High-throughput DNA sequencing, 184–188,197–202
Hirudin, 251
His-tag, 274–275
HIV *see* human imunodefiency virus
HIV-1
    regulation of gene expression, 333–334
    tat, 333–334
    see also human imunodefiency virus
Homoduplex, 447g
Homologous, 447g
Homologous recombination, 448g
Homology, 220, 448g
Homozygous, 448g
Horizontal transfer
    pathogenicity islands, 27
    plasmids, 26
Human genome, 23
Human Genome Project, 86, 418, 448g
Human growth hormone, 251
Human immunodeficiency virus, 309, 438
Huntington's disease, 30, 424, 425
Hybridization, 448g
    colony, 127, 129, 131
    dot blot, 125, 427, 429
    in diagnostic tests, 423
    northern, 326–328, 354
    plaque, 127
    Southern, 125, 427
    zooblot, 125, 167, 168
Hydrophobic, 448g
Hydroxyl radical footprinting, 345

I
Ideogram, 29
    band numbering, 30
In vitro, 448g
In vitro fertilisation, 142, 426
In vivo, 448g
Inclusion bodies, 261–262
Independent assortment, 159
Inducer, 448g
Insertional inactivation
    of antibiotic resistance genes, 53–54
    of the *lacZ* gene, 54–56
Insulin, 14, 126, 250, 251, 418
Interferon, 251
Interspersed transposon-derived repeats, 13
Intron, 11, 102–103, 210, 216, 448g
Intron-exon boundary, 215
Inverse PCR, 148
Inverted repeats, 143
Isoschizomer, 51, 448g

IVF *see in vitro* fertilization

K
Karyotype, 28, 29, 448g
Kilobase, 448g
Klenow fragment, 65, 181
Knock-out, 448g

L
Labeling
    DNA probes, 127, 130–131
    end labelling, 320–321
    flourescent labels, 129
    labels detected immunologically, 128
    nick translation, 130
    radioactive labels, 128
    random prime labelling, 130
lacZ, 55, 58, 96, 295–297
    blue-white selection, 55, 58, 96
    gene fusion studies, 295
    insertional inactivation, 55–56, 58, 96
    promoter probe, 337
    translational fusion with, 295–297
Lambda, 448g
    see also bacteriophage λ
Library *see* DNA library
LINE *see* long interspersed elements
Linkage, 159, 163, 448g
Linker, 106, 448g
Liposome-mediated transfer (lipofection), 272, 339
Locus, 448g
Log phase, 448g
Long interspersed elements, 15
LTR retrotransposon, 13

M
M13 *see* bacteriophage M13
Macro-array *see* DNA array
MALDI-TOF MS *see* matrix-assisted laser desorption/ionization time-of-flight mass spectrometry
Map based cloning, 159–160, 448g
Marker rescue, 448g
    see also complementation
Mass spectrometry, 358–361
Matrix-assisted laser desorption/ionization time-of-flight mass spectrometry, 358–359
MC1R, 416
Melting temperature (Tm), 66–67, 448g
Membrane proteins, 293
    mapping, 293–297
Mendel, Gregor, 159
Mendelian, 449g
Mendelian inheritance, 142, 384, 417

Metallothionein, 381, 400–401
Methioine sulphoxamine (MSX), 272
Methotrexate (MTX), 272
Microarray *see* DNA array
Micro-injection, 449g
Micro-satellite DNA, 13, 412
Mini-satellite DNA, 13
Minisequencing, 420
MLPA *see* Multiplex ligation-dependent probe
    amplification
Model organism, 20
    Arabidopsis thaliana, 20
    Caenorhabditis elegans, 20
    Chimpanzee, 20
    Danio rerio, 20
    Drosophila melanogaster, 20
    Escherichia coli, 20
    Mus musculus, 20
    Rattus norvegicus, 20
    Sacchromyces cerevisiae, 20
Molecular beacon, 429–431, 437, 438
Molecular chaperones, 262
Molecular clock, 240
Molecular diagnostics
    bacterial disease, 437–438
    cancer, 439–441
    inherited disease, 422–436
    viral disease, 438–439
Moloney murine leakemia retrovirus, 385
Monosomy, 30
Motif, 234–237, 449g
MS/MS *see* tandem mass spectrometry
Multiple cloning site, 49, 449g
Multiple displacement amplification, 436
Multiple sequence alignment, 233–235, 239
Multiplex ligation-dependent probe
    amplification, 427, 428
Multiplex PCR, 414, 420, 431, 449g
Mus musculus, 20
Muscular dystrophy, 424
Mutagen, 449g
Mutagenesis, 449g
    in determining whether an ORF is functional,
        281
    signature-tagged, 152–156
    site-directed, 304–309
    using T-DNA, 405–406
Mutation scanning, 432–435
Myogenin, 400

N
N terminus, 449g
N50 length, 196
NAT *see* nucleic acid tests
National DNA Database, 416

NCBI *see* National Center for Biotechnology
    Information
National Center for Biotechnology Information,
    228
NDNAD *see* National DNA Database
N-glycosylation, 265, 268
nirB, 337, 346, 349
Non-coding DNA, 10, 16
    impact on genome analysis, 16
Non-homologous recombination, 284, 381–383,
    449g
Nonsense suppression
    use in chromsome jumping, 166
Northern analysis, 326–328, 354
*Not*I
    use in chromsome jumping, 164
Nuclear run-on, 330–334
Nuclear transfer, 401–403
Nuclease protection, 319–323
Nucleic acid tests, 438
Nucleosomes, 340–341
Numerical chromosomal abnormality, 30

O
Oligo (dT)-anchor primer, 324–325
Oligonucleotide, 449g
    allele-specific, 427
    calculating melting temperature, 67
    degenerate, 126–127, 132
    linkers and adaptors, 106
    molecular beacon, 429–431, 437, 438
    primers for PCR, 66–68
    primers for primer extension, 324–325
    primers for sequencing, 175–179
    probes, 124–127
    TaqMan®, 431– 432, 438
Oncogene, 371–372, 449g
Oncomouse, 371–372, 449g
One-hybrid screens, 351–353
Open reading frame, 208–214, 213, 279–280, 449g
Operon, 449g
ORF *see* open reading frame
Origin of replication, 36–37
Orthologue, 231, 233, 449g

P
Paralogue, 231, 449g
Partial digest, 90, 449g
PCR, 449g
    see also polymerase chain reaction
Periplasm, 449g
Pfam, 225
Phage *see* bacteriophage
Pharmacogenomics, 449g
Phenotype, 419, 449g

Phenylketonuria, 421, 450g
phoA
    translational fusion with, 295, 421
Phylogenetic tree, 239–241
Phylogeny, 450g
Pischia pastoris, 239–242
Plaque, 94
Plasmid, 450g
    see also vector
Plasmid, copy number *see* vector
Plasmid, F plasmid *see* vector
Plasmid, R plasmid *see* vector
Polycation-mediated transfection, 339
Polygalacturonase, 376–377
Polylinker *see* multilpe cloning site
Polymerase chain reaction, 64–77
    allele specific PCR amplifiction, 423
    cloning products of, 74–76
    fluorescent, 413, 425
    fluorogenic 5′ exonuclease assay, 431–432
    for gene cloning, 86
    for quantitation, 76
    in diagnosis of infectious disease, 437–439
    in diagnosis of inherited disease, 423–432
    in forensic analysis, 412–415
    in sequencing, 185–188, 201–202
    inverse PCR, 148
    library screening, 133–134
    multiplex, 414, 420, 431
    primer specific amplification, 423
    primers, 66,72–73
    quantitative real-time reverse transcriptase-
        polymerase chain reaction (QRT-PCR), 76,
        327–329
    reaction, 68–72
    real-time PCR, 76
    signature-tagged mutagenesis, 154, 155
    site-directed mutagenesis, 308
    TaqMan®, 431–432, 438
    temperature cycling, 71
Polymorphism, 18–19, 77, 160, 161–163, 167,
    425, 450g
Polynucleotide kinase, 107, 320
Polyploid, 24, 450g
Positional cloning, 450g
    see also map based cloning
Pre-implantation diagnosis, 423, 426, 436, 450g
Prenatal genetic diagnosis, 423, 426, 450g
Primer, 450g
    see also oligonucleotide
Primer extension, 318–319,320, 324–326, 419–420
Primer specific amplification, 423
Primer walking, 182–183, 191, 192
Probe
    Allele Specific Oligonucleotide (ASO), 427

cloned DNA fragment, 127
degenerate, 126, 132
DNase I hypersensitivity assay, 340–341
expression library screening, 353
Fluorescent In Situ Hybridisation (FISH),
    434–435
gene identification, 148, 155, 162, 165, 167
immobilized, 331,334
labeled RNA, 322–323
molecular beacon, 429–431, 437, 438
molecular diagnostics, 437–439
multiplex ligation-dependent probe
    amplification, 427, 428
northern analysis, 326, 328
nuclear run-on, 331–334
nuclease protection assay, 321–323
oligonucleotide, 66–67, 124–127
random hexanucleotide, 108, 130, 328, 329,
    332, 356, 357
screening DNA libraries, 123–127, 450g
TaqMan®, 431–432, 438
Transcriptomics, 354–355
ProDom, 225
Profile, 237, 450g
Prokaryote, 450g
Promoter, 210, 253, 316–318, 450g
    alcohol dehydrogenase, 269
    alcohol oxidase (AOX1), 270
    ara, 253, 254
    cauliflower mosaic virus 35S, 392–393
    ferritin, 272
    for production of proteins from cloned genes,
        253–259, 269, 271
    gal-1, 269
    heat shock gene, 271, 388
    HIV, 333–334
    lac, 253–256
    lambda P_L, 255–258
    nirB, 337, 346, 349
    pBAD, 253, 254
    T7, 257–258
    tac, 255–256
    techniques used to study promoter function,
        318–350, 354–356
    triose phosphate isomerase (TPI), 269
    ubiquitin, 271
Promoter probe, 334–335, 376–377, 450g
Prosite, 225, 237
Protease
    as examples for sequence analysis, 226–239
    cleaving translational fusions, 264
    factor Xa, 263
    in *E. coli,* 261
Protease – *contd*
    TEV protease and TAP tagging, 303–304

Protein
  addition of "tags" to, 264, 274–275
  aggregation of, 261
  disulfide bonds in, 263, 265
  N-glycosylation of, 265–266
  production from cloned genes, 249–277
  purification, 274–275
  redox state of, 263, 265
Protein A, 301–304
Protein expression
  in *E. coli*, 252–265, 267
  in insect cells, 273
  in mammalian cell cultures, 270–273
  in transgenic organisms, 372–374
  in yeast, 268–270
  proteomics, 356–361
Protein family, 232, 235
Protein fingerprinting, 358–359
Protein fusions *see* fusion protein
Protein identification, 358–359
Protein sequencing, 359–361
Protein turn-over, 358
Protein-protein interactions, 297–304
  detecting with co-immunoprecipitation,
    301–302
  detecting with TAP-tagging, 302–304
  detecting with two-hybrid methods,
    297–301
Proteins, expression of, 249–277
Proteolytic fingerprint, 358–359
Proteome, 450g
Proteomics, 356–361, 450g
Pseudogene, 16
Pseudomonas aeruginosa, 149, 171, 209, 241
Pull-down assays, 301–304, 353
Pyrosequencing, 197–202

Q
QRT-PCR *see* quantitative real-time reverse
    transcriptase-polymerase chain reaction
Quantitative real-time reverse transcriptase-
    polymerase chain reaction (QRT-PCR), 76,
    327–329

R
RACE *see* rapid amplification of cDNA ends
Rapid amplification of cDNA ends (RACE),
    323–325
Rattus norvegicus, 20
Reading frame, 208–211, 450g
Real-time PCR, 76
  see also quantitative real-time reverse
    transcriptase polymerase chain reaction
Recombinant, 450g
Recombination, 450g

Recombination, homologous, 159, 282–284,
    382–383, 398–403
Recombination, non-homologous
  use in constructing gene deletions, 382–383
Recombination, site-specific, 382–383, 403–405
Red hair, 416
Regular expression, 235, 450g
Regulation of gene expression, 315–361
Repeat sequence
  plants, 23
Repetitive DNA, 9, 13–15, 23, 31, 87, 90, 93, 156,
    163, 411–415
Reporter gene, 335–337, 376–377, 450g
Reporter gene assays, 264, 326, 334–338,
    347–349, 376–377, 379
  deletion analysis, 347–350
  example of use, 337
  promoter probes, 334–335
  reporter proteins, 335–337
  site-directed mutagenesis, 347–350
  vectors, 338
Reporter protein, 264, 295, 334–337, 450g
Repressor, 316, 339, 347–349, 451g
Resistance gene, 156–159
Restriction endonuclease, 451g s *see also*
    restriction enzyme
Restriction enzyme, 38–40, 47–51, 451g
  4bp recognition site, 49
  6bp recognition site, 47
  blunt ends, 51
  compatible ends, 49
  frequency of recognition sites, 47–48
  function, 38–40
  isoschizomers, 49
  naming, 38
  partial digestion, 90
  rare cutters, 51
  recognition sequences, 48
  sticky ends, 39, 40–41
  type II, 38
  use in gene cloning, 38–40, 47–51
Restriction fragment, 451g
Restriction fragment length polymorphisms, 125,
    161–163, 451g
Restriction map, 58–59, 451g
Retrovirus, 27
Reverse northern dot-blot analysis, 332–334,
    354
Reverse transcriptase, 27, 103, 318, 327, 357
RFLP *see* restriction fragment length
    polymorphisms
Ribosome binding site, 209, 210, 258–259
RISC *see* RNA-induced silencing complex
RNA interference, 287–290, 374, 375, 451g
  small interfering RNA, 287

RNA ligase, 325
RNA mediated inhibition *see* RNA interference
RNA stability, 330
RNAi *see* RNA interference
RNA-induced silencing complex, 287–288
RNase, 319

S
S1 nuclease, 319
sacB, 285
Saccharomyces cerevisiae, 20
    cloning for expression in, 268–270
    genome wide survey of protein localisation,
        291–293
    genome wide survey of gene function,
        284–285
    promoters, 269
Salmonella typhimurium, 153, 239
SARS *see* severe acute respiratory syndrome
Satellite DNA, 13
Screening, 86, 117–135
    complementation, 119–120
    detecting a DNA sequence, 120
    differential, 132
    expression library, 353
    immunological, 120
    marker rescue, 120
    PCR, 133–134
SDS-PAGE, 121, 259
Segregation, 451g
Sequence comparison
    pair-wise alignment, 217– 220, 223
Sequence logo, 235, 236
Sequence-tagged sites, 195
Severe acute respiratory syndrome, 203, 439
Sex chromosome, 9
Shigella dysenterie, 239
Shine-Dalgarno sequence *see* ribosome binding
        site, 451g
Short interspersed elements, 15
Short tandem repeat, 412
Shotgun sequencing, 183–184,189–197
Shuttle plasmids, 269
Sickle cell disease, 422, 424, 428
Signal sequence/signal peptide, 451g
Signal sequences, 266
Signature tag, 153–155
Similarity, 220–222
    search, 224
Simple sequence repeats
    micro-sattelite, 13, 14
    mini-sattelite, 13, 14
    satellite, 13, 14
    trinucleotide repeat, 13, 14
SINE *see* short interspersed elements

Single nucleotide polymorphism, 18–19, 161,
    419, 424
Single nucleotide primer extension, 420
Single-stranded conformational polymorphism,
    432–433, 440
SNF1/SNF4 interaction, 298–299
SNP *see* single nucleotide polymorphism
Southern blot, 125, 340
Splicing, 12, 216
SSCP *see* single-stranded conformational
        polymorphism
Start codon, 209, 210, 451g
Stop codon, 208, 210, 451g
STR locus, 413
STS *see* sequence-tagged sites
Sub-cloning, 151
Substitution
    conservative, 220
    matrix, 221
Suicide plasmid, 149, 286, 451g
Supercoil, 451g
Supermouse, 381–384
SwissProt, 224, 225
Synteny, 21

T
T7 DNA polymerase, 181
Tandem affinity purification, 302–304
Tandem mass spectrometry (MS/MS), 359–361
TAP-tagging *see* tandem affinity purification
Taq DNA polymerase, 65, 72, 181
Tat, 333–334
T-DNA, 391–393
Telomere, 100, 101, 451g
Template, 64, 177, 179, 451g
Terminal deoxynucleotidyl transferase (TdT),
    324–325
Terminator *see* transcription terminator
TEV protease, 303
Thymidine kinase gene, 398
Ti plasmids, 391
Tissue plasminogen activator (tPA), 251
Tn5, 149
Trait, 159, 370, 451g
Transcription, 315–318, 451g
Transcription startpoint, determination, 318
    by nuclease protection, 319–323
    by primer extension, 318–319
    by RACE, 323–325
    comparison of techniques, 325
Transcription terminator, 210, 258, 392–393,
    451g
    nos, 392–393
Transcription, determining the level, 326–338
Transcription, global studies, 353–357

Transcription, identifying regulatory proteins, 350–353
Transcription, identifying regulatory regions in DNA, 338–350
Transcriptomics, 353–357, 451g
   probe design, 355
Transfection, 272, 339
   see also transformation
Transformation, 36, 42–44, 339, 451g
   by biolistics, 272, 395
   by electroporation, 44, 272, 339
   by lipofection, 272, 339
   by micro-injection, 381, 388
   by particle bombardment, 272, 395
   of *E. coli*, 42–44
   of mammalian cells, 272, 339
   of plants, 391–393
   stable germ-line transformation, 380
Transgene, 451g
Transgenic, 451g
Transgenic organism, 365–407
   biopharming, 373–374
   disease models, 371–372
   problems with, 396, 397
   production, 378–397
   reporter genes, 376–377
   turning genes off, 374–376
   used for protein production, 372–374
   uses in agriculture, 368–371
Translation, 210, 211, 451g
Translational fusions, 263–264
   use in mapping membrane protein topology, 294–297
   use in two-hybrid methods, 297–301
Transposase, 143, 157
Transposition, 13–15, 144
Transposon tagging, 144–149, 172
Transposons, 13–15, 143, 382–383, 451g
   autonomous, 143
   Hermes, 388–390
   non-autonomous, 143
   non-replicative, 144
   replicative, 144
   use in gene tagging, 144–159
   use in signature tagged mutagenesis, 154
TrEMBL, 225
Trinucleotide repeat, 13, 424–427

Trisomy, 30
Two-dimensional gel electrophoresis, 121, 358–359
Two-hybrid screens, 297–301, 353, 452g

U
UniProt, 225

V
Vaccines
   edible, 373, 390–394
Vector, 452g
   BAC, 100–101, 102
   bacteriophage λ *see* bacteriophage λ
   cosmid, 98–100, 446g
   expression, 96, 109, 252–253
   F plasmid, 37
   for expression in mammalian cells, 271
   for expression in yeast, 268–269
   for gene expression, 252–259
   for preparing RNA probes, 322–323
   for reporter gene assays, 338
   from retroviruses, 385–386
   insertion vector, bacteriophage, 95,105
   plasmid, 7, 36–37, 53–58, 450g
   plasmid copy number, 25, 57
   R plasmid, 37
   replacement vector, bacteriophage, 96–98
   suicide, 286
   YAC, 100,101
Vitamin A deficiency, 370

W
Western blot, 281
Whole genome amplification, 435–437
Wild type, 452g

X
Xenotransplantation, 402, 452g
Yeast artificial chromosome (YAC), 452g
Yeast one-hybrid screen *see* one-hybrid screens
Yeast two-hybrid *see* two-hybrid screens
YEps, 268

Z
Zooblot, 125, 167, 168, 452g